# Daring French Explorations

## EDITOR'S NOTE

*Daring French Explorations* is first and foremost the brainchild of a passionate collector, based on a selection of extracts from travel journals and other works written in the eighteenth and nineteenth centuries that describe the circumnavigations undertaken by fifteen French explorers. This material, including the quotations that punctuate the book, is reproduced as it appears in the original works; any adjustments (to punctuation, spelling, etc.) are for the sake of readability only. For example, the modern equivalents of certain place names may be indicated in square brackets to facilitate understanding of the course traveled by the explorer. Units of measurement are retained as they appear in the source texts without conversion.

Otherwise, this book voices what the sailors themselves experienced—what they felt and what they saw—expressed in their own words, which must be read in their historical context and not taken to represent the views of the author and publisher. Most of the illustrations in this book are taken from historical, hydrographic, botanical, and zoological atlases. Some captions appear in italics, as they are drawn from the original title of a document, hence the variable spellings of certain place names.

The routes followed by each explorer are shown in sixteen maps custom made for this book from hydrographic atlases, from the explorers' accounts, and from other historical documents published at the time (logbooks, letters, articles, speeches, etc.). The background of the map is inspired by an 1820 engraving by Ambroise Tardieu representing the voyage of the corvette the *Favorite*, to which part of the Antarctic coastline has been added. The original reflects Western knowledge of the surface of the Earth at that time and includes various small, recently discovered islands of no relevance to the present work, which therefore are not featured here. Unfinished lines indicate coasts that had been sighted by navigators, but not yet mapped. A bibliography and list of illustrations at the end of the book provides details for all the written and illustrative material used in its production.

Hubert Sagnières

FOREWORD
Edward Duyker

# Daring French Explorations

## TRAILBLAZING ADVENTURES AROUND THE WORLD: 1714–1854

FEATURING
**BOUGAINVILLE, LAPÉROUSE,
DUMONT D'URVILLE, AND MANY MORE**

Flammarion

## BIOGRAPHY

"Discover, explore, share" could be the motto of Hubert Sagnières, a born explorer who cut his teeth crisscrossing the South Pacific islands with his family in the footsteps of Bougainville and Lapérouse.

In 1991, as head of a leading French industrial group, he moved to Canada with his wife and children and developed a passion for the Far North that took him on more than twenty winter expeditions to the Canadian Arctic. More recent years have seen him on forays into the most far-flung islands of the Indonesian Archipelago, driven by a thirst for discovery that led to fascinating encounters with the Dayaks of Borneo and the Mentawai people on Siberut Island.

With this same passion for travel and adventure, Hubert Sagnières created this exceptional compilation of extracts drawn from antiquarian books on exploration, bringing together the essence of the great explorers of bygone times in their own worlds. He meanwhile continues to crisscross the planet, as eager as ever to discover all those places that have acquired mythical status over time.

All of the royalties from this work will be donated to the humanitarian society Karuna-Shechen, founded by Matthieu Ricard in 2000.

**https://karuna-shechen.org**

“Storms, fog, and snow will be a nuisance from time to time. When they are, think of those who have been there before you and simply tell yourself, ‘What they could do, I can do too.’”

Antoine de Saint-Exupéry, *Wind, Sand and Stars*, 1939
French writer and aviator

“Dream, Dare, Do, and Expect the Unexpected.”

Mike Beedell
Canadian explorer and photographer

For Margaux, Paul, Marco, Luca, Adèle, Maïa, Félix and Lir of course,
and my future grandchildren,
may the spirit of adventure drive everything they do in life!
But also for my children who accompany me on my expeditions,
Fanny and Louis, Amandine and Xavier,
Marigona and Yves, Alice and Luc, Bénédicte and Damien.
But mainly, yes mainly ...
For Anne.

# FOREWORD

**Edward Duyker** – Fellow, Australian Academy of the Humanities

Over the centuries, perceptive private individuals have been responsible for the preservation of rare books and manuscripts that might otherwise have been lost to posterity. Many great university and national libraries have had their genesis in such collections and they constitute part of the memory of humanity. Serious scholars are likely to owe a debt of gratitude to one or more collectors in their chosen field. Catalogs of private libraries more often appear when collections are offered for posthumous sale. Generous living collectors will sometimes enable private access to rare titles or lend them for exhibitions. Others might facilitate the publication of transcriptions or facsimiles. Bibliophile Hubert Sagnières has done something more unusual here; he has offered the public a beautifully illustrated collection of extracts from sixteen precious works in his private collection. Together, they tell the story of French exploration from the early eighteenth century to the middle of the nineteenth century—almost entirely within the pre-photographic age and through to the end of the age of sail. The works of expedition draughtsmen and artists—sometimes gifted amateurs—were engraved and, in later years, splendidly lithographed in color.

The vivid selection of images and accounts presented here draws fresh attention to expeditions that ushered many parts of the globe into the human imagination. Hubert Sagnières' library is clearly more than the sum of its individual parts. The methodical dedication of the private collector encompasses breathtaking range. For many readers in the English-speaking world, the manifold voices, images, and maps from these French voyages will be a revelation. While these expeditions had differing political and scientific objectives, virtually all contributed to an expansion of cartographic knowledge. On one voyage after another, the exploration teams would compare longitudinal determinations and enable precision on the location of islands, capes, reefs, and continental points. Expedition commanders received orders to investigate the assertions of speculative cartographers, based on the journals, charts, and sometimes very vague reports of earlier navigators. One can see the gradual development of scientific method in these voyages. The naturalists who sailed with them did not merely contribute to the cataloging of creation, or the gathering of potentially useful plants and animals. Many of their biogeographical and ecological observations helped lay the foundations for later evolutionary biology.

Similarly, geological specimens and magnetic intensity studies in different parts of the globe contributed to modern earth sciences. Those who began to measure seawater temperatures at depth and to record meteorological data expanded the frontiers of nascent oceanography and climate science. Their findings continue to provide us with baseline data that is of value in understanding the great challenge we all now face: global warming. In recording vocabularies and cultural information during their landfalls, they offer us a window on our shared humanity. The published narratives, atlases, and scientific studies they produced were often masterpieces of the printer's art; the evidence is before us in this present volume. We can all be grateful for their extraordinary endeavors.

Plate 20, George Wolfgang Knorr, 1766.

# EXPLORERS IN THE AGE OF SAIL: IN THEIR OWN WORDS

**Hubert Sagnières**

*"We have travelled the world, crisscrossed the seas and studied archipelagos and continents. We have returned with our heads and hearts full of the great things that we have studied; and we have no wish to keep to ourselves all that has caught our eyes and our imaginations.*

*Each of us deserves an equal share in the praise or blame of the reader, for these pages are a truthful reflection of the phenomena that we have pursued, driven by our ardent thirst for study and curiosities. The reward we claim for our zeal is but the faith that attaches to every work of conscience.*

*Yes, our stories are true; yes, they are true in detail and true in general; and yes, the manners and customs we describe are true to the manners and customs of those people with whom we mixed, whether hostile or amenable to civilization.*

*To lie is a disgrace for any narrator and never more so than for the man who writes the histories of nations. Lying for him is an act of cowardice. Because every thing is joined in that long succession of periods and ideologies that go to form kingdoms and empires; every thing is linked and every thing is coordinated. Today is the consequence of yesterday, and the year that is commencing is the reflection of the year just ended.*

*It is in that spirit that we offer you this book, as a work entirely devoid of futilities and containing, we trust, some useful truths for those who wish to learn the history of the peoples scattered upon our globe. Revealed in its pages is that incessant struggle of civilization against barbarism, the former carried forward by the three great powers that govern the world: gold, bronze, and religion."*

Élie Le Guillou (1806–1860), *Voyage Autour du Monde de l'Astrolabe et la Zélée, sous les ordres du contre-amiral Dumont d'Urville, pendant les années 1837–1840.*

The "truth" as told by Élie Le Guillou is that of a nineteenth-century Frenchman who served as ship's surgeon on Jules Sébastien César Dumont d'Urville's perilous expedition to the Antarctic Ocean.
It was a time when ships with evocative names like *Bonite* and *Vénus* were dispatched to conquer the waves in pursuit of geographic and scientific knowledge, claiming new territories as they went, spying on other nations, and acquiring wealth for the benefit of the world's then-expanding empires. It was a time that belongs to a bygone past and a bygone historical context, when recently discovered populations were by turns viewed with friendship, admiration, amazement, and condescension—a time when people still spoke of "peoples hostile or amenable to civilization ..."

The nineteenth century also saw the emergence of that emblematic figure, the explorer. Yet the personalities of the individuals who embarked on these around-the-world adventures were very varied, as were their objectives and motivations. In this book, I invite sixteen of them to speak for themselves: fifteen Frenchmen and one Frenchwoman who undertook these extraordinary voyages in the period 1714–1854. To that end, I have drawn from ancient writings collected over the years, seeking to share the personal experiences of the authors with a wider audience: tell their stories as they told them, based on their own logbooks and correspondence, or the books they published on their return. Reading the original versions of their stories was an honor and a joy for me, and also an incredible thrill. Touching and smelling the old paper added a sensory dimension that seemed to bring the author back to life. It was as if in reading their words and viewing their illustrations I somehow resuscitated the personality behind the writings.

The extracts presented here are selected from thousands of pages of explorations and observations of the world: an ancient world but no less an echo of our twenty-first century, many of the issues raised by these navigators remain just as relevant today.

It was in the period 1519–1522 that the Portuguese explorer Fernand de Magellan (1480–1521) and his Spanish counterpart Juan Sebastián Elcano (1476–1526) completed the first circumnavigation of the globe under sail, with a few French sailors on board. But the first official, around-the-world voyage on behalf of France would have to wait another quarter of a century, for the expedition led by Louis de Bougainville in 1766, which was financed by King Louis XV. Over the course of those 247 years, while more than twenty circumnavigations of the globe were completed, most of them were conducted on an unofficial basis, often by Frenchmen and usually for commercial purposes. It is this absence of any legal context together with the astronomical profits made by the shipowners that explain why no record was ever made of these voyages. The South Seas were then in the hands of pirates and buccaneers bringing back gold and silver from Peru.

Other sailors did of course sail the seas in the course of those 140 years between the first and the last of the great French maritime expeditions; but none of them published his adventures.

One notable example is Jean-Baptiste Cécille (1787–1873) aboard the *Héroïne*, who defended French whalers, rounded the two capes and accomplished other extraordinary feats. But fortunately for posterity, we do possess written records of sixteen circumnavigations completed by French navigators (and women passengers) in the period 1714–1854, some based on the accounts or logbooks of the explorers themselves, others drawn from relevant publications.

On their to return France after years of challenging navigation, some of these intrepid folk were welcomed as heroes; others were marginalized because the government had meanwhile changed and nobody knew what to do with them. Many returned with their holds full of riches and discoveries; others came back empty-handed, were called before the council of war or fell into obscurity. Some indeed are better known outside France, in those countries where they landed and which they mapped.

Glorified or not, France owes all these explorers a debt of gratitude for having risked their lives to reconnoiter hundreds of islands, coasts, bays and maritime passages; for collecting thousands of specimens of fish, insects and mammals; and for advancing the cause of navigation, hydrography, astronomy, physics and more besides. They should also be remembered for their willingness to meet with unknown populations, to observe their ways and sometimes even to live with them. The accounts of their adventures are surprisingly contemporary, and display an open-mindedness that is often quite remarkable. What is more, it is clear from the wealth of intelligence they brought back that they already foresaw the impact of the actions of Western governments on the rest of the world.

It was in 1714 that the first Frenchman to recount his around-the-world adventures set sail from Cancale, and it was in 1854 that the last French three-masted circumnavigator returned to the port of Toulon. Twenty years later, Jules Verne published *Around the World in Eighty Days*. Exploring the oceans had become mundane and there was nothing left to learn about the world and its people. Henceforth explorers would concentrate on the North and South Poles.

But let us not forget these great explorers of the eighteenth and nineteenth centuries–these men and women who dared to venture to the other end of the Earth. This book salutes them, together with their crews and the two thousand sailors who accompanied them. Over the course of a century and a half, they accomplished extraordinary feats that nourrished the imaginations of millions of children and adults and inspired countless careers. May their adventures always make us dream, and foster a spirit of adventure in anyone who thinks China or the North Pole may be discovered at the click of a mouse.

# Le Gentil de La Barbinais

**1692–1731**

AT THE DAWN OF THE REIGN OF THE THEN FIVE-YEAR-OLD KING LOUIS XV, A SAINT MALO NATIVE PRONE TO SEASICKNESS SET SAIL FOR REGIONS STILL LARGELY UNKNOWN. HIS NAME WAS LE GENTIL DE LA BARBINAIS. HE WAS DRIVEN BY AN UNSPOKEN DESIRE TO SHAKE HIS OWN PREJUDICES. ALONG WITH HIS THIRST FOR DISCOVERY, A LONGING TO SEE THE WORLD'S LANDSCAPES AND THE MEN AND WOMEN WHO LIVED THERE, AND THE POSSIBILITY OF MAKING HIS FORTUNE IN THE PROCESS. THIS WAS, AFTER ALL, BOOMTIME FOR MARITIME TRADE. BUT ONE THING IS FOR SURE: HE DID NOT EXPECT TO BECOME FAMOUS.

Plate 55, George Wolfgang Knorr, 1766.

# VOYAGE AROUND THE WORLD BY LE GENTIL DE LA BARBINAIS

**1714–1718**

**A daring and curious adventurer, La Barbinais was the first Frenchman to embark on a circumnavigation of the globe and the first to record his discoveries in writing, complete with all of those emotions aroused by a journey full of surprises that took him all the way to Peru and China …**

On August 30, 1714, La Barbinais embarked on the *Vainqueur*, a 230 ton merchant ship belonging to his uncle, Julien Bourdas. With twenty-two cannon and a crew of seventy men, the vessel set off to trade in South America, defying the ban on such voyages imposed by King Louis XIV on January 18, 1712. Everybody could see that this was a risky venture but for the time being at least shipowners in Saint-Malo saw it only as an opportunity to get rich. Buccaneers and other pirates were returning from Peru with their holds filled with gold and silver seized from the Spanish—why not La Barbinais? So it was that driven by the winds of temptation La Barbinais embarked on his around-the-world voyage.

Off the coast of Cancale, France, the winds carried the *Vainqueur* to the Canary Islands and thence to Brazil. At this point she made her first crossing of the equator, stopping south of Rio on December 12 to reprovision and prepare for Cape Horn, which she rounded in January 1715 despite a first mutiny on board. The ship then continued northwards along the Chilean coast, reaching Concepción Bay after four months at sea where La Barbinais found more than forty French ships that had defied the ban on trading in South America, each more miserable than the last. On going ashore, he was deeply shocked by the violence of Spanish colonization and quickly set sail again for Valparaiso, then in the throes of a native uprising, before coasting to Coquimbo. The fortnight he spent in Chile, staying in Arica, taught him a lot about the local way of life—enough to know that he liked the country but that business was bad. He then headed for Peru, where he stayed for seven months, visiting a number of places and venturing inland all the way to Lima, only to discover that trade there was even more deplorable than in Arica. Then came two months of indecision, our young *Malouin* (St Malo native) meanwhile mingling with the Jesuits whose flagrant and wholly accepted libertinism astonished him. The local authorities were no less reprehensible, being riddled with corruption and plundering the country to line their own pockets.

On March 4, 1716, La Barbinais left Peru for China, this time sailing aboard the *Jupiter*: a 350 ton vessel commanded by Captain Maurice Béven, with 24 cannons and one 106 men on board. She had left Bayonne on July 17, 1714, also in defiance of the royal ban. They crossed the Equator in March, skirted Formosa [Taiwan] on June 25, and entered Amoy Bay [Xiamen] a few days later. It was then that things got complicated, the *Jupiter* needing to reprovision and the Mandarins were having none of it until a certain Father Laurearty stepped in and persuaded them to allow the local people to supply the ship with food. La Barbinais and Béven then spent long weeks in negotiations—months of talks that so exhausted our young explorer that he eventually retreated to a monastery on the small island of Colomsu where he immersed himself in the history of China, and its takeover by what were then known as the Tartars. He became fascinated by the customs and traditions in China and spent Christmas with the bonzes on the island.

The *Jupiter* left China in February 1717, loaded with provisions, to begin its return voyage to France. But no sooner had she left port than the ship struck a reef and ended up on its side. Refloated thanks to the ingenuity of Chinese sailors, the ship set course for Cochinchina, south of present-day Vietnam, and the Strait of Malacca. Sailing between Sumatra and Bangka, she passed the coast of Batavia [Jakarta] at the end of March but did not stop for fear of encountering the Dutch. Not that this prevented La Barbinais from venturing ashore on the island of Java and marveling its wonders ...

After the Sunda Strait, in early April 1717, the *Jupiter* set sail for Bourbon Island [La Reunion]. There, La Barbinais had the chance to catch his breath after his Chinese adventures, and stayed for five months, waiting for the best season to sail even farther south toward the Cape of Good Hope. Rounding the Cape in October, the vessel sailed up the Atlantic to Brazil, which she reached in November, gradually drawing closer to French shores. But her troubles were far from over, starting with a second mutiny that forced La Barbinais to stop over in Rio for several months, with many sailors imprisoned.

Then, in early April 1718, the *Jupiter* arrived in Spain to learn that its owner had gone bankrupt and that she herself could be seized by the Spanish at any moment for having visited Peru illegally. The new owners therefore ordered La Barbinais to go to Genoa, which he did ... but on foot. He liked Italy so much that after his testing time at sea he wondered whether he wouldn't be better off staying there. In the event, he remained until August 1718 when he returned home to Saint-Malo.

In July 1724, La Barbinais recorded his adventures in the form of sixteen long letters addressed to an imaginary correspondent. Unlike those adventurous navigators who came after him, he clearly took great pleasure in recounting the habits and customs of the people he met and describing the places where they lived. He later moved to Santo Domingo, where he made his fortune, before settling in Nantes, where he died on December 30, 1731.

► View of the Cape of Good Hope.

**GENERAL REFERENCE MAP**
For the Voyage of the ***VAINQUEUR*** and the ***JUPITER***
Led by Monsieur Le Gentil de la Barbinais
1714–1715–1716–1717–1718

Western Longitude, Paris Meridian
150
120
90
60
30
0
Paris Meridian
Melville Is.
Lancaster Strait
Baffin
Bay
Davis Strait
GREENLAND
Iceland
SWEDEN
Baltic Sea
Bering Strait
Cook's River
Port des Français
Lituya Bay
Kodiak Is.
Sitka
Alaska
Queen Charlotte's Is.
Vancouver Is.
Aleutian Islands
Nootka
Hudson Strait
Hudson
Bay
Cape Farewell
LABRADOR
CANADA
NORTHERN
BRITISH ISLES
London
Paris
Brest
FRANCE
Aug. 1714
St Lawrence River
Quebec
Newfoundland
NORTH
AMERICA
NORTHERN OCEAN
UNITED STATES
Acadia
New York
Washington
ATLANTIC
Azores Is.
Toulon
ITALY
SPAIN
Madrid
MEDITERRANEAN SEA
Strait of Gibraltar
San Francisco
Monterey
Bermuda Is.
New Orleans
San Antonio
California
MEXICO
Gulf
of Mexico
Florida
Madeira Is.
BARBARY
MOROCCO
Canaries Is.
OCEAN
Tropic of Cancer
Mazatlán
Lucayan Is.
Hawaii Is.
Sandwich Is.
San José
Mexico City
Santo Domingo
Caribbean Sea
Jamaica
Acapulco
Guadeloupe
Martinique
Cape Verde
SENEGAL
AFRICA
GUINEA
Panama
COLOMBIA
Cayenne
GUYANA
Equinoctial Line
Gallapagos Is.
Paita
PERU
SOUTH
AMERICA
Ascension Is.
Marquesas de Mendoza
Feb. 1716
Lima
CONGO
Navigators Is.
Pisco
Bahia
BRAZIL
Nov. 1714
St Helena Is.
Maouna Is.
Tahiti
Society Is.
Arica
Nov. 1715
Friendly Is.
Tropic of Capricorn
Rio Janeiro
Traitors Is.
Mangareva
Pitcairn Is.
C. Frio
Easter Is.
Coquimbo
May 1715
CHILE
St Catherine's Is.
SOUTH
Valparaiso
Montevideo
Mar. 1715
Buenos Aires
Rio de la Plata
Concepción
Cape of Good Hope
ATLANTIC
Chatham Is.
PATAGONIA
OCEAN
Falkland Is.
OCEAN
Strait of
Magellan
Tierra del Fuego
Is. de Los Estados
Cape Horn
Jan. 1715
Sandwich Is.
Antarctic Circle

When I left France, Sir, you engaged me to write you with details of every thing that might happen to me in the course of my journey. I would most earnestly request that you have recourse to a world map where you will see at a glance all the countries that you would have me describe. You believe, and you may not be wrong, that it is enough to be a traveler to love storytelling: therein, however, lies not my wont and I would not be keeping my word were it not that the honor and esteem in which I hold you oblige me to give you the satisfaction you have demanded of me. Moreover, do not expect to read pompous accounts and flowery descriptions for I am no orator. I will simply recount, as sincerely as I may, all of the things that I remarked.

LE GENTIL DE LA BARBINAIS,
MARCH 1724

"In 1714, a Frenchman by name of La Barbinais le Gentil set sail on a particular vessel to engage in trade off the coast of Chile and Peru. From there he traveled to China where, having rested for a year at diverse trading posts, he embarked on a vessel other than that which had brought him thither, and returned to Europe, having assuredly made a personal voyage around the world but not such a voyage around the world as could be said to have been made by the French nation."

EXTRACT FROM *VOYAGE AUTOUR DU MONDE*
BY LOUIS-ANTOINE DE BOUGAINVILLE, 1771
*(A VOYAGE ROUND THE WORLD, PERFORMED BY ORDER OF HIS MOST CHRISTIAN MAJESTY, IN THE YEARS FROM 1766 TO 1769).*

**AUGUST 1714**
CANCALE, FRANCE

*We left the Bay of Cancale* on August 30 in the year 1714 aboard the *Vainqueur*. Buffeted by headwinds we were forced to put in at the Isle of Sark, at a distance of three leagues from Guernsey. Having waited three days for the return of fair weather, the while embarking several oxen and other provisions, we set sail on September 4. It matters not to you, Sir, what wind was blowing at the time; suffice to say that it was favorable to those wishing to travel to the Canary Islands. Our intent was thither to hove to and purchase provisions that our hasty departure from France and the capacities of the Isle of Sark had prevented us from so doing. After ten days under sail, I noticed with pleasure that I was becoming familiar with the sea and that my stomach did no longer pay the main its habitual ransom.

**NOVEMBER 1714**
ILHA GRANDE, BRAZIL

*Ilha Grande* measures fourteen leagues in circumference and lies in the Tropic of Capricorn, within two leagues of the American continent. It is a land of eternal springtime. I had the curiosity to stride out alone to the top of the mountain opposite the harbor where we had dropped anchor. After a good deal of trouble caused by the thickness of the woods and the constant rain, I succeeded in reaching the top. I lost my way on the return journey and descended by an uncertain path. Fatigue led me to consider curiosity as an incommodious passion. Orange and lemon trees grow on this island solely to afford umbrage for monkeys and crocodiles. The most unpleasant creature is a small worm that insinuates itself beneath the fingernails and toenails; there it gorges itself and grows by degrees fatter, setting off a painful itching.

Though we were sorely wanting in food and water, we durst not go to Rio Geneiro [Rio de Janeiro], and made do with such provisions as we could obtain from dwellings on dry land. The Portuguese were still smarting from the damage lately inflicted upon them by the René Duguay-Trouin squadron—the wound was yet still bleeding.

**DECEMBER 1714**
ILHA GRANDE, BRAZIL

*You know, Sir,* that as a consequence of the King's orders and conventions between France and Spain, those wishing to take up arms in behalf of Peru are obliged to keep their enterprise secret. Our shipowner had commissioned an English vessel, in the name of an Englishman who was to carry the title of captain aboard the ship but exercise none of the functions inherent therein. This precaution also obliged us to hire English sailors, who

*The first compliment they paid us was to congratulate us on coming to swell the numbers of the wretched.*

were almost equal in number to their French counterparts. France and England being rarely of one accord, squabbles broke out every day and in accordance with a natural preference (which should have been tempered by politics) we shewed an undue partiality that led us to take the part of our Frenchman. The English swore revenge and resolved to kill all of those officers and sailors who had so displeased them. A young Guernsey man warned the captain of the plot afoot and the latter, being a man of honest disposition, advised us of the same forthwith.

**APRIL 1715**
BAHÍA DE CONCEPCIÓN, CHILE

*Finally, Sir, after six months at sea* and a thousand fatigues owing to want of food and water, we caught sight of the mountains, followed shortly afterwards by the Bay of Concepción in Chile. We are hardly at peace here, Sir, having encountered hitherto nothing but misfortunes and contretemps that seem to follow one after another. Assuredly, if the French Crown knew what it costs those who venture into these waters in defiance of its orders, far from punishing them, it would have sympathy for their madness and perhaps praise their zeal in purging the kingdom of an infinity of fabrications, which they come here to barter for money but always at a considerable loss to themselves.

We were not expecting to find in the Bay of Concepción such a large company of people from our own country, nor to hear the sad tidings which they rattled off on our arrival. Some roundly cursed us, others droned on about the sorry state of their affairs to the point of tedium. In a word, confusion reigned. There are currently forty French ships in these waters. For all the love that I bear my country and loathe as I am to speak ill of her, I must nevertheless admit that there is none more often led astray by ambition or less suited to trade with the Indies than France. And such

▲ View of the town of Talcahuano and the Port of Concepción Bay.

TERRE FERME
PANAMA
Ancien Perou
I. des Perles
Popayan
Pasto
Cali
Cartage
S. Sebastien
S. Fe Bogota
Velez
S. FE
Castille d'Or
PARIA
Tamara
Acamacori
Meta R
Cataries
Paria ou Orenoque R
Mines d'or
P. de Bonaventure
I. des Gorgones
I du Coq
Catames
R de S Jago
Provinci dos Emeraudes
Rio Calliente
Madrigal
Tusa
la Culata
Ligne Equinoctiale
Otavalo
Quito
Contiega
Zumague
Los Quixos
AUDIENCE DE QUITO
Bracamaros
Canalla
R. des Amazones ou Oreghano
Riviere d'or
Maragnon ou Amazones
Negro R
I. et Marais dittes Lagunas des Amazones
Guamares
Arcabucos
Portovieyo
I. de Puna
V. R. Tumbes
Poecoi R
Chira R
Pasta
Tumibanba
Ayabuca
Paltas
Garrochamba
Caxas
Caparra
S. Jaques de Montagnes
Tangarata
Valladolid
Caxamalca
Motupe
Colique
Litibamba
Chicapoyas
Levante
Guamacuco
PROVINCES DES AMOZONE
MER DE PEROU
P. Malabri
Truxillo
Chimo
S. Jaques des Vallées
Santa
Guanueo ou Leon
Chincacoa
Atamillos
Condesugo
Chupas
Tovet de la Bavanca
Bonbom
Tambo ougaura
Vallee de Lima
V. Los Ries
V. Pachacama
V. Chinca
Lobo Isle
Paviacaca
Xauxa
AUDIENCE DE LIMA
Guamango
Vilacinga
Guaitara
Mala
Andagui
R. Abancay
Cusco
Urcos
Carabaya
Lumicaca
Nasca
Catabamba
R. Apurima
Limatamb
Xaquixaguana
Chinchero
Sandes
Madera
la Grande R
PARTIE DE
LA MER DE
SUD
Villahormosa d'Arequipa
Xuli I.
Collao
Titicaca
Incaya
Inguarma
Chuquito
Hayohayo
Guavina
Pavia
Laraix
Chuquiaqua
Chicoana
Caribes
Mulobamba
Antropophage
R la Plata
Coponata
LOS CHARCAS
Hicaos
Vallee de Terapaca
Pica
Lac Paria
Pocona
la Plata
Potosi
la Plata ou Grande
Ste Croix de la Vieille Montagne
Tropique du Capricorne
Vallee d'Omaguaca
295
300
305
310
315

is the view held by other European peoples. Is it not, in fact, a wanton waste of pecuniary resources to send forty vessels to Peru when ten would suffice?

**JULY 1715**
ARICA, PERU [CHILE SINCE 1880]

*You cannot ignore, Sir,* that the Spanish conquest of this country was other than a violent usurpation, with religion as its imaginary pretext. Following many battles, the Spaniards subjugated Peru and Chile and sealed their new conquest with the blood of several million people—a weak and timid people unfamiliar with the art of the war. Those spared the wholesale slaughter were granted an apparent liberty that was in fact nought but a diluted form of slavery. Thus were the Indians placed under the yoke of their new masters, the while secretly decrying their loss of freedom and subjected to inhuman treatment by the Spaniards.

**SEPTEMBER 1715**
PISCO, PERU

*I left Pisco* on September 4 for the province of Chincha, whose capital today is a little Indian village that bears the name of the province. The village was formerly a powerful city with some two thousand families in its bosom. The province was likewise home to several million inhabitants but today stands deserted, with scarcely five hundred families remaining. It is easy to see from this how many were slaughtered by the Spaniards, who freely admit that their victory was sealed in the blood of their victims.

The Spaniards, dazzled by the riches before their eyes, soon sought a plausible pretext to eliminate the Indians and their king. Religion supplied the pretext. A bishop who preached the virtues of the Gospel to king Atabalippa, [Atahualpa] an Inca with scant regard for what he did not understand, was then the herald of war. The Indians were set upon and put to the sword; Atabalippa himself was overthrown, clapped in irons and held prisoner in the Spanish encampment.

Peru became the theater of a bloody war and avarice, which from the beginning had produced such sad catastrophes, and reaped consequences still more grim in the time that followed.

**DECEMBER 1715**
LIMA, PERU

*I finally arrived in Lima* after nine days of walking during which I suffered every imaginable incommodity and fatigue. My face had such a brown complexion that I could easily have been mistaken for one of our Indian guides. Mosquito bites rendered me unrecognizable. I learnt on arrival that trade was as disadvantageous in the capital as in the other ports of Peru. I found several vessels ready to return to France, having exchanged French goods for silver bullion at a fifty percent loss.

*I will pass over in silence, Sir, all the cruelties that the Spanish continued to inflict on the Indians.*

◄ **Facing-page:**
Map of South America.

◄ **Left:**
Lima Cathedral, Peru.

*There are in the city of Lima fifteen monasteries for females, whose debauchery is so great ...*

The parlors of female monasteries are perpetually full of priests and clergy who stay there all day long, doing and saying God knows what. I would blush and scruple to give details of what I have seen and heard on this subject. The fairer sex in this country enjoys unbridled license and glorifies debauchery.

### JUNE 1716
### AMOY [XIAMEN], CHINA

— *We were engaged for two days* loading the provisions given to us by the Viceroy: one hundred chickens, forty ducks, four oxen, six sheep, eight pigs and vegetables. On June 7, with a favorable east-northeast wind in our sails, we departed Lima in the company of three other vessels. On the twenty-ninth day of the month the mountains of China came into view and a number of fishing boats drew up alongside us. Shewing no fear, the fishermen presented us with fresh fish, the while making all manner of gestures beyond our understanding. We could only surmise that they wished to dissuade us from going to Amoy. Had we but acted on that supposition we would not be in the predicament in which we find ourselves today.

We entered Amoy Bay at four o'clock in the evening and dropped anchor opposite the chief temple on the island, within two leagues of the port and the town itself.

### JULY 1716
### AMOY [XIAMEN], CHINA

— *We were awoken by a strident fanfare* issuing from two sampans or warships that came to drop anchor near our vessel—a clanging of cauldrons and bronze basins that made our heads spin and amused us not in the least. The *Houpou* Mandarin, which is to say the Collector of Customs for the island of Amoy, had sent out his sampans, supposedly in our honor but in reality to observe us and prevent the inhabitants of Amoy from bringing us their merchandise. This is the habitual practice in all Chinese ports regarding foreign vessels.

On July 2, ten of our party went ashore. We attended at the house of a Chinese man calling himself a Christian, who had prepared a magnificent breakfast for the occasion. He had sent for ten sedan chairs mounted on bamboo poles that were borne on the shoulders of bearers. Accompanied by all the Mandarin's interpreters, we commenced our visits by waiting on the *Houpou* Mandarin.

Facing-page:
*Vue de la ville d'Agana (Isles Marianne)*
(view of the town of Agana, Mariana Islands).

Below:
*Gondole de Mandarin*
(Mandarin's gondola).

Bottom, left:
*Manière de consulter le sort*
(a way of divining destiny).

Bottom, right:
*Famille de pêcheurs*
(fishing family).

Pages 26-27:
*View of the Tchin-shan, or golden island in the Yang-Tse-Kiang, or Great river of China.*

**NOVEMBER 1716**
AMOY [XIAMEN], CHINA

*All the vessels in Canton* are preparing to depart, Sir, whilst we, deceived and betrayed by the Chinese, run the run risk of wintering here and missing the season of return to Europe. All travelers agree that China is a country which abounds more than any other in every thing that may contribute to render life convenient and even delicious.

The Chinese face should not be judged by the portraits they send us on their screens and fans. Ordinary people, which is to say those that necessity exposes to hard work and the ardors of the sun, have a slightly weather-beaten complexion, particularly in the south of the empire. They take care to grow their facial hair and they never cut their nails. I saw Chinese merchants with nails one inch long and more.

*... that it seems they took the cloth to engage more in worldly practices than to escape them.*

**DECEMBER 1716**
AMOY [XIAMEN], CHINA

*Their manner of dress is very simple.* Rich people wear three silk tunics without linings: a short undertunic and two calf-length tunics that cross over. Their drawers are likewise of very fine silk. The Chinese find our own clothes surprising. They say they are too close-fitting and too short, and that a body thus encumbered cannot be free and available. I sincerely believe they are right.

Let me now tell you about Chinese women. Their lot can only arouse your compassion if you judge by your preconceived notions. Banned from civil society, forced to please an often disgusting and always jealous husband, they live in a perpetual state of reclusion and those joyful nuptials that bring so much freedom to our own ladies serve only to rob them of what little freedom remains to them. It is surprising that a people displaying so much passion for the female sex should have so little regard for women. In possessing women, Chinese men seek only to satisfy their senses; but even the most brutal love has its subtleties, and I am oftentimes astonished that Chinese men should heap such woe on the very women who make them happy.

**JANUARY 1717**
AMOY [XIAMEN], CHINA

*Chinese men have a peculiar taste* in women. They seek out a form of beauty that has nothing to do with Nature even if the latter produces in China the same bodies as it produces in Europe. When a girl reaches the age of three, her feet are broke and the toes folded back underneath the sole. Nitric acid is then applied to burn the flesh, and the foot is bound in place with several bandages until the bones are set. The pain lasts a lifetime, making walking difficult and giving women a most ungracious gait. I have yet to grasp the reasons behind this custom. I daresay even the menfolk cannot explain it. For my part I imagine that their forbears were more jealous still and introduced the custom to make women still more submissive and dependent on men. Some people say it was a punishment meted out to women who tried to shake off their husband's yoke. Whatever the case, they suffer their incommodity with a smile, and seek to make their foot as small as possible. They are right for therein lies their dowry and all their inheritance.

**FEBRUARY 1717**
AMOY [XIAMEN], CHINA

*And so, Sir, what think you now* of the Chinese? Do you see something barbaric in their customs, something repugnant to humanity? Nay, doubtless you do not. But Chinese they remain, and because thousands of nautical miles separate us, we regard them here in Europe as a savage people, lacking both manners and morals. For myself at least, I am cured of my prejudices and secretly congratulate myself on having freed myself from this self-deception.

Before I flew the nest, the French were my heroes, and I regarded all other European nations as barbarians by comparison. This fostered my self-esteem and as a Frenchman I could claim the same superiority that I attributed to my country. I was accustomed since childhood to hearing Indians called barbarians. Yet the first Indians I saw had all the appearance of men, and even of men who were more robust and inured to work than we are.

Bottom:
Chinese mode of transport.

Facing-page:
Interior of a Mandarin's house.

*You must know that Chinese people calling themselves Christian do so to curry favor with Europeans in the hope of preferential treatment.*

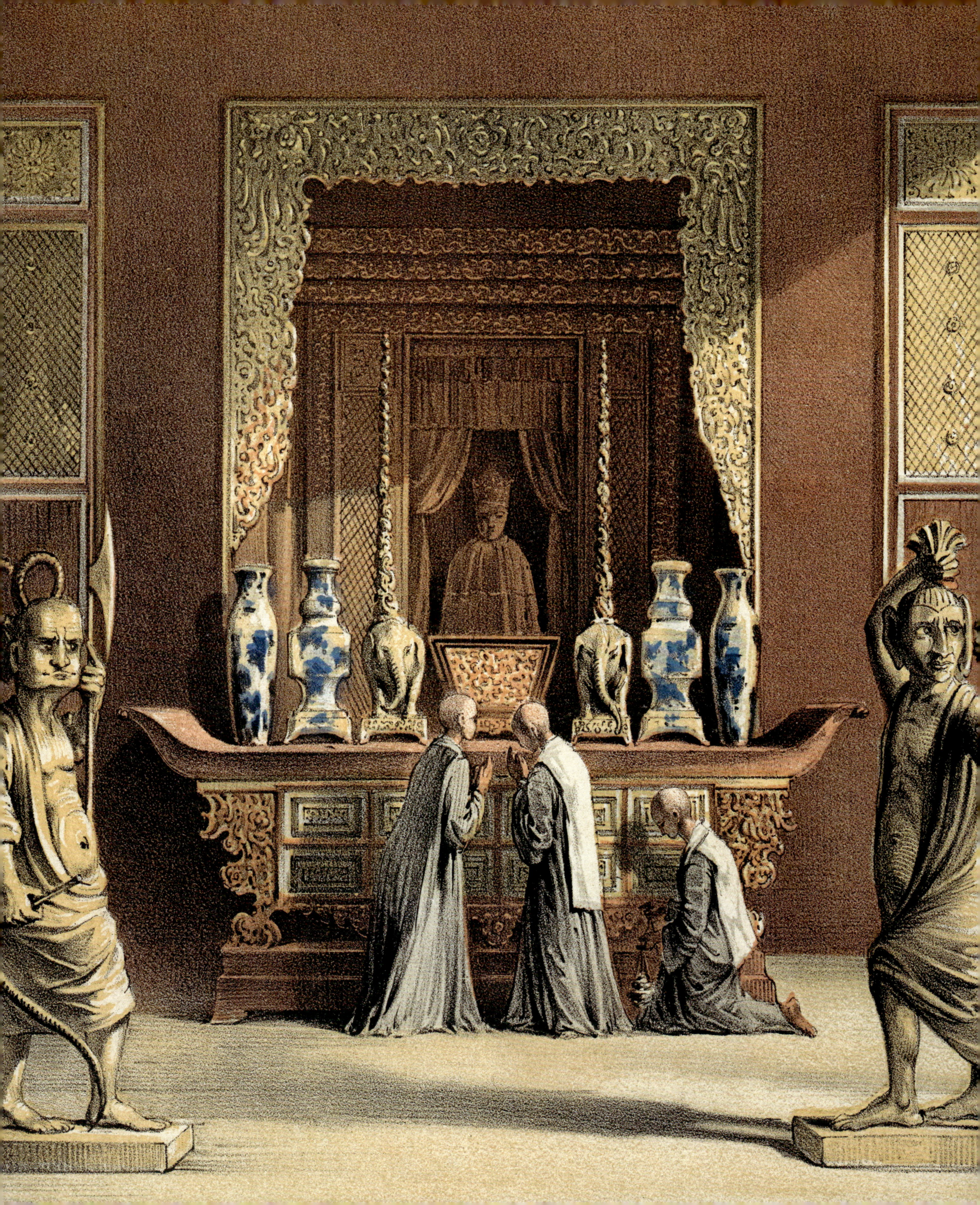

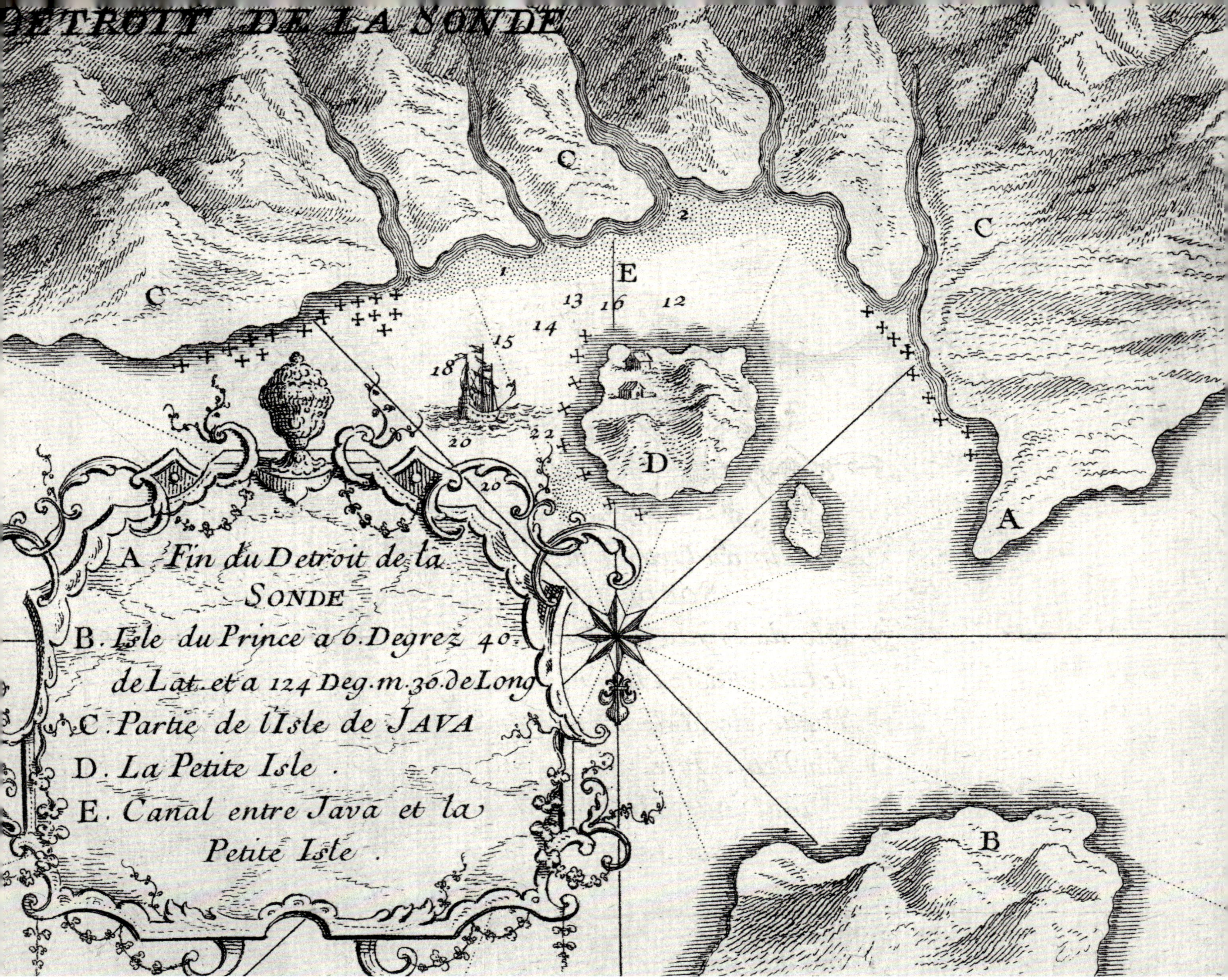


*Having seen off the Portuguese and the Spanish, the Dutch have made themselves masters and sole arbiters of the spice trade.*

**MARCH 1717**
SUMATRA, DUTCH EAST INDIES

*Sumatra Island appears* covered with trees all the way to the shore and irrigated by rivers. I will say nothing more about the island, Sir, which is all too well known today through the accounts rendered by the Dutch. We dropped anchor in the embouchure of the Sunda Strait. Being so close to Batavia, it would have been only natural that we should put into a port where abundance reigns and trade flourishes, making this port the richest and the handsomest in all of the Dutch East Indies. However, we durst not think of coming ashore there lest we should suffer some insult at the hands of the Dutch, who are very jealous in regard to trade. They can barely tolerate that other European people should use the Straits of Sunda. They have acquired such a redoubtable empire in these waters that they believe they can commit any act of their chusing with impunity. I am oftentimes astonished that the French, the English, the Spanish and the Portuguese have not yet sought to avenge the injuries inflicted on them by this ambitious nation. The Dutch have fortified themselves in such a manner that it is well-nigh impossible today to see them off without all the aforesaid powers joining forces to that end. And so it is more to our liking to seek succor from barbarians than to beg for help from such intractable folk.

### FEBRUARY 1718
AT SEA, OFF THE COAST OF BRAZIL

*The Lord has confounded our prudence* and overturned our projects. The whims of fortune once again conspire to separate us from our dear motherland. Brazil is a country of corrupt habits, where brazen-faced men know no shame and their womenfolk rival them in debauchery.

The convents serve as retreats for streetwalkers. I know not, Sir, whether to expound on their licentiousness; it seems preferable to remain silent about their crimes and there being no virtue in them worthy of praise, at least to conceal their vices to avoid scandalizing the Church by revealing the iniquities of her ministers.

### APRIL 1718
GALICIA, SPAIN

*We were in ignorance of the state* of European affairs. It was put about in Brazil that Spain was vastly re-arming and prudence demanded that we seek intelligence before entering any port. Our merchandize from China barred us from entering our own ports; our voyage to Peru gave the Spaniards a kind of right to confiscate our vessel.

The director arrived from Bayonne within a month and brought orders that we should proceed to Genoa. Never were orders so short-sighted as these, for bringing silk to Italy was tantamount to bringing water to the sea. But all the shipowners having gone bankrupt during our voyage, their interests in the vessel had been ceded to their creditors. I proceeded overland to France by way of Spain. I admit that I would prefer a thousand times to travel in the mountains of Peru where one finds more attractions and, dare I say, more humanity. I staid three days in Bayonne then arrived in Marseille where I took ship for Genoa.

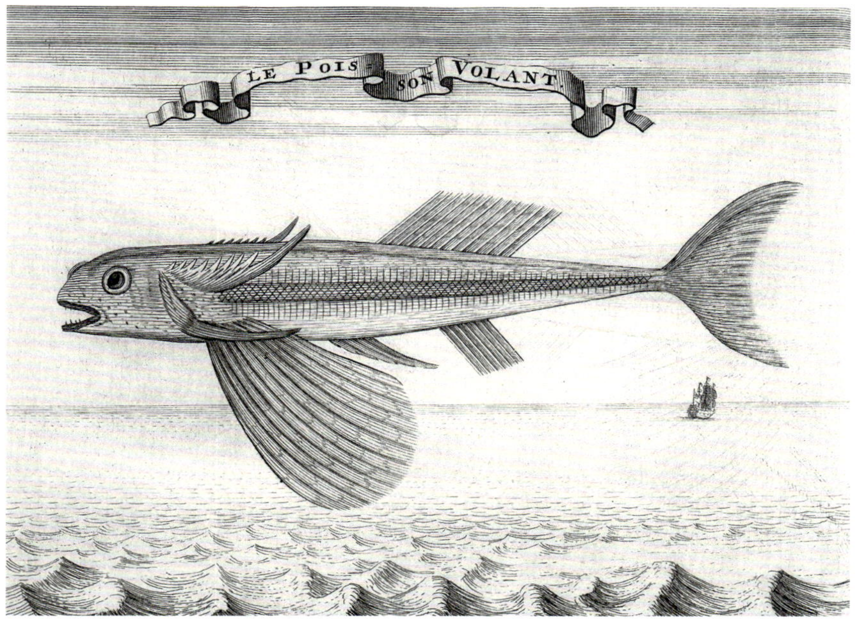


Top:
Porpoises and flying fish encountered on the crossing.

Facing-page:
Map of Sunda Strait in 1715.

*The clergy openly traffic in women and are known rather by the names of their mistresses than their own.*

### JULY 1718
GENOA, ITALY

*I have been here for nearly one month.* My business concluded, I may return to France, or I may remain in Italy for a few years. Fortune, which I traveled so far to find without success, may yet lie hidden here waiting for me.

There is no one doubts the utility of travel, even to the point of braving the sea if only the perils were less common; but the risks are so vast as to quash even the firmest of resolutions. The imagination conjures up nothing but innumerable tasks, an involuntary diet, interrupted slumbers, tempests, pitfalls, &c. But what is life without risk and the necessity of confidence? Our prejudices get the better of our courage and our resolve. A man born in the heart of Paris, whose longest trips are from Paris to the village of Saint Cloud, may tremble on the river Seine, but were he but born by the sea he would not tremble in the midst of the ocean. I do not advise such timid souls to under take long voyages. But I do dare to demand of them (as I demand it of you) that they listen most attentively to the d-tailed accounts rendered to them of the rarities and customs in those countries that they want the courage to visit themselves. It is only too common to see these idlers censure what they do not understand and what surpasses their sphere of understanding.

# Louis-Antoine
# de Bougainville

**1729–1811**

FOR LOUIS-ANTOINE DE BOUGAINVILLE THERE WAS NO QUESTION OF BEING LEFT BEHIND WHEN THE ENGLISH RESUMED THEIR MARITIME EXPLORATION AT THE END OF THE SEVEN YEARS' WAR. ON THE CONTRARY, WITH THE BLESSING OF KING LOUIS XV, HE WOULD UNDERTAKE A GREAT VOYAGE OF DISCOVERY THAT WOULD MAKE THIS OFFICER OF THE KING AND EXPERIENCED SAILOR THE FIRST TO SAIL AROUND THE WORLD ON BEHALF OF FRANCE. WITH A STOWAWAY WHAT'S MORE, AND A WOMAN AT THAT. LONG BEFORE THE CURRENT WAVE OF FEMINISM, THIS ENDEARING GENTLEMAN WAS ALREADY CHAMPIONING THE RIGHTS OF WOMEN.

Plate A, George Wolfgang Knorr, 1766.

# VOYAGE AROUND THE WORLD ABOARD THE *BOUDEUSE* AND THE *ÉTOILE*

**1766–1769**

**At the age of thirty-seven, Bougainville ventured beyond Cape Horn, observed the Patagonians, discovered Tahiti and, being a true Man of the Enlightenment, nurtured the myth of the noble savage. His portrayal of Tahiti as a paradise on Earth would strike the imaginations of readers for generations to come.**

Born in Paris on November 12, 1729, in the reign of King Louis XV, Bougainville was destined to become one of France's most famous maritime explorers. Before that, however, he was a brilliant mathematician, author of a treatise on integral calculus (*Traité de Calcul Integral)* at the age of 25, an advocate to the Parliament of Paris and First Secretary to the Embassy in London. In short, Louis-Antoine de Bougainville was destined for extraordinary things.

Just as the prestigious Royal Society opened its doors to him, the Seven Years' War broke out. Bougainville, ever keen to see action, left for Canada to help General Montcalm prevent the English from getting their hands on Quebec. After fighting in every battle, winning some and losing others, he was wounded and returned to France. Though he would later sing the praises of the "primitives" of Oceania, he found nothing to admire in the American Indians who slit his men's throats at night. The fact that the English were allied with the Indians also made the whole conflict horribly unbalanced, but King Louis XV remained deaf to his entreaties to send reinforcements. Only the Marquise de Pompadour sympathized, donating two hundred thousand piasters of her own money to the cause—generous but far from enough. Without additional funding, the French would lose the game but Bougainville returned to Quebec anyway and kept fighting to the bitter end. On September 13, 1759, Montcalm was killed on the Plains of Abraham in Quebec City and Bougainville was taken prisoner by the English. The following year, on September 7, 1760, he negotiated the French surrender in Montreal. And with that, France kissed goodbye to Quebec.

Though released on parole by his captors, Bougainville wanted revenge. At his own expense, he set out to conquer the Falkland Islands with a handful of Acadians who felt as he did. In 1764 he seized the Falklands in the name of France, only to return to Paris to be ordered by King Louis XV to return them forthwith to Spain. And just to add insult to injury, Bougainville was entrusted with the negotiations. The time had clearly come to put into action a plan he had been hatching for a very long time ... With everything in readiness, including a set of instructions to himself, he informed the King a few months before departure that, yes, he would hand over the Falklands to the Spanish but that he would continue from there to make a full circumnavigation of the globe, with the aim of taking possession of new territories for France. He chartered a frigate, the *Boudeuse*, and a cargo ship, the *Étoile*, surrounded himself with the best sailors and, most important, enlisted the services of renowned naturalist: Philibert Commerson.

After a false start and a severe storm, Bougainville eventually left Brest in December 1766, heading for Buenos Aires and Montevideo to negotiate the details of the Falklands handover with the Spanish. On April 1, 1767, the French flag was raised over the islands and the assembled Frenchmen were told, in a letter from the King read by Bougainville, that they could choose either to remain in the Falklands under Spanish rule or return to France. His mission accomplished, Bougainville could now take the *Boudeuse* around Cape Horn and thence across the Atlantic. But his ship took on water and the *Étoile* fell foul of the Spanish within moments of joining the expedition. For Bougainville, there was nothing for it but to remain in Brazil while his boats were repaired—several months that were put to good use by botanist Commerson and his assistant Jean Baré, who discovered an unknown flower they named Bougainvillea.

A year after leaving France, in December 1767, the *Boudeuse* reached the Straits of Magellan. Bougainville encountered the Patagonians of Tierra del Fuego and finally sailed into the Pacific Ocean in search of Easter Island ... which he didn't find. But on March 23, the Tuamotu Islands were sighted, followed by Tahiti on April 4. Bougainville stayed only nine days, but what a stay! The whole of Paris would revel in his idyllic descriptions of what he called Nouvelle-Cythère, with its "noble savages" and its "wild women", and from which he brought back a young Tahitian volunteer, Aotourou. It was in Tahiti that Jean Baré would turn out to be Jeanne Barret, a young woman of twenty-six who was thus the first woman to circumnavigate the globe. Commerson's "housekeeper" for several years, she had disguised herself as a man to accompany him on this trip, women being strictly forbidden on board royal ships. When Bougainville found out, he agreed to let her stay and it was she who replaced Commerson when he developed an abscess on his leg. Jeanne scoured the island collecting botanical samples (some of her illustrations are preserved in the Jardin des Plantes in Paris) and mysterious herbs to make antiscorbutic infusions. In addition to flora, she had a passion for insects and fish, which she dried on board after eviscerating them, inventing taxidermy in the process. In 1773 she returned to France after Commerson's death and inherited his estate. At Bougainville's request, King Louis XV dubbed her a *femme extraordinaire*, making her famous beyond France's borders. When later she married, she insisted on keeping her own name and property, something no woman had ever done before.

Meanwhile, in 1768, the *Boudeuse* and the *Étoile* continued their westward journey. On behalf of France, Bougainville claimed the Navigator Islands [Samoa], the Great Cyclades [Vanuatu], and discovered a beautiful island in New Guinea [Papua New Guinea] that he named Bougainville. The ships then sailed to unknown climes, enduring hurricanes and finding a haven of peace in New Brittany in Port Praslin Bay; and finally, after an essential stopover in Batavia [Jakarta], rounded the Cape of Good Hope and headed for France.

On their return in March 1769, Aotourou spent a year in Paris with Bougainville while the latter wrote up his voyage. Published two years later, the work spoke to the imagination of French and English readers alike. Philosophers likewise thrilled to Bougainville's tales of discovery, Diderot in particular, writing his *Supplément au Voyage de Bougainville* in 1772. Bougainville for his part, still nursing his grudge against the English, now had a new travel project in mind: to head for the North Pole and discover the Northwest Passage before his enemies across the Channel did. And just as before, when everything was ready he presented his project to the King ... who refused. A big mistake, because twenty years later, the English would be calling the shots in the Arctic. King Louis XVI, in his turn, preferred to finance Jean-François de Galaup de Lapérouse, and in 1775 dispatched Bougainville to fight in America in the War of Independence. And just as before, he was wounded and taken prisoner.

In 1781 he married and became a father, to a son named Hyacinthe. In the French Revolution, despite his loyalty to the King, he continued to serve, never mind that he was arrested in the Terror. Napoleon rehabilitated him with full honors in 1799, appointing him a Senator, then a Grand Officer of the Legion d'Honneur, Count of the Empire and finally President of the Council of War. Bougainville meanwhile developed a passion for plants and flowers, especially roses. Stricken with dysentery, he died in Paris on August 31, 1811 at the age of eighty-one. He is buried in the Pantheon in Paris.

► The Patagonians of Tierra del Fuego.

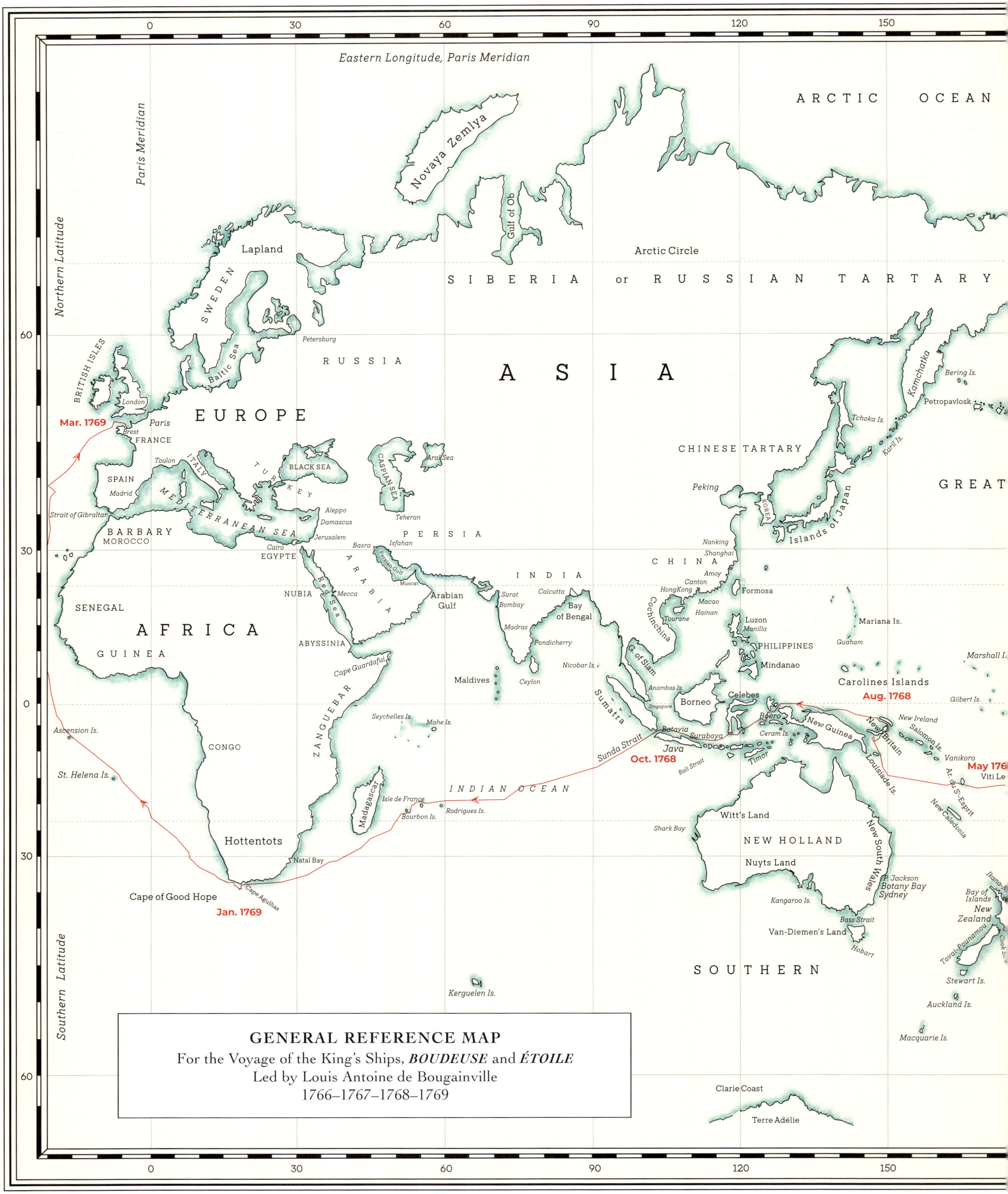


**GENERAL REFERENCE MAP**
For the Voyage of the King's Ships, ***BOUDEUSE*** and ***ÉTOILE***
Led by Louis Antoine de Bougainville
1766–1767–1768–1769

150
120
90
60
30
0
Western Longitude, Paris Meridian
Paris Meridian
Melville Is.
Lancaster Strait
Baffin
Bay
Davis Strait
GREENLAND
Iceland
Bering Strait
Hudson
Bay
Hudson Strait
Cook's River
Port des Français
Lituya Bay
Kodiak Is.
Sitka
Alaska
Cape Farewell
BRITISH ISLES
SWEDEN
Baltic Sea
Queen Charlotte's Is.
LABRADOR
CANADA
NORTHERN
London
Vancouver Is.
Aleutian Islands
Nootka
St. Lawrence River
Newfoundland
Nov. 1766
Paris
Brest
FRANCE
NORTH
AMERICA
Quebec
Acadia
UNITED STATES
ATLANTIC
Toulon
ITALY
SPAIN
Madrid
NORTHERN OCEAN
San Francisco
New York
Washington
Azores Is.
Strait of Gibraltar
MEDITERRANEAN SEA
Monterey
Bermuda Is.
New Orleans
San Antonio
California
MEXICO
Gulf
of Mexico
Florida
Madeira Is.
BARBARY
MOROCCO
Canaries Is.
OCEAN
Tropic of Cancer
Mazatlàn
Lucayan Is.
Sandwich Is.
Hawaii Is.
San José
Mexico City
Santo Domingo
SENEGAL
Caribbean Sea
Jamaica
Acapulco
Guadeloupe
Martinique
Cape Verde
AFRICA
GUINEA
Panama
COLOMBIA
Cayenne
GUYANA
Equinoctial Line
Gallapagos Is.
SOUTH
AMERICA
Paita
PERU
Lima
Marquesas de Mendoza
Ascension Is.
CONGO
Navigators Is.
Maouna Is.
Tahiti
Society Is.
Friendly Is.
Traitors Is.
Apr. 1768
Mangareva
Pitcairn Is.
Tropic of Capricorn
Pisco
Arica
BRAZIL
Bahia
Sep. 1767
St Helena Is.
Rio Janeiro
C. Frio
Easter Is.
Coquimbo
CHILE
St Catherine's Is.
SOUTH
Valparaiso
Feb. 1767
Montevideo
Buenos Aires
Rio de la Plata
Concepción
Cape of Good Hope
ATLANTIC
Chatham Is.
OCEAN
PATAGONIA
June 1767
Strait of Magellan
Falkland Is.
OCEAN
Tierra del Fuego
Is. de Los Estados
Dec. 1767
Cape Horn
Sandwich Is.
Antarctic Circle

An epistle to His Majesty

Sire,

The voyage I am about to recount is the first of its kind ever undertaken by a Frenchman and executed in His Majesty's vessels. It has already introduced the world to the face of the Earth and determined its dimensions. Those of your subjects to whom this important discovery was entrusted were selected from among the most illustrious French savants. The intrepid Magellan, sailing under the auspices of a King who knew something about men, served to open the way and overcame the difficulties; despite the fate that deprived him of the pleasure of returning in his original vessel to Seville from whence he had departed, nothing could rob him of the glory of being the first to circumnavigate the Earth. Heartened by his example, English and Dutch seafarers discovered new lands and enriched Europe by the enlightenment they brought. But such primacy of primogeniture in matters of discovery does not deprive French seafarers of their rightful claim to a share in the glory attached to these brilliant but arduous enterprises. Several regions of America were found by courageous subjects of Your Majesty's kingly ancestors. Since that time, various reasons as much internal as external in nature appear to have suspended the taste and activity of the nation in this regard. Your Majesty wished to take advantage of the privilege of peace to uncover geographical knowledge useful to humanity. Under your auspices, Sire, we embarked on a career where trials of all sorts awaited us at every turn; patience and zeal were not wanting on our part. The story of those efforts is what I dare to present to Your Majesty here; your approbation will ensure its success.

I AM, WITH THE MOST PROFOUND RESPECT,
YOUR MAJESTY'S VERY HUMBLE AND OBEDIENT SERVANT,
DE BOUGAINVILLE

**DECEMBER 1766**
**BREST, FRANCE**

*By the month of February 1764,* France had established its first settlement on the Falkland Islands. Two years later, the islands were claimed by Spain and its rights being recognized by the King, I received orders to travel thither, transfer the colony to Spanish authority and thence proceed to the East Indies, sailing across the South Sea between the Tropics. For this undertaking, I was given command of the frigate *Boudeuse*, armed with twenty-six twelve-pounder cannon, and I was to be joined at the Falkland Islands by the fluyt *Étoile* bringing the provisions necessary for our long navigation and destined to follow me throughout the remainder of our campaign.

My general staff comprised 11 officers and 3 volunteers, seconded by a crew of 203 ordinary seamen, marine officers, soldiers, cabin boys and servants; the campaign was led by Monsieur the Prince of Nassau-Siegen with the King's consent. On the fifth of the month, at four o'clock in the afternoon, the middle of Ouessant Island [Ushant Island] stood in the north-northeast quarter of the compass, at a distance of some five and a half leagues; thence did I establish our point of departure on the *Neptune français*, a reference that I have ever used in the course of sea voyages.

**JANUARY 1767**
**RIO DE LA PLATA, ARGENTINA**

*On the thirty-first day of the month,* at eleven o'clock in the morning, we dropped anchor in the Bay of Montevideo, in four fathoms of water over a seabed of soft, black mud. The two Spanish frigates destined to take possession of the Falkland Islands had been at anchor here for one month. Their commander, captain Don Philippe Ruis Puente, had been appointed governor of these islands. We proceeded together to Buenos Aires to work in concert with Governor General Don Francisco Bucarelli on the measures required to cede the colony that I was to hand over to the Spanish.

Monsieur the Prince of Nassau and myself had traveled upriver from Buenos Aires in a schooner, but because returning the same way would have required sailing upwind, we crossed the river above the colony of Saint-Sacrement and traveled overland for the remainder of the journey to Montevideo where our frigate had remained.

**APRIL 1767**
**FALKLAND ISLANDS**

*On the first day of April, I handed over* our colony to the Spanish who took possession by raising the Royal Standard of Spain, this flag being honored by a twenty-one-gun salute fired on land and from the ships. I had read the French inhabitants of this nascent colony a letter from His Majesty consenting to their remaining there under the dominion of a Catholic king.

A few families took advantage of this opportunity; the others, together with the colonial authorities, were embarked on the Spanish frigates, which set sail for Montevideo on the morning of April 27. For my part, I was compelled to remain in the Falklands to await the *Étoile,* without which I could not continue my journey.

## MAY 1767
### SOUTH AMERICA

*The Indians were so* cravenly submissive to their priests that they permitted themselves to be flogged in punishment for public sins, and actively sought punishment for private sins.

The priest would rise at five o'clock in the morning, say mass, have breakfast, pray in accordance with the liturgy of the hours, work with his ministers then visit the seminary, schools and workshops. If he went out, it was on horseback in the company of his entire retinue. He dined at eleven o'clock, alone or with his curate, conversed until midday and dozed until two o'clock. He remained shut indoors until praying the rosary, after which there would be conversation until seven o'clock at night; then did he take supper. By eight o'clock he would normally be abed.

The people by contrast would be put to work at eight o'clock in the morning, either in the fields or in the workshops, and the *corregidores* saw to it that this routine was strictly adhered to. The women spun cotton. At half past five in the evening every one gathered to pray the rosary and to kiss the hand of the priest. Such deadly tedium provides a sufficient explanation for what we were told: that the Indians quit this life with no regrets and die having never lived. Indians who chanced to fall sick rarely recovered. When asked whether they were afflicted by the prospect of death, they said that they were not, and they said it with conviction. It must come as no surprise therefore that when the Spaniards

**Bottom:** Map of the Falkland Islands in 1760.

**Pages 42-43:** Houses on the Falkland Islands in 1767.

*By converting the Indians to Catholicism, the Spaniards civilized wild men and made themselves masters of a vast country.*

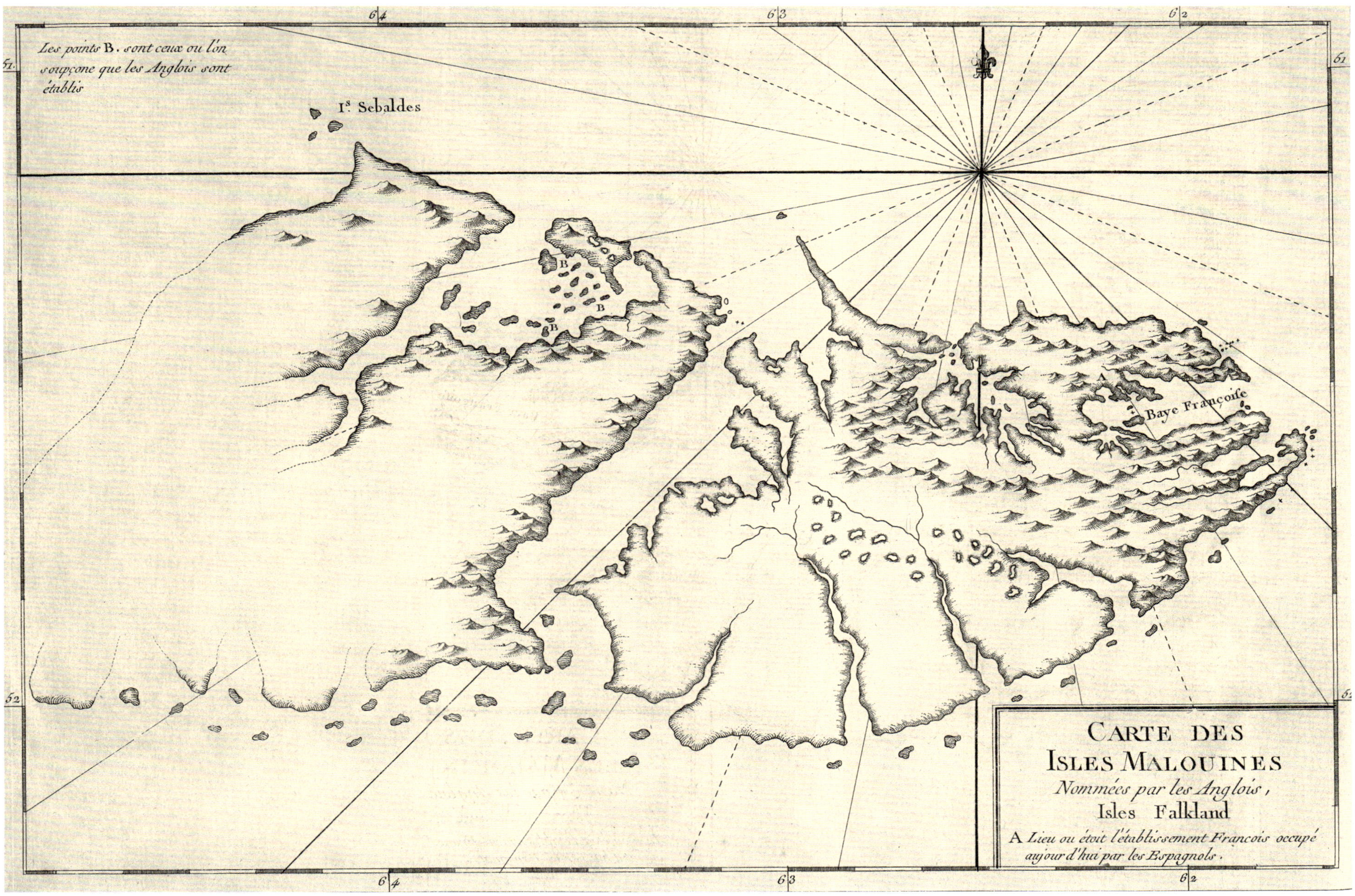

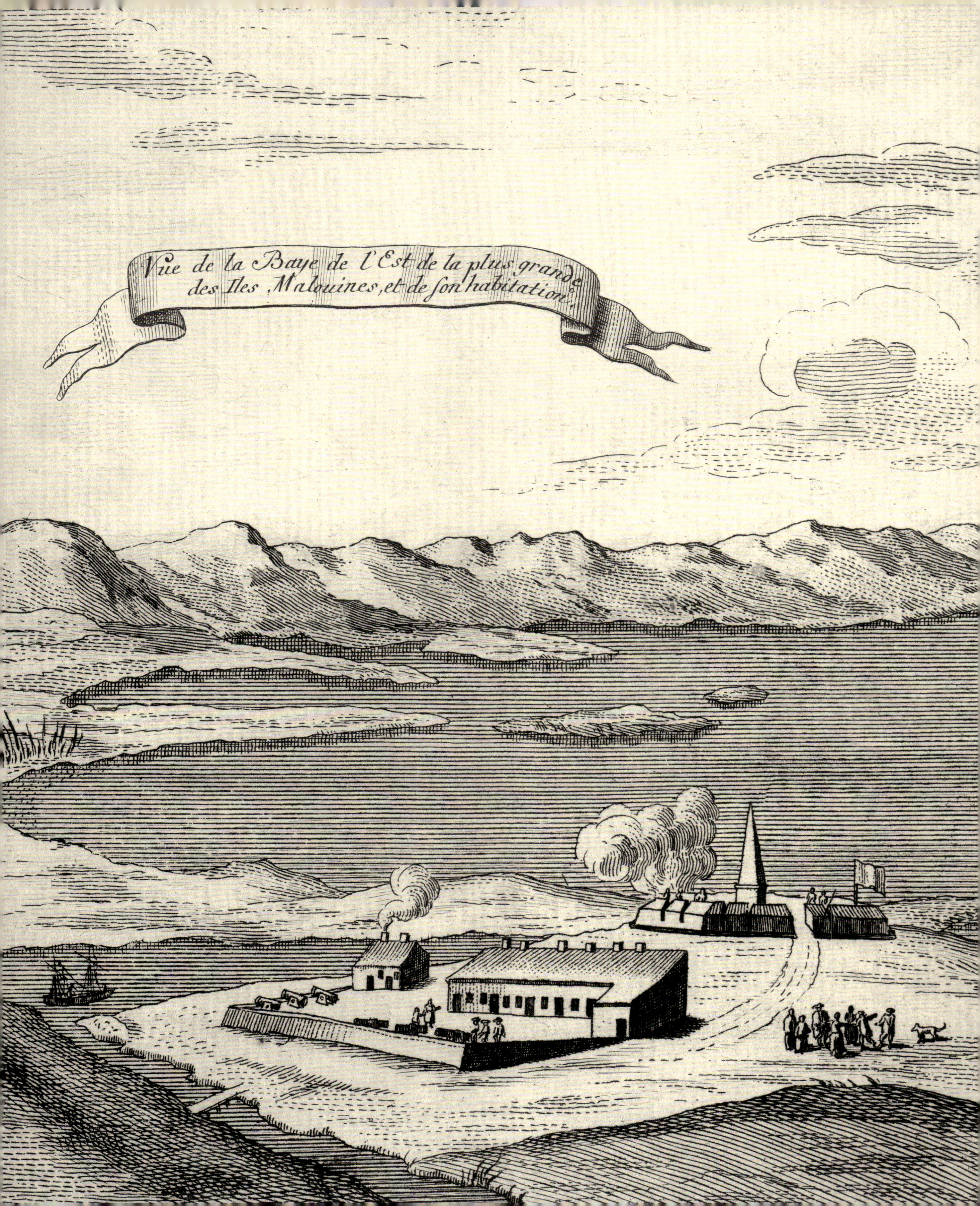
Vue de la Baye de l'Est de la plus grande
des Iles Malouines, et de son habitation.

entered a mission, this great people, subjected to monastic rules, evinced the greatest desire to force its closure. Furthermore the Indians as the Jesuits saw them were men of a kind who would never rise above the intelligence of children; the life they led deprived these overgrown children of the joyfulness of little ones.

**DECEMBER 1767**
TIERRA DEL FUEGO

*This morning, the Patagonians,* who throughout the night had kept a fire burning at the bottom of Possession Bay raised a white flag on a hill and we responded by raising our ensign. These are without doubt the Patagonians seen by the *Étoile* in the month of June 1766 and the flag they raised was the one given to them by Monsieur Denys de Saint-Simon as a gesture of alliance. The care they have taken to preserve it betokens a gentle people, faithful to their word or at least grateful for the presents that are offered to them.

We could also clearly distinguish, when we were in the gully, some twenty men on Tierra del Fuego, wearing animal skins and running at full tilt to follow our course along the coast. Now and then they even appeared to be gesturing at us, as if beckoning us to approach them. According to the reports received from the Spaniards, the people that inhabit this part of the Land of Fire have none of the cruel ways intrinsic to most aboriginals. These are robust, well-fed Americans with taut sinews and firm, solid flesh. Their countenance is neither hard nor displeasing and is often quite pretty. Their face is round and somewhat flat; their eyes are bright and their teeth extremely white, though likely too big for Parisian tastes; their hair is long and black and tied in a topknot.

*What struck me as prodigious were their enormous stature, large heads and dense limbs.*

**APRIL 1768**
**TAHITI, FRENCH POLYNESIA**

*From the fires that we rejoiced to see* burning all along the coast, we knew that the island was inhabited. As we ran under full sail toward the shore, we noticed a pirogue approaching from the open sea and advancing toward the coast using its sail and paddles. It passed us from astern and joined an infinity of other canoes that hailed from all parts of the island and raced before us. One of them preceded the others and was propelled by twelve naked men who presented us with banana tree fronds, and from their gestures it was clear that these were meant as an olive branch. We responded with such signs of friendship as came to mind; then did they draw up alongside the vessel and one of them, remarkable for his enormous head of hair that stood up in spikes, extended together with his token of peace a piglet and a bunch of bananas. We accepted his gift, which he attached to a cord that we threw to him. We gave him bonnets and handkerchiefs, presents by way of introduction that were intended to betoken our alliance with these people.

As we approached the shore, we found ourselves increasingly surrounded by islanders: pirogues full of women whose pleasing features might rival those of the greater number of their European counterparts, and with bodies quite unmatched for beauty. They began by dallying with us from their boats, but with a naiveté that betrayed their awkwardness; either Nature has everywhere embellished the fairer sex with an ingenuous timidity, or even in these countries where the Golden Age of Love still reigns supreme, women appear not to want what they desire the most. The menfolk were simpler and quickly expressed themselves more freely. They pressed us to choose a woman, follow her ashore and become acquainted with her in a fashion that their unequivocal gestures made only too explicit. In the face of such a spectacle, what hope was there that four hundred young French seamen who had not seen a woman for six months could be kept working?

**APRIL 1768**
**TAHITI**

*I suggested to the chief of the district* in which we were staying that we make a French-style garden and sow different seeds, a proposal that was received with joy. A short while later, Ereti saw to the fencing and the clearing of the plot of land selected by our gardeners. I saw to the digging of the soil, our garden tools drawing much admiration. We have reason to believe that these plantings will be well tended, for the people here appear to love husbandry and I believe could easily be trained to make the most of what is assuredly the most fertile soil in the universe.

The deed of acquisition was engraved on an oak plank, which I buried near the shed together with a tightly sealed bottle containing the names of the officers on the two vessels. I did the same for all the lands we discovered over the course of our journey.

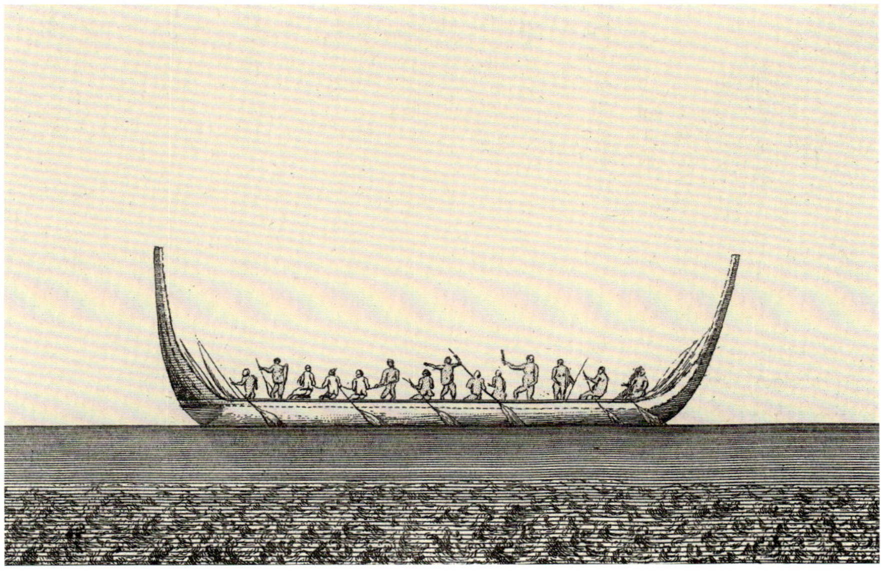

Facing-page:
*Bougainville abordant à Taïti* (Bougainville landing on Tahiti).

Top:
*Canot de l'Isle Taiti à la voile* (Tahitian sailing canoe).

Above:
*Canot sauvage de l'Isle Choiseuil* (Choiseuil Island, native canoe).

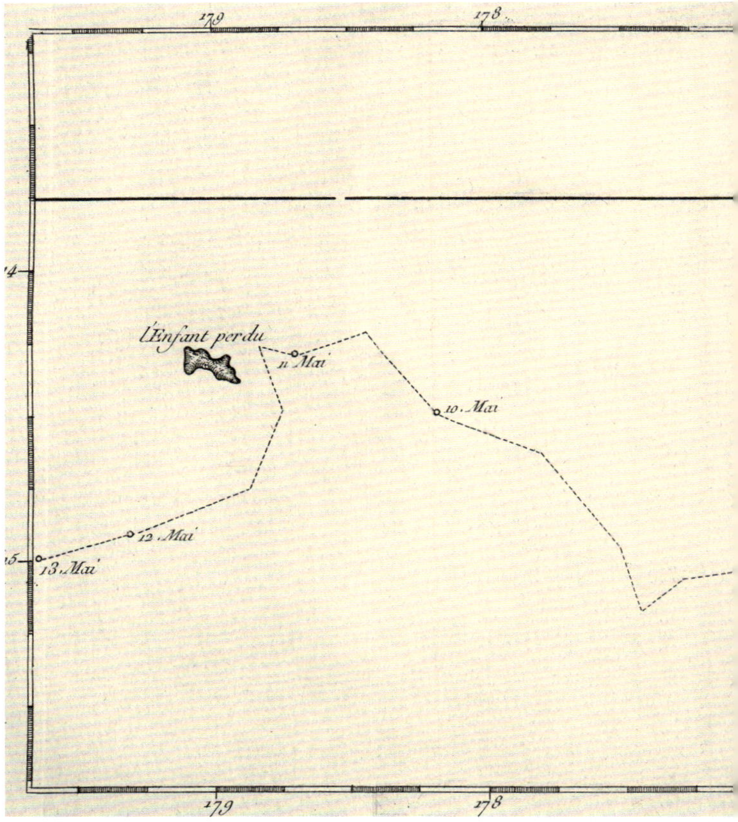


*The fact is that it is exceeding common in capital cities to find people who ask questions ...*

**APRIL 1768**
TAHITI

*A short while later, Ereti's giant pirogue* drew up alongside us loaded with all manner of refreshments. Also in the boat were Ereti's wives, together with the same islander who had come aboard the *Étoile* on the day we landed. Ereti took him by the hand and presented him to me, the while giving me to understand that Aotourou, for so was he called, wished to follow us, and entreating my consent thereto. He then presented him to all the officers, each one in his turn, saying that he was entrusting his friend to his friends and recommending him in all earnestness. Thus did we take our leave of these good people and I was not the less surprized by the chagrin they expressed at our departure than I was by the affectionate trust they shewed on our arrival.

**APRIL 1768**
TAHITI

*I shall end this chapter* by attempting to justify myself, for I find myself obliged to use this term, to justify myself I say, for having taken advantage of Aotourou's good nature in making him undertake a voyage that he assuredly did not expect to last so long. I spared neither money nor care to render his stay in Paris agreeable and useful. He remained there eleven months, during which time he evinced no boredom. There was a great keenness to see him, a gawking curiosity that served only to give false ideas to persons born to scoff—men who have never set foot outside the capital, of shallow culture and mistaken beliefs of all kinds, who judge solely by their own prejudices yet proffer rigid opinions that allow of no inflection. "Really!", some exclaimed, "you mean to say that in this man's country they speak neither French, nor English nor Spanish?" What answer could I give? Yet it was not the surprising nature of such a question that rendered me speechless. I was used to such interrogation, knowing as I did that on my arrival several of those very people who are allegedly erudite had argued that I had not circumnavigated the world since I had not been to China. Others, trenchant critics to a man, formed and spread a decidedly threadbare idea of our poor islander on the grounds that after a two-year stay in France he could scarcely speak a word of French. Do we not see on an everyday basis, said they, Italians, Englishmen and Germans, for whom a one year's stay in Paris suffices to learn French? I should perhaps have answered that the Tahitian, being indolent in mind and body and possessing only those ideas relevant to a simple and narrow community, would have to create, so to speak, a world of new ideas in order to arrive at the correct choice of French word. But making this point would have taken a few minutes and I have invariably noticed that when plied with questions as I was, by the time I was disposed to satisfy my esteemed interlocutor with a reply he would no longer be at hand to hear it.

However, though Aotourou tortured what few French words he could sputter, he went out every day on his own, and scoured the city without ever losing

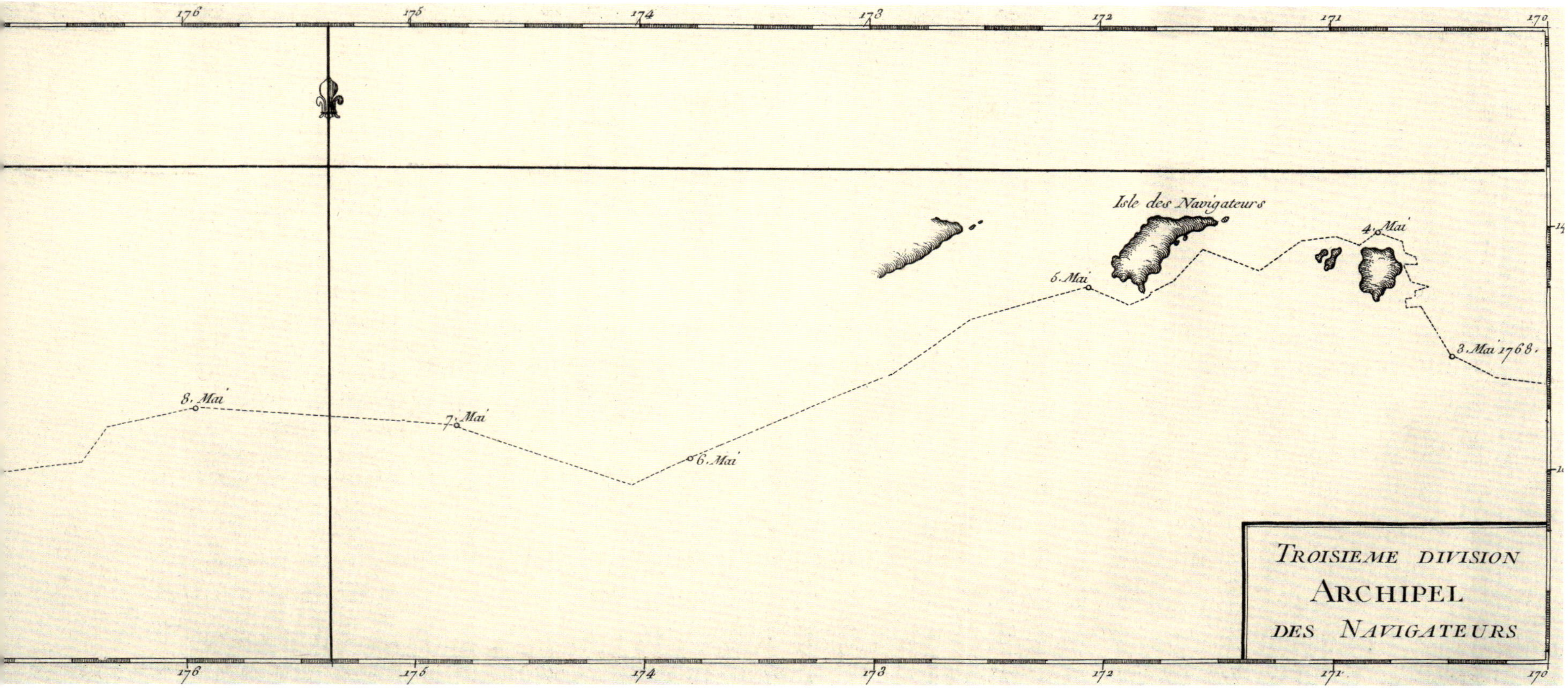


*... not as curious souls seeking instruction but as judges preparing to pass sentence.*

his way. The only one of our spectacles to his liking was the opera, for he had a passion for dancing. He was perfectly informed of the dates of performances, which he attended alone, paying at the door like every one else, and his favorite place was in the corridor. Out of the great number of people who desired to see him, he always remarked upon those that were kind to him, with a grateful heart that never forgot them.

He departed in the month of March 1770, embarking at La Rochelle on the vessel *Brisson*, which was to take him to the Île de France [Mauritius]. The minister ordered the governor and intendant of the Île de France to convey Aotourou back from thence to his home island.

**MAY 1768**
**TAHITI**

*For some time now,* there had been a rumor circulating aboard both vessels that Monsieur de Commerçon's man servant [Philibert Commerson], Baré by name, was a woman. All the signs pointed to the rumor being well-founded, not least Bare's build, pitch of voice, absence of beard growth and the pains he took to avoid changing his linen or relieving himself in the company of others. That being said, how could this indefatigable Baré possibly be seen as a woman? This, the seasoned botanist whom we had seen follow his master in his naturalist forays across the snow fields and mountain glaciers of the Magellan Straits? This, the person who would carry all the supplies necessary for those arduous walks—food, weapons and plant notebooks—with a strength and courage that saw him dubbed "my beast of burden" by his mentor? It took an event in Tahiti for the suspicion to be confirmed. Monsieur Commerson went ashore to collect plants, Baré following behind him with the notebooks under one arm. Scarcely had he set foot on dry land than he was surrounded by Tahitians crying that this was a woman and wishing to welcome her in the traditional Tahitian manner.

When I went aboard the *Étoile*, a tearful Baré admitted that she was a girl. She told me that in Rochefort she had deceived her master by turning up dressed like a man just as they were embarking; that she had formerly worked as a footman for a Genevan in Paris; and that, born an orphan in Burgundy, she had decided to disguise herself as a man after losing a court case that had left her penniless. She added that she had embarked in the knowledge that the ship would be circumnavigating the world, an adventure that piqued her curiosity. She would be the first, and it is fair to say that she always comported herself on board with flawless wisdom. Forsaking the tranquil occupations of her sex, she dared to brave the fatigues, dangers and other exertions that may reasonably be expected on an excursion of this nature.

**Top, left:**
*Canot des îles des navigateurs à la voile* (Navigators' Islands, sailing canoe).

**Top, right:**
Route taken by the *Boudeuse* through the Navigators' Islands.

## MAY 1768
### NAVIGATORS ISLANDS [SAMOA]

*On May 3, almost at daybreak,* we discovered a new land in the northwest, within ten or twelve leagues of our position. I had the rudder set to steer toward the windward side of the northern tip of this land with the intention of reconnoitering. On the morning of May 5 we perceived this new land to be a beautiful island of which we had only glimpsed the peaks the previous day. It was interspersed with mountains and vast plains covered with coconut palms and an infinity of other trees. By ten o'clock in the morning we had taken cognizance of another land in the west but our view was subsequently obscured by clouds. Indeed, to judge from the pirogues that we had seen sailing out to sea, there were likely other islands in these parts, suggesting that these lands formed a chain extending along the same meridian.

*These lands formed a chain, we called them the Archipel des Navigateurs (Navigators Islands.)*

Top:
*Vue de l'isle d'Oubi depuis la baye d'eau douce dans l'île Batchian* (view of Obi Island from the freshwater bay on Batchian Island).

Above, center and bottom:
*Indiens de Moa, Jamna et autres isles adjacentes avec une de leurs pirogues* (Indians from Moa, Jamna and other neighboring islands, with one of their canoes); *Papous de la Nouvelle Guinée avec une de leurs pirogues* (Papuans in one of their canoes).

## SEPTEMBER 1768
### BURU, MOLUCCAS, DUTCH EAST INDIES, INDONESIA

*Scarcely had we dropped anchor* than two unarmed Dutch soldiers, one of whom spoke French, came aboard to inquire on behalf of the factory director what reasons had brought me to this port, knowing as we surely did that its entrance was exclusively reserved for the vessels of the Dutch East India Company. I sent them back with one of my officers to inform the director that our need for provisions had obliged us to enter the first port we encountered, without regard to the treaties that forbid foreign vessels from putting into the ports of the Moluccas, and that we would leave directly he had provided us with the succor we so urgently required. The two soldiers returned a short while later to inform me of an order signed by the governor of Amboine (Ambon), to whom the resident of Buru is directly answerable, by virtue of which the latter is expressly forbidden to allow foreign vessels into his port. The resident meanwhile bade me furnish him with a written declaration of the reasons for my putting into his port, in order that he might submit this as justification to his superior, to whom he would convey the line of conduct incumbent upon him on receiving us here. It was a reasonable request to which I acceded. From that moment forward, we had no further difficulties. The resident, in compliance with the policy of the Dutch East India Company, put on a brave face and offered us what he could as liberally as if he had been master in his own house.

Facing-page:
View from the harbor of Bourou Island.

Pages 50-51:
View of the port of Zamboanga, with Fort Pilar in the background, Philippines, 1776.

Vüe du Fort

P. Sonnerat del.

A La Maison du Gouverneur
B l'Eglise qui est dans la Citadelle
C Tour pour les Decouvertes
D Tour de l'Hopital
E Le Ruisseau qui se jette à la Mer

Avril Sculp.

F Debarquadaire
G La Corvette le Necessaire
H Une Galere Espagnole
K Une Somme Chinoise
M La Flutte l'Isle de France

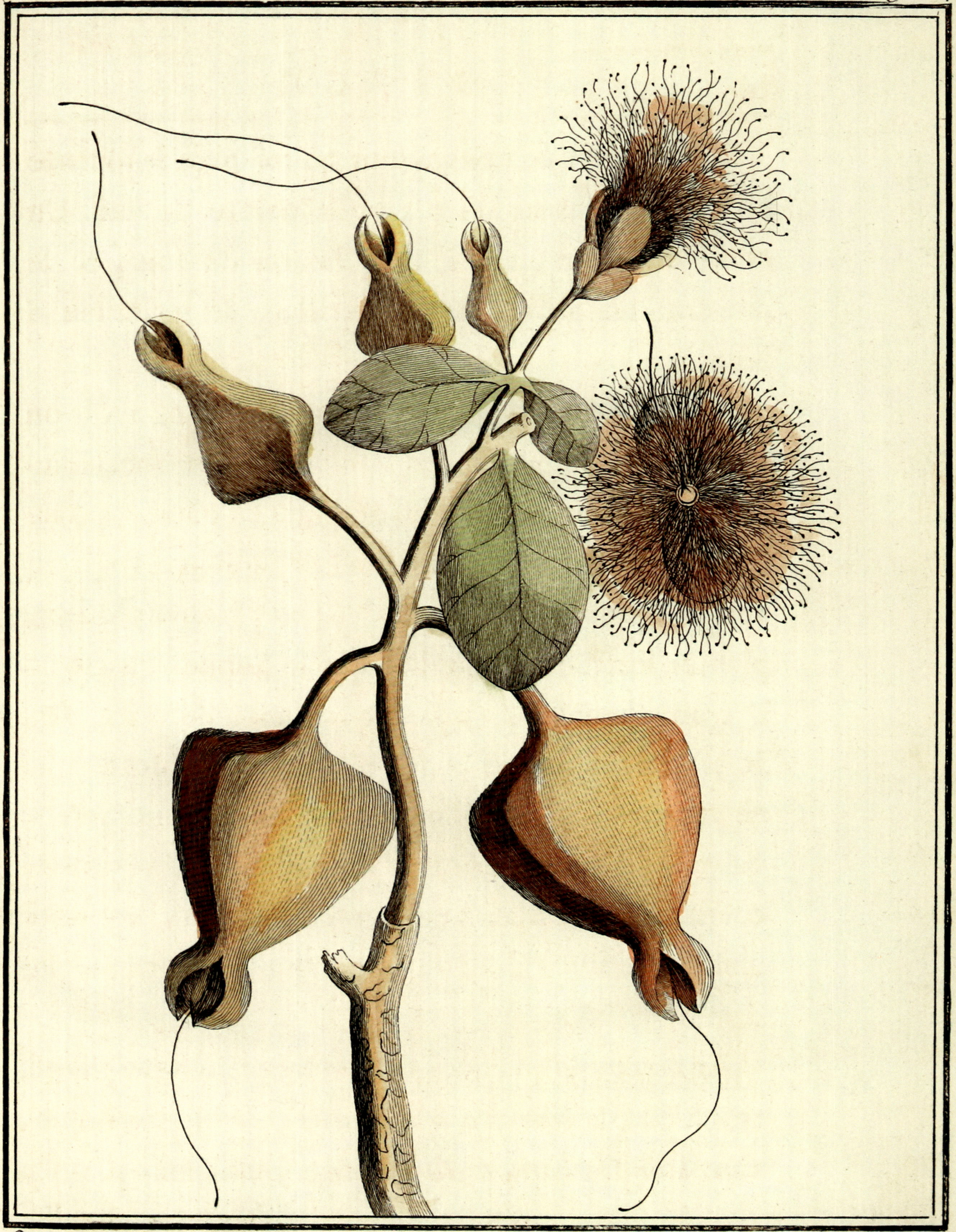

P. Sonnerat, del...

therese Martinet Sculpsit

Le Bonet quarré ou la Commerçona.

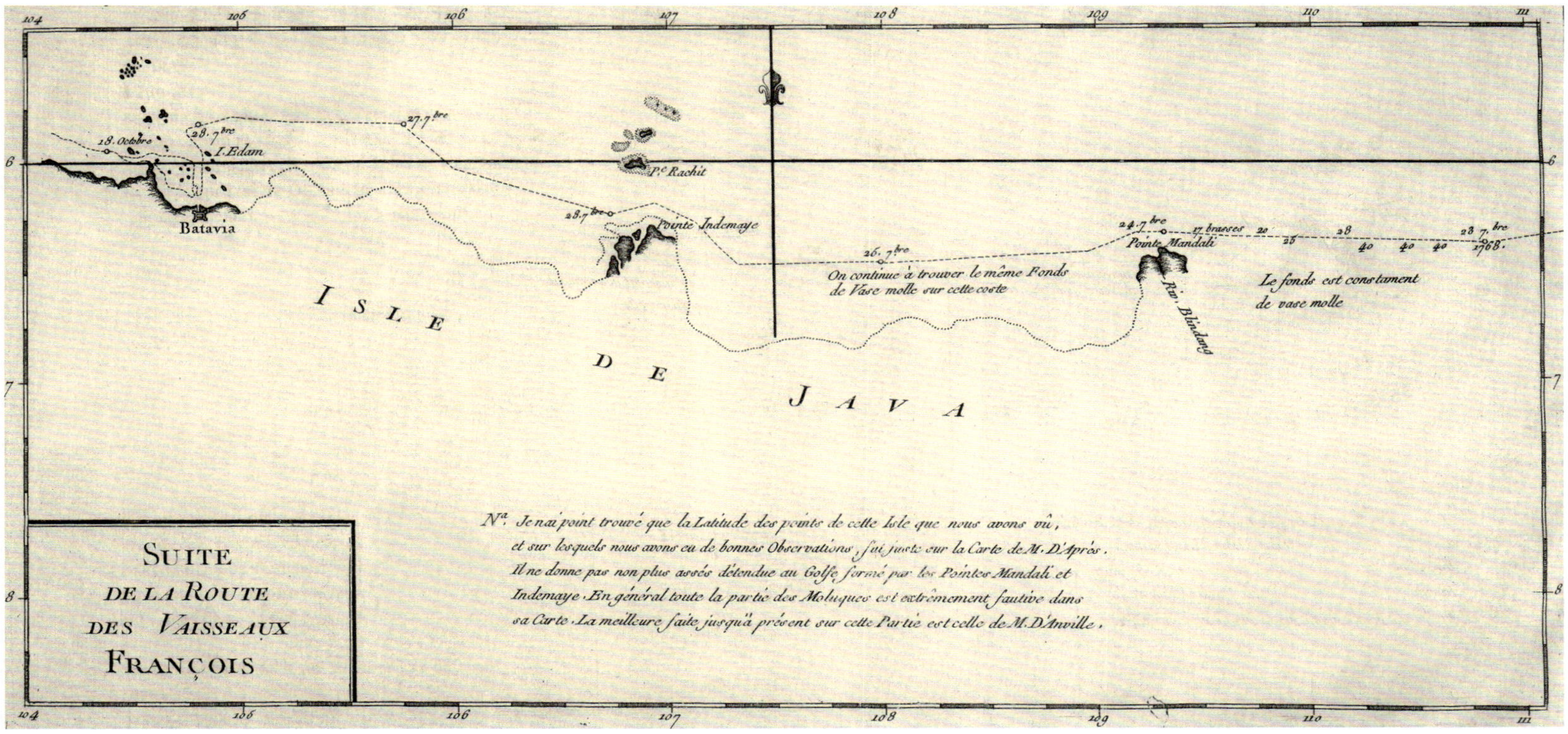


*An inhabitant of Batavia was flogged and banished to a deserted island for having shown an Englishman a map of the Moluccas.*

**OCTOBER 1768**
BATAVIA [JAKARTA] DUTCH EAST INDIES

*We never tired of* promenading in the environs of Batavia. Even the most seasoned city dweller would be impressed by the magnificence of the city outskirts. They encompass a wealth of grand houses and splendid gardens, all maintained with that good taste and attention to cleanliness that strikes the visitor to Dutch territories. I have no hesitation in saying that they surpass in beauty and in opulence those of our largest French cities and come close to the magnificence of the outskirts of Paris.

Equally striking is the luxury that obtains in Batavia: nearly all the house interiors reflect a sumptuousness and discerning taste that testify to the wealth of the residents.

**NOVEMBER 1768**
BATAVIA [JAKARTA] DUTCH EAST INDIES

*After the Dutch* had seen off the Spanish and Portuguese, a victory owed to clever maneuvering, courage and patience, they fully appreciated that to establish a monopoly of the spice trade it was not enough merely to have expelled their European competitors from the Moluccas. The great number of islands rendered their safekeeping almost impossible. The Dutch East India Company had even more to fear from the purloining of tree seedlings and their successful cultivation elsewhere. It therefore decided to destroy spice trees wherever possible, leaving only those trees on the smallest and most easily defensible islands. It was at this point that everything was reduced to fortifying these precious deposits, bribing sovereign rulers where necessary to secure their consent to destroy the very sources of the spices on which their revenue depended. Thus does the Dutch East India Company subsidize the King of Ternate and a few other princes of the Moluccas to the tune of twenty thousand rijksdaalders per annum. In places where it has been unable to convince the sovereign to consent to the burning of his plants, the Company has burnt them regardless, taking advantage of its superior might, or purchased the leaves of the trees on a yearly basis, knowing that after three years of such defoliation the trees would die, something of which the Indians are likely ignorant.

By means of this device, the harvesting of cinnamon is confined to Ceylon, the culture of nutmeg to the Banda Islands and the culture of cloves to Ambon, with cloves being disallowed in Banda and nutmeg being disallowed in Ambon. These deposits produce more than enough to meet world demand. Holland's other factories in the Moluccas exist for the purpose of keeping out other nations, prosecuting the search for spice trees and burning them, and provisioning those islands where they are grown. Furthermore, all the engineers and seamen engaged in this practice are obliged, on leaving their work, to relinquish their maps and plans and to swear that they have kept none back.

Facing-page:
*Le Bonet quarré ou la commerçona*, New Guinea plant (*Commersonia*).

Top:
Map showing the route taken by the *Boussole* around Indonesia.

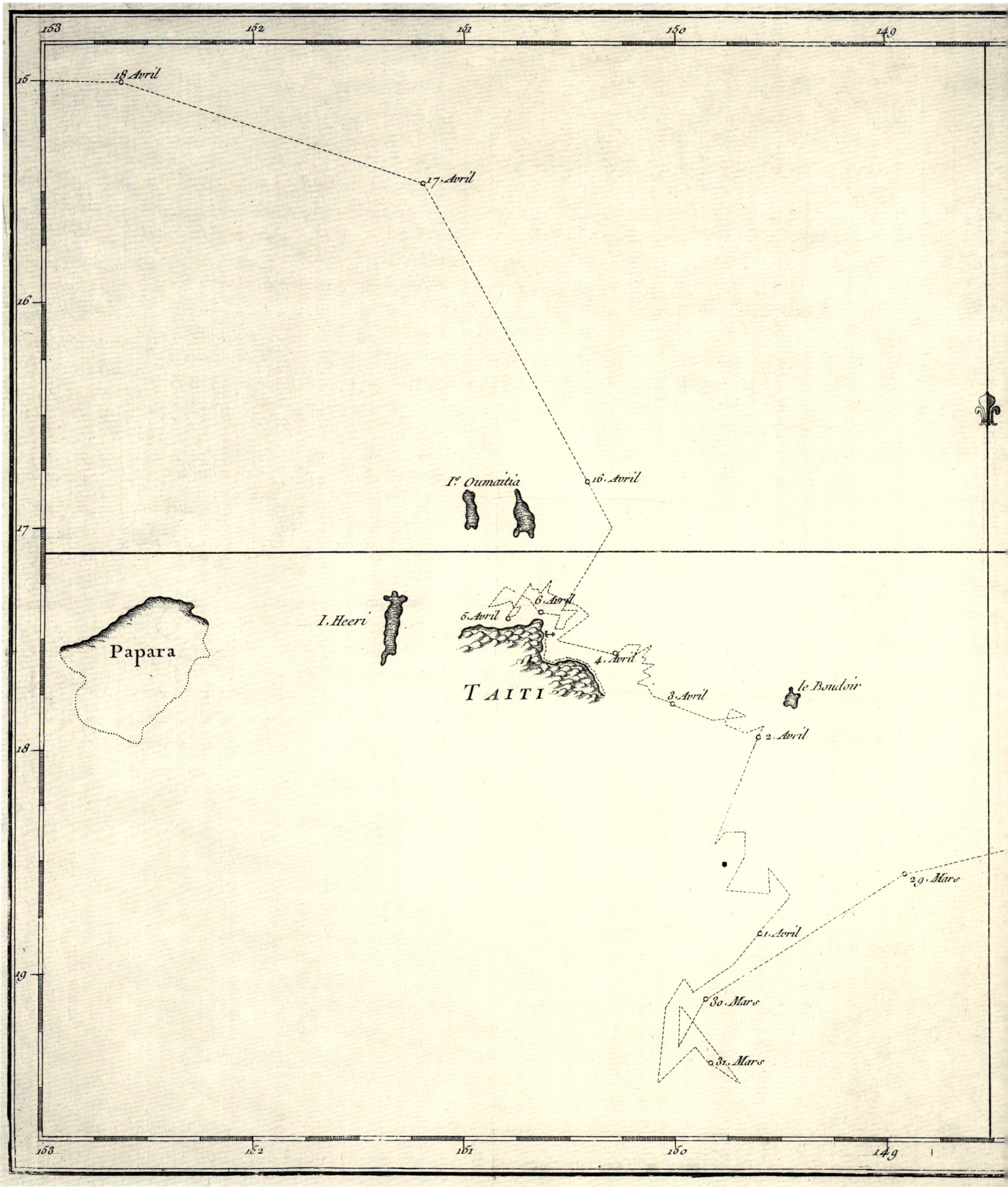

153
152
151
150
149
15
16
17
18
19
18 Avril
17. Avril
16. Avril
Ie Oumaitia
I. Heeri
Papara
6. Avril
5. Avril
4. Avril
3. Avril
2. Avril
1. Avril
TAITI
le Boudoir
29. Mars
30. Mars
31. Mars

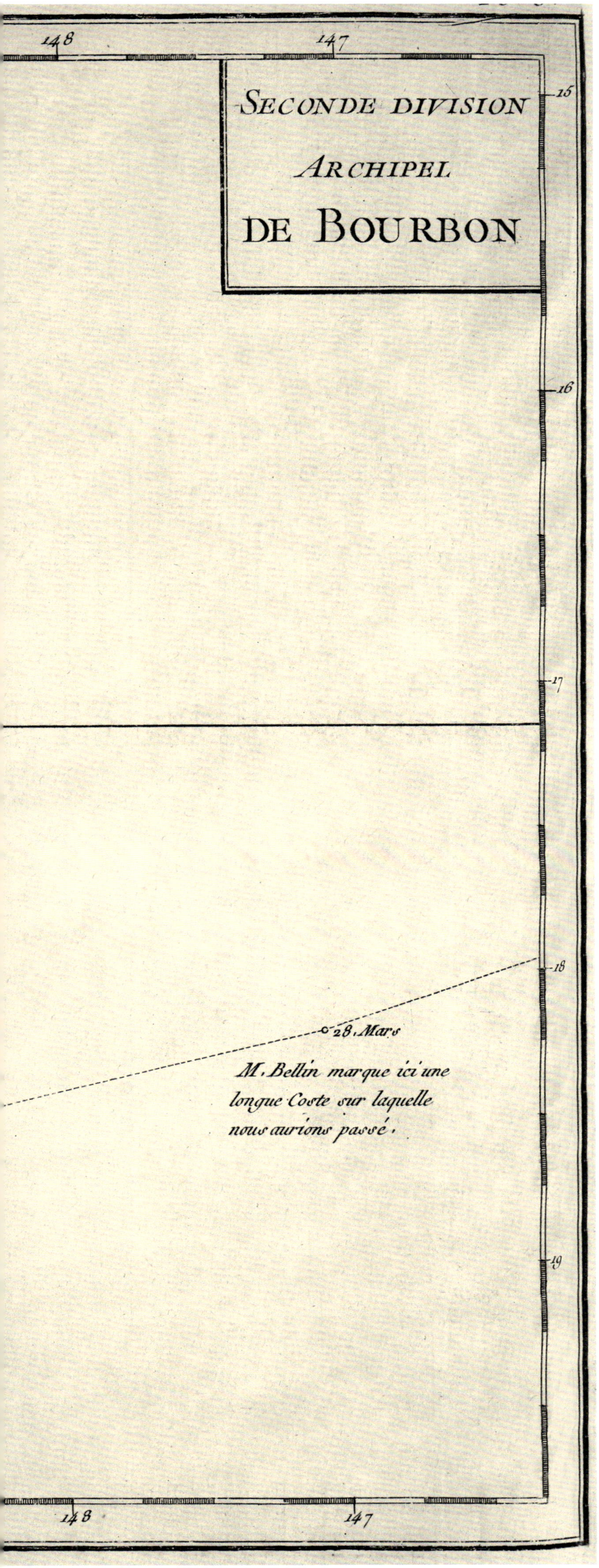


**JANUARY 1769**
CAPE OF GOOD HOPE

*At nine o'clock in the morning* we dropped anchor in the Cape, at the northern end of the anchorage. There were fourteen large vessels of different nationalities already there, and several more arrived during our stay. Monsieur Philip Carteret had departed from there on the Feast of the Epiphany. We greeted the town with a fifteen-gun salute, which was returned in like manner.

We had every occasion to praise the governor and the inhabitants of the Cape of Good Hope, who strove to combine the useful with the pleasant. I shall not pause to describe this place, which as every one knows is under the direct control of Europe and certainly not a dependency of Batavia, neither in terms of military nor civil administration, nor in terms of employment. Indeed, to have been employed in the Cape is enough to preclude employment in Batavia.

**MARCH 1769**
SAINT-MALO, FRANCE

*Determined to enter the port of Brest,* I was beating to windward, with variable southwest winds veering northwest, when, on the morning of March 15, I was informed that the mizzen mast was close to breaking beneath the hounds. I therefore changed course downwind, heading for the port of St Malo that was then our closest place of refuge. We entered therein on the afternoon of March 16, 1769 having lost but seven men in two years and with four months lapsed since our departure from Nantes.

◄ Map charting the *Boussole's* arrival in Tahiti, April, 1768.

PIERRE MARIE FRANÇOIS DE PAGÈS
WAS NO ORDINARY SAILOR, NOT LEAST BECAUSE
HE OFTEN TRAVELED ON FOOT.
HE WAS ALSO AN ETERNAL OPTIMIST
WHO HAD NO OTHER THOUGHT THAN TO BE USEFUL,
WHETHER TO THE NAVY, TO FRANCE
OR HUMANITY AS A WHOLE.
A GREAT ADMIRER OF NATURE IN ALL ITS BEAUTY,
PAGÈS WAS CURIOUS ABOUT EVERYTHING,
SPONTANEOUS AND ENDLESSLY ENTHUSIASTIC.
EAGER TO DO GOOD AND WITH
NOTHING BUT THE KNAPSACK ON HIS BACK,
HE SET OUT ALONE TO EXPLORE
THE WORLD AND ITS PEOPLE, TO OBSERVE,
TO WONDER AND TO ENJOY HUMAN ENCOUNTERS.

# Pierre Marie François de Pagès

**1740–1792**

Plate C1, George Wolfgang Knorr, 1766.

# VOYAGE AROUND THE WORLD BY LAND AND SEA

**1767–1771**

**On a stopover in Haiti, Pagès decided to desert. Aged twenty-six at the time, he then crossed the United States and explored India, on foot, on horseback, by canoe or riding on a camel, to become one of the first romantic adventurers of the modern world.**

Pagès was a contemporary of Bougainville and Lapérouse but he had quite a different approach to travel. He was born in Toulouse on October 16, 1740, at a time when King Louis XV still held onto his crown, and Louisiana and Quebec remained French. Attracted by adventure and the great outdoors, he left the cradle of his family to join the navy at the age of seventeen, embarking on the *Triton* at the height of the Seven Years' War. He was wounded in action and promoted to the rank of ensign. He then joined the East India squadron in Haiti to defend the West Indies against pirates. Ten years later, Pagès had had enough. Bored with his monotonous existence, he repeatedly asked the Ministry of the Navy for leave, which was just as repeatedly refused. Pagès was infuriated and took the decision that would change his life: he deserted.

In 1766, Pagès disembarked without authorization at Cap-Haïtien. His plan was to reach China and India by the western route, i.e. across America and the Pacific Ocean, and seek passage along the coast to the north. The fact that no-one had ever done this before and that unlike all other explorers he had no financial backing didn't worry him. Everything was very simple for Pagès. He would travel modestly, moving from village to village, mingling with communities, working with his hands, bartering—in short, doing anything necessary to earn his crust. He would eat whatever he was given and whatever came his way. In the process, he would learn and grow in understanding, justifying an optimism that was matched only by his naivety. A wide-eyed innocence that even his fiercest critics found disarming.

From Haiti, Pagès travels up the Mississippi by canoe to New Orleans, where he is shocked by the attitude of the French colonists towards the local population, whose side he takes. Wherever he goes, he seeks out the "simple ways of the people" and the "unspoiled charms of nature." Everything is a source of wonder, whether the beauty of a river, the prodigious height of the trees or the quality of the water. He waxes lyrical about "nature in its adolescence, its pleasing forms as yet unchanged by man's handiwork." Our eighteenth-century explorer was an environmentalist before environmentalism existed. He then hires a local guide and travels across Texas to Mexico City, where once again he is enchanted by everything he sees, with the notable exception of the treatment meted out to the Indians by the missionaries. The negative effects of colonialism were already only too obvious to Pagès, who came to regard all the non-European peoples he met as beautiful, kind, generous, cheerful and helpful, and very much at risk of being corrupted by foreigners.

In 1768, after crossing America, Pagès arrived in the port of Acapulco. Seeing a ship bound for the Philippines, he climbs aboard on impulse, still penniless but prepared to work as a carpenter to pay for his bed and board. In Manila and Batavia (Jakarta), true to his desire to mingle with the locals, he lives, sleeps, eats and drinks like his native hosts, working alongside them. He surprises even himself by not falling ill, unlike all those other foreigners who live European-style.

Having landed in China only to be refused entry, our man shrugs it off and moves on to northern India where he arrives in December 1769. This time it is the Brahmins who capture his heart and he lives and dresses like them throughout his stay. Traveling up the Gulf of Oman, he then arrives in Basra in April 1770 and joins a group of camel riders to continue his journey. When one day he is robbed, he owes his survival to the generosity of the Bedouin. He crosses Iraq, Syria and reaches Israel, where the end of his journey is nigh and he boards a ship for France.

Pagès arrived in Marseille on December 5, 1771, at the end of a four-year voyage that had left him exhausted, destitute and unrecognizable: dressed Oriental style and sporting an impressive beard he was mistaken for a Turk. He managed to reconnect with an old friend, the viscount du Bouchage, who recognized him as a former officer of the King and succeeded in convincing the new Minister of the Navy, in Paris, of the usefulness of Pagès' observations and discoveries. King Louis XV's clemency did the rest. Reinstated in 1772, Pagès rejoined the navy with the rank of ensign, and even received all his back pay. In 1773, he joined Yves Joseph de Kerguelen's second expedition to the South Pole. The adventure was very much to his taste, unlike the man, about whom he says conspicuously little in his memoirs.

In 1776, taking his cue from Bougainville and with the King's permission, he set sail on a whaler bound for the North Pole. He explored the Arctic, came within 180 leagues of the Pole and mapped Spitsbergen. But having done all that, though aged only forty-two, he was no longer in good health and left the navy in 1782, after writing an account of his voyages and earning the gratitude of King Louis XVI in the form of a generous captain's pension. With that, he moved to Haiti, where he acquired a plantation and was murdered ten years later, on January 11, 1792, in a slave revolt on his farm. So much for the myth of the noble savage ...

► A young man from Malaysia.

Pl. XVI.

T. Heaphy del.t A. Cardon fecit

**A MALAY,**
native of Bencoolen.

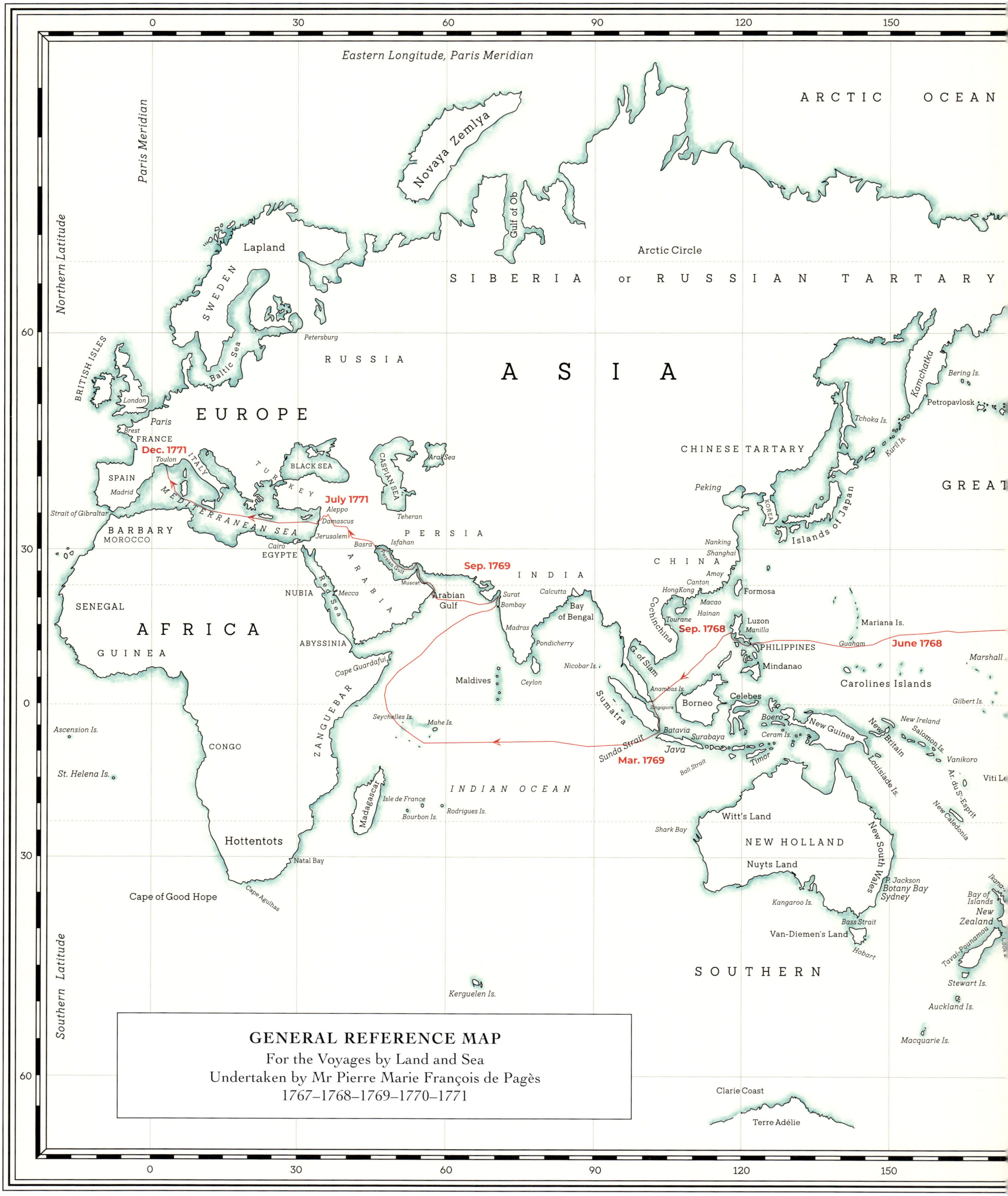
GENERAL REFERENCE MAP
For the Voyages by Land and Sea
Undertaken by Mr Pierre Marie François de Pagès
1767–1768–1769–1770–1771
Eastern Longitude, Paris Meridian
Paris Meridian
Northern Latitude
Southern Latitude
0
30
60
90
120
150
ARCTIC OCEAN
Novaya Zemlya
Gulf of Ob
Arctic Circle
SIBERIA or RUSSIAN TARTARY
Lapland
SWEDEN
Baltic Sea
Petersburg
RUSSIA
ASIA
BRITISH ISLES
London
Paris
Brest
FRANCE
EUROPE
Dec. 1771
Toulon
ITALY
SPAIN
Madrid
Strait of Gibraltar
MEDITERRANEAN SEA
TURKEY
BLACK SEA
CASPIAN SEA
Aral Sea
July 1771
Aleppo
Damascus
Jerusalem
Teheran
PERSIA
Isfahan
Basra
Persian Gulf
Muscat
BARBARY
MOROCCO
Cairo
EGYPTE
NUBIA
Red Sea
Mecca
ARABIA
Arabian Gulf
Sep. 1769
INDIA
Surat
Bombay
Calcutta
Bay of Bengal
Madras
Pondicherry
Ceylon
Maldives
Nicobar Is.
SENEGAL
AFRICA
GUINEA
ABYSSINIA
Cape Guardafui
ZANGUEBAR
CONGO
Ascension Is.
St. Helena Is.
Seychelles Is.
Mahe Is.
Madagascar
Isle de France
Bourbon Is.
Rodrigues Is.
INDIAN OCEAN
Hottentots
Natal Bay
Cape Agulhas
Cape of Good Hope
Kerguelen Is.
CHINESE TARTARY
Peking
KOREA
Islands of Japan
CHINA
Nanking
Shanghai
Amoy
Canton
HongKong
Macao
Hainan
Formosa
Cochinchina
Tourane
Sep. 1768
Luzon
Manilla
PHILIPPINES
Mindanao
G. of Siam
Sumatra
Anambas Is.
Singapore
Borneo
Celebes
Batavia
Surabaya
Java
Sunda Strait
Mar. 1769
Bali Strait
Timor
Boero
Ceram Is.
New Guinea
Kamchatka
Bering Is.
Petropavlosk
Tchoka Is.
Kuril Is.
GREAT
Mariana Is.
Guaham
June 1768
Marshall
Carolines Islands
Gilbert Is.
New Ireland
New Britain
Salomon Is.
Vanikoro
Louisiade Is.
Ar. du St-Esprit
New Caledonia
Viti Le
Witt's Land
Shark Bay
NEW HOLLAND
Nuyts Land
New South Wales
P. Jackson
Botany Bay
Sydney
Kangaroo Is.
Bass Strait
Van-Diemen's Land
Hobart
Bay of Islands
New Zealand
Tavai-Pounamou
Stewart Is.
Auckland Is.
Macquarie Is.
SOUTHERN
Clarie Coast
Terre Adélie

150
120
90
60
30
0
Western Longitude, Paris Meridian
Paris Meridian
Melville Is.
Lancaster Strait
Baffin
Bay
GREENLAND
Davis Strait
Iceland
Bering Strait
Cook's River
Port des Français
Lituya Bay
Kodiak Is.
Sitka
Alaska
Queen Charlotte's Is.
Aleutian Islands
Nootka
Vancouver Is.
Hudson
Bay
Hudson Strait
Cape Farewell
SWEDEN
Baltic Sea
BRITISH ISLES
London
LABRADOR
CANADA
NORTHERN
Newfoundland
St-Lawrence River
Quebec
Acadia
Brest
Paris
FRANCE
NORTH
AMERICA
UNITED STATES
New York
Washington
ATLANTIC
Toulon
ITALY
SPAIN
Madrid
MEDITERRANEAN SEA
NORTHERN OCEAN
San Francisco
Monterey
Azores Is.
Strait of Gibraltar
Bermuda Is.
Dec. 1767
New Orleans
San Antonio
Madeira Is.
BARBARY
MOROCCO
California
MEXICO
Gulf
of Mexico
Florida
Canaries Is.
OCEAN
Tropic of Cancer
Mazatlàn
Lucayan Is.
Hawaii Is.
Sandwich Is.
San José
Santo Domingo
SENEGAL
Mexico City
Caribbean Sea
June 1767
Acapulco
Jamaica
Guadeloupe
Martinique
Cape Verde
AFRICA
Apr. 1768
Panama
GUINEA
COLOMBIA
Cayenne
GUYANA
Equinoctial Line
Gallapagos Is.
0
SOUTH
Paita
PERU
Ascension Is.
Marquesas de Mendoza
AMERICA
CONGO
Lima
Navigators Is.
Bahia
St Helena Is.
Pisco
BRAZIL
Maouna Is.
Tahiti
Society Is.
Arica
Friendly Is.
Tropic of Capricorn
Rio Janeiro
Traitors Is.
Mangareva
Pitcairn Is.
C. Frio
Easter Is.
Coquimbo
CHILE
St Catherine's Is.
SOUTH
Valparaiso
Montevideo
Buenos Aires
Rio de la Plata
Concepción
Cape of Good Hope
ATLANTIC
Chatham Is.
OCEAN
PATAGONIA
Falkland Is.
OCEAN
Strait of
Magellan
Tierra del Fuego
Is. de Los Estados
Cape Horn
Sandwich Is.
Antarctic Circle

This is neither the work of a man of letters nor a scholar, it is my work: a simple narrative by a friend of Man and Nature who makes no pretense of profound observation, merely to sincere and sensible observation. This sensibility of soul to certain facts and certain manners will at times elicit reflections for which I crave the reader's indulgence. Perhaps some may be forthcoming from the reader who has the goodness to take an interest in these arduous journeys, undertaken by the writer in the hope and desire of being useful to his country and to satisfy his heart.

YOUR MOST HUMBLE AND OBEDIENT SERVANT,
PAGÈS
SANTO DOMINGO, BARADERES DISTICT,
MAY 10, 1780

**JUNE 1767**
**SANTO DOMINGO** [HAITI]

*Different circumstances* relating to my particular views and my service in the Navy led me to form a plan to acquaint myself with the Indian Ocean and travel there by way of a westerly route. Thereafter, I had a mind to travel across China to Tartary, on the borders of the Kamchatka Sea. My object was to scour the northern coasts in search of the Northern Sea Route, using an approach that seemed to me quite simple. I wished to acquaint myself with the customs and ways of life of the northern peoples, then adopt those customs so as to spend as long as possible following those peoples in their activities. Thus would I take myself from village to village all along the coast line. By keeping to this route I would surely find a passage toward the north of Siberia, or satisfy myself as to its impossibility if the continuity of these coasts should lead me to North America. This second part of my project could not be realized for I was unable to procure the means necessary to travel across China.

Such was the situation of my heart on that day in 1766 when in the performance of my duties I departed from Rochefort on the coast of Santo Domingo. I embarked on a French boat bound for New Orleans, awaiting the fruits of Providence, my own courage and patience and the simplest way of living possible; relying on habitual practice of a hard life to render more bearable the fatigues and such manual labors as circumstance might force upon me. I was ready for everything and nothing could have surprized me.

**JULY 1767**
**LOUISIANA** [USA]

*I cannot overstate the excellence* of the inhabitants of Louisiana, who embrace the life of primitives for the sake of making their fortune, even to the extent of dressing like them and undertaking the most arduous tasks for an often-insignificant profit. Some go fowling along the coast, wandering thereabouts in a muck-sweat; others venture four or five leagues inland to hunt bear and deer, whose skins they bring back together with the fat and smoked meats; still others go deep into the forests to work the timber of cedar, cypress and maple trees, in which this colony does a flourishing business with the American Islands.

The manners of these primitives led me to believe that they were primitives in name alone and that we had called them primitives solely by virtue of their simple customs and a way of life that is hard by comparison with ours. I particularly admired their steadfast composure and serenity, their never taking that keen or anxious interest in success or failure as we do here in Europe.

**SEPTEMBER 1767**
BETWEEN LOUISIANA AND TEXAS [USA]

*Thus did I set off, alone* having enlisted neither traveling companion nor servant. There were times on that brief journey, fifty leagues at most, when I could not sufficiently admire the decrees of Providence. When, resting on the bear skins that did duty for a mattress, with my horse tethered to some bushes and grazing alongside me, I pictured myself alone in these vast forests. I led a life not unlike that of a primitive, for my best and most transportable provisions were a little sun-dried meat and roasted corn flour.

To the shame of the French settlers in Louisiana, I am obliged to report that they have implanted their vices in the very bosom of these innocent families. In Nachitoches I saw a very becoming young girl emerge from the depths of the forest and offer herself, not to just anyone but only to those proffering gifts albeit of trifling value. So it is, as I have often remarked, that commerce and association with foreigners do instill an idea of luxury in these innocent peoples, making them want necessities that Nature knows not, the while breeding a worm of corruption that would never otherwise have shew itself.

*I spent my time admiring the beauty of this country: for the first time, I saw these men that we call "primitives" as men.*

**NOVEMBER 1767**
SAN ANTONIO, TEXAS [USA]

*In the environs of this factory* are four missions situated within two or three leagues of one another along the river. Each one is managed by two Franciscan brothers who keep families of savages taken captive in the war and whom they have christened and united in matrimony. Each mission supports seven or eight families, including the women and children, and puts them to work for the benefit of the mission. In terms of temporal life, the rules applying in these missions are very much the same as those obtaining in the Jesuit establishments of Paraguay, where the Indians are nevertheless less restricted than here.

The Spanish take no chances when fighting these arrow-wielding Indians. They cover the head with a helmet, and the body with a cassock made of three or four deer skins stitched together, which is proof against any arrow. Where there are but a few savages and the Spanish horsemen are sure of victory, the latter only open fire in the last resort; rather they lasso their quarry like horses, chasing them and throwing the rope around their necks and feet in such a way that they fall at full length and are rendered defenseless.

**FEBRUARY 1768**
MEXICO CITY, MEXICO

*The road to Mexico City* is very fine and admirably well maintained; measuring at least one hundred feet in breadth and one league in length. Within the city are public gardens, beautiful walks and large and lovely hostelries, as in all of the towns hereabouts, but with few comforts as the bedrooms are unfurnished and no board is provided. The Indians here engage in painting and sculpture with quite some success, their ecclesiastical works being particularly striking. Goldsmiths are among the city's premier artisans, creating imposing handiworks that exhibit taste and proficiency.

In the time of my sojourn, the Inquisition, which is particularly severe in Mexico City, had various persons publicly flogged, among them two women who were the victims of a ridiculous superstition. They were accused of inflicting wounds on their enemies by means of certain invocations, then healing the affected parts using a sort of doll that they possessed for that purpose, and kept hung around their necks.

◄ Pages 64-65:
*Dépècement de la baleine* (dismembering the whale).

▲ Top:
Extract from the engraving, *Vue de l'aiguade de l'isle de Tinian* (view of the watering place on Tinian Island).

► Pages 68-69:
Mausoleum of a Brahmin widow who immolated herself on the funeral pyre of her dead husband.

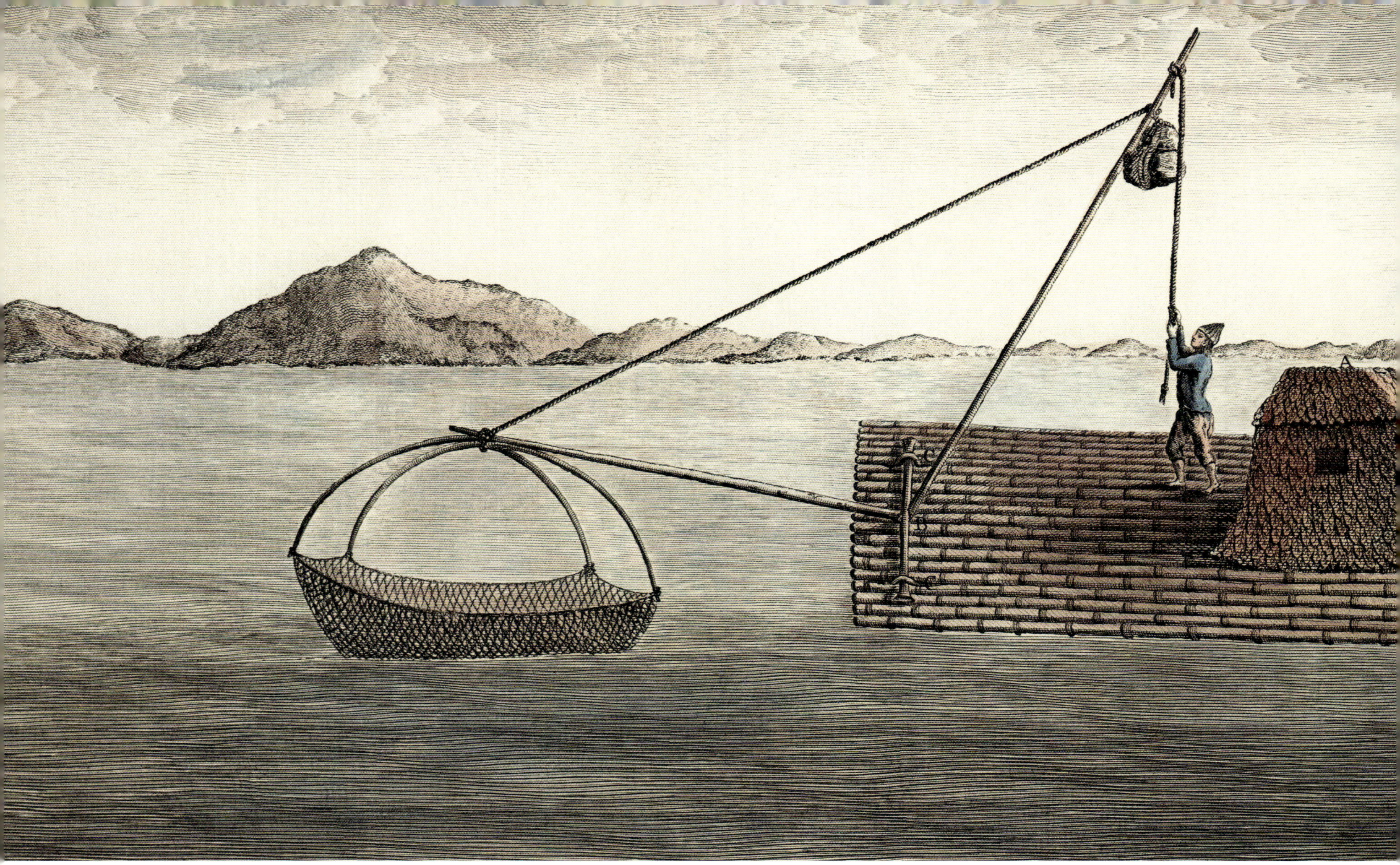

*They tie-up the savages and escort them to the missions where, by dint of gentle treatment, hunger and the women ...*

The whip is but the prelude to the punishment, always inflicted the moment the accused are found guilty and before the final sentencing. The punishments of the Inquisition are revered as being pleasing to the Almighty. According to the Spanish version of Christianity, those who go astray shall be punished and not led back into the paths of righteousness.

**APRIL 1768**
**AT SEA, IN THE PACIFIC OCEAN**

*On April 2, 1768 we left Acapulco* for Manilla with a northwest wind in our sails. It is difficult to describe the confusion that reigned aboard the ship, even if I was informed that it was mild by comparison with the disordered state of the sailing from Manilla to Acapulco. This was a vessel of just five hundred tons which in addition to the crew, carried exiles, women, monks, merchants, officers of every rank both military and judicial and a great many clerks and ship's officers. These officers are not mariners, their commissions being purchased at every sailing with the promise of high salaries and trading opportunities. Navigation is understood only by the pilot, who ranks as a senior officer. He is the one who maneuvers the ship and sets its course. I contrive to eat with one such a few days before our departure, having not had the time to lay in provisions in Acapulco like the others—on this vessel every one must purchase his own food and water and every one eats alone. There was not even a cauldron for the common use of the crew, though some had one of their own. This brings about a great confusion due to the number of servants, which exceeds the number of masters as certain seamen have as many as two servants apiece.

**SEPTEMBER 1768**
**LUZON ISLAND, PHILIPPINES**

*Conduct at the court of Spain* and in its colonies led me to reflect on certain matters. That esteem, probity, impartiality and confidence that unite the hearts of those who profess the same religious beliefs are powerful reasons indeed to combine their temporal interests. Carefully selected missionaries with a true sense of devotion must therefore be key to winning the confidence of new citizens. Civil Government must protect and assist them, the while setting limits that keep them from fanaticism and ambition.

Le Pélican rose de l'Isle de Luçon.

Pic surnommé le Cardinal de l'Isle de Luçon.

*... they are given as wives, it is hoped to tame them; only then are they catechized and christened.*

It is likewise useful to establish total equality between the individuals of conquered countries and those of the sovereign nation, each according to his abilities and faculties; the same rewards and the same employments must be dispensed in every kind of state.

This is not the rule in our colonies where slavery is the established practice. The Europeans who have settled there disdain manual labor as debasing and have become bourgeois, using natives to toil in their place and even in place of animals. We Europeans have invented a form of slavery unmatched anywhere in the world for the sheer savagery of its punishments.

**MARCH, 1769**

**BATAVIA [JAKARTA], DUTCH EAST INDIES, INDONESIA**

*On March 7, 1769 I left Manilla* for Batavia sailing aboard a Spanish schooner. During my four-months sojourn in Batavia and its environs, I could not help but notice how insubstantial are the Dutch settlements there in comparison with those of the Philippines. From the time the Dutch were first established in these parts, far from integrating the Indians they have succeeded only in forfeiting their goodwill, and must now resort to extreme measures to safeguard trade, be it by trickery, force or sweetness.

I could not help but compare the stability of these possessions with those of the Philippines. Is it not surprizing that in the time this country has been a Dutch possession, its Dutch masters have never been at peace in their remote factories? They are constantly at war and constantly the victors, victory being but a Hydra from which are born new wars. At the time of writing, they had just concluded their wars in the Moluccas and in Malacca; they were waging another major battle east of Java; and they were preparing an attack on Sumatra against the King of Palimban [Palembang].

◄ **Facing-page:** *Machine pour pêcher dont se servent les indiens de l'isle de Laïçon* (fishing contraption used by Laicon Island Indians).

▲ **Top:** Pink pelican (left); Luzon Flameback, Philippines (right).

*There were many Christians in this village, with a priest and a church; foreign believers are as free to worship here as they are in France.*

The harshness of the Dutch is at least as great as that of the Spanish. Every body knows the abasement to which they subject the Indians under their control, and the carnage they inflicted upon the Chinese when they rebelled.

**DECEMBER 1769**
NORTH INDIA

*Intent on acquainting myself* with the Maratha caste, I had a set of traditional clothes specially made for the purpose and set off six days after my arrival in the company of a Maratha guide. On the morrow, having traveled barely ten leagues, I arrived in Navsari, a town of medium size where they make cotton canvas; here, there is a Maratha fort together with charming pagodas, gardens and parterres. I was struck by the tameness of the various animals that frolicked among us. The trees were full of birds that did not flee as we approached, of monkeys and squirrel-like creatures that cavorted above our heads, jumping nimbly from bough to bough and on the rooftops. The other quadrupeds being of a similarly sweet disposition, we felt no fear in their presence. Such is the happy effect of the tradition of these people, who never kill any animal!

From Damun [Daman] onward there are Christians everywhere, the region having been formerly conquered by the Portuguese and only passed into the hands of the Maratha, who tolerate all religions, some 30 years since.

**JANUARY 1770**
BOMBAY [MUMBAI], BRITISH INDIA

*I learnt that a vessel* of the French East India Company, the *Indien*, had dropped anchor in Surate. Wishing to take this opportunity to write home, I made my way to Danou [Daman] where the transporting of mail to Surate presented no difficulty. I arrived after five days on the road, revisiting Bassein [Vasai-Virar] along the way where I was struck anew by the refined simplicity of the inhabitants, the while conscious of the subtle variations arising from the different religions.

Those solid conclusions that I had reached on the Island of Samar were further confirmed by the ways of living and thinking of the Brahmins, who I emulated in every way excepting that of religion. I dwelt in gardens where I led a soft and unchanging existence, surviving solely on rice, fruits and greens that I picked and prepared for myself as had long been my custom. Two pieces of cotton sufficed for my daily wear, one

Top:
*Vue de l'Église des Hollandais à Batavia* (view of the Dutch Church at Batavia).

Facing-page:
Luzon islanders, Philippines.

*It was tempting to think that the softness of the Indians came from their abstaining from blood and meat.*

around my waist and the other over my shoulders. I had let my beard grow in the manner of the Greatest Brahmins and I often walked like them, bareheaded and barefoot. Added to this, I slept on matting in the local manner, with a restfulness that only the coolness of the night could procure.

**APRIL 1770**
BASSORA [BASRAH], ARABIA

*We set sail for Bassora* on April 20, 1770, piloted and escorted to the mouth of the gulf by an English sailing vessel armed with cannon and destined to rid the coast of pirates. I was tolerably well treated by the Moors aboard, whose soft and peaceful nature accorded with my own. Like all town-dwelling Mussulmans, they were fanatical about their faith but I refrained from interfering in their observances. They were free to pray and read alongside me, and I was careful only to place myself behind them when they knelt toward Mecca as is their wont. They extended the same affability to Gentiles, Christians, Jews and Mussulmans alike, which pleasing custom led me somewhat to reconsider the unfavorable view that their imperious demeanor had first given me of this religion and all those who professed it. I discerned that its righteous principles, albeit very strict as concerns morality and wholly derived from the patriotic beliefs of their founder, tended to persuade its disciples that they are superior to non-believers; but that sweet charity of the Moors was also owed to their Asian character and manners, and to the precepts of the Mahometan religion.

They seemed angry to have been brought to play a part in European pursuits and ways of thinking whose dire consequences they had become aware of only too late. I sought to impress them with declarations of glory, honor and service to the State; but they knew only the glory, honor and duty that resides in righteous action and simple, charitable works. In the end I knew not who was right, me or them, but I could see that I had not convinced them despite their earnestly wishing it to be so.

The Moors asked me why the French did not think simply; why they could not remain peacefully at home but rather must go running to the ends of the Earth to gather money and spend it; and why they spread discord wherever they went?

*Associating with strangers implants an idea of luxury, kindling unnatural desires and hatching the worm of corruption.*

**SEPTEMBER 1770**
DAMAS, SYRIA

— *I was very well received every where* I went. The inhabitants of these mountains appeared to me of a noble simplicity. In truth, there is none of the luxury and wealth found in Europe and what wealth exists is more equally distributed. In consequence, extreme poverty is rarer than in many parts of France. It seems to me that there can be no better remedy for human misery than to live in the manner of these mountain folk, who possess only what is necessary, but that in abundance. Idleness I would venture to say must not be considered as the ultimate reward for Man's labors; a modest recompense for his exertions is a powerful enticement to keep working and earn an honest living by so doing. Such a society, to be sure, produces neither scholars nor particularly amiable people, just happy men naturally inclined to do their duty, to take care of their families and be good citizens. To my mind such people are worth more than all the idle rich who corrupt the morals of their own societies but profit them not one whit.

**MARCH 1771**
SYRIA

— *In the beginning, I found this treatment* of women repellent, the while remarking certain similarities between the customs of the American aboriginals, the Arabs and the Asians. There are discernible likenesses between these different peoples, which is the more remarkable for their being so remote from one another. In America, the native carries only his rifle while the woman carries the luggage; in Asia the same custom obtains. The American native neither converses with his wife nor includes her in his gatherings. The same is so for Syrian and Asian men. The Arab rides on his donkey, while the woman follows him on foot carrying a large bundle on her head. The American native oftentimes sits at ease in his pirogue while the women row with all their might. I found it generally surprizing that such remote peoples, one so ancient and one likely so recent, should be so remarkably similar, and that we should have such opposite manners despite being equally distant from these different peoples.

Facing-page:
Detailed map of François de Pagès' travels through Palestine.

Bottom:
*Pondichery. Femme du peuple* (working woman in Pondicherry).

Pages 74-75:
Detailed map of François de Pagès' travels in the East.

Women are never the objects of conversation and are paid no attention in the streets. Their dwelling places are sacred and entry is forbidden to men. Merely saluting a woman would be a disgrace. The Europeans attribute these customs to male jealousy; but it is rather an excess of respect for the honor of women, which the Asians hold in far higher regard than do we Europeans.

Women are undivided in their interests and seek only to amuse themselves. In consequence they are freer and more sociable; gardens, baths and tombs are their public places and they also attend on each other in their respective apartments.

**DECEMBER 1771**
MARSEILLE, FRANCE

*We set sail anew,* arriving six days later in Tunis where we dropped anchor, the ship's owner having business to attend to there. I was very well received by the Consul. I had occasion to see several Mahomedans in whose character I discerned a softness comparable to that of the Bedouin Arabs of Basra and Muscat and quite unlike the harshness of the Turks in Syria.

Though we were ready for sea, contrary winds prevented our landing in Sardinia until November 27 and kept us there for two days, at anchor in the Gulf of Palma (Porto Palmas). It was with renewed pleasure that I encountered in this place so close to France the remains of a rustic simplicity that always made me yearn for the life our forefathers.

We made ready for sea once again and skirted the coast of western Sardinia, giving a wide berth to the Corsican coast and landing six days later in Pomègues, an island in the gulf of Marseille where ships are put under quarantine. On the morrow, December 5, 1771, I disembarked at the Marseille infirmaries to begin my period of forced isolation and thanked God for having brought my long journey to such a happy conclusion.

*Women enjoy themselves as much, if not more, in their own company, than our own European ladies do in their societies.*

**JULY 1771**
SAINT JOHN OF ACRE [ACRE]

*My sojourn in this part of Asia* having been exceeding long, I proposed to regain Europe by way of Saint-John of Acre, a port little used by Marseille merchant ships.

We set sail for Marseille at the end of the month of August 1771, heading for the island of Cyprus. After sailing along its coast, we found ourselves turning northward due to the westerly and therefore contrary winds that prevail in summer in these parts. We were looking for the coast of Karamania, hoping to encounter northerly winds, as turned out to be the case. Lack of water caused us to put into an outer anchorage in the southern part of the Island of Rhodes. But the Turks eyed us with suspicion, believing us to be obtaining supplies for the Russians and making us sufficiently wary to beat a hasty retreat.

We pursued our route by way of the canal of Candia [Heraklion] then headed for Malta, where we dropped anchor on October 15. Several companions of mine were also moored up there, aboard French frigates, and I was pleased to find that their affection for me had not waned with absence.

CARTE D'UN
DEPUIS LA MER MÉDITE
DANS L
Pour servir aux Voyages au Tou
M. de PAGÈS, Chevalier de l'O
des Vaisseaux du ROI, et Corres
Route de M. de Pagès.
Villages Arabes Bâtits.
Ruines de Ville, ou Châteaux.
TURQUIE
Alexandrette
Chypre
Alep
Orfa
Diarbekir
Merdin
Mosul
CURDISTAN
Euphrate Fl.
Ruines de Palmyre
Ana
Bagdad
Tripoli
Barut
Damas
S. Jean d'Acre
Jerusalem
Gaza
Mesquet-Aly
Korna
Curdes
Bassora
Ispal
L'EMPIR
Neubandja
Rajian
ARABIE DESERTE
ARABIE PÉTRÉE
EGIPTE
Suez
Route des Pelerins
Route des Pelerins de la Perse
Karah
Sekiale
Hagiar
El Thoalabian
Hail
Feid
Anizé
Vizal
Maaden al Nocra
Kliaibar
GOLFE DE PERSE
I. Baharem
I. Samac
Pays du Cheikr d'Elcatif
Elkatif
Goda
Mascalat
Tropique
Médine
Jambo
Possessions du Cherif de la Mèque
la Mèque
Taret
Padla
PARTIE DE L'ARAB
HEUREUSE
Cette Partie est un
Nageran
Tebalé
Sadun-rai
Giorash
Darboni
Gibezan
PAYS D'ADEN
Sanaa
Mareb
Mouab
Damar
Zebid
Traza
Betlafakih
I. Dalak
AFRIQUE
PARTIE DE LA MER ROUGE
Sassare
Cap Fortu
30
35
40
45
50
25
20
15

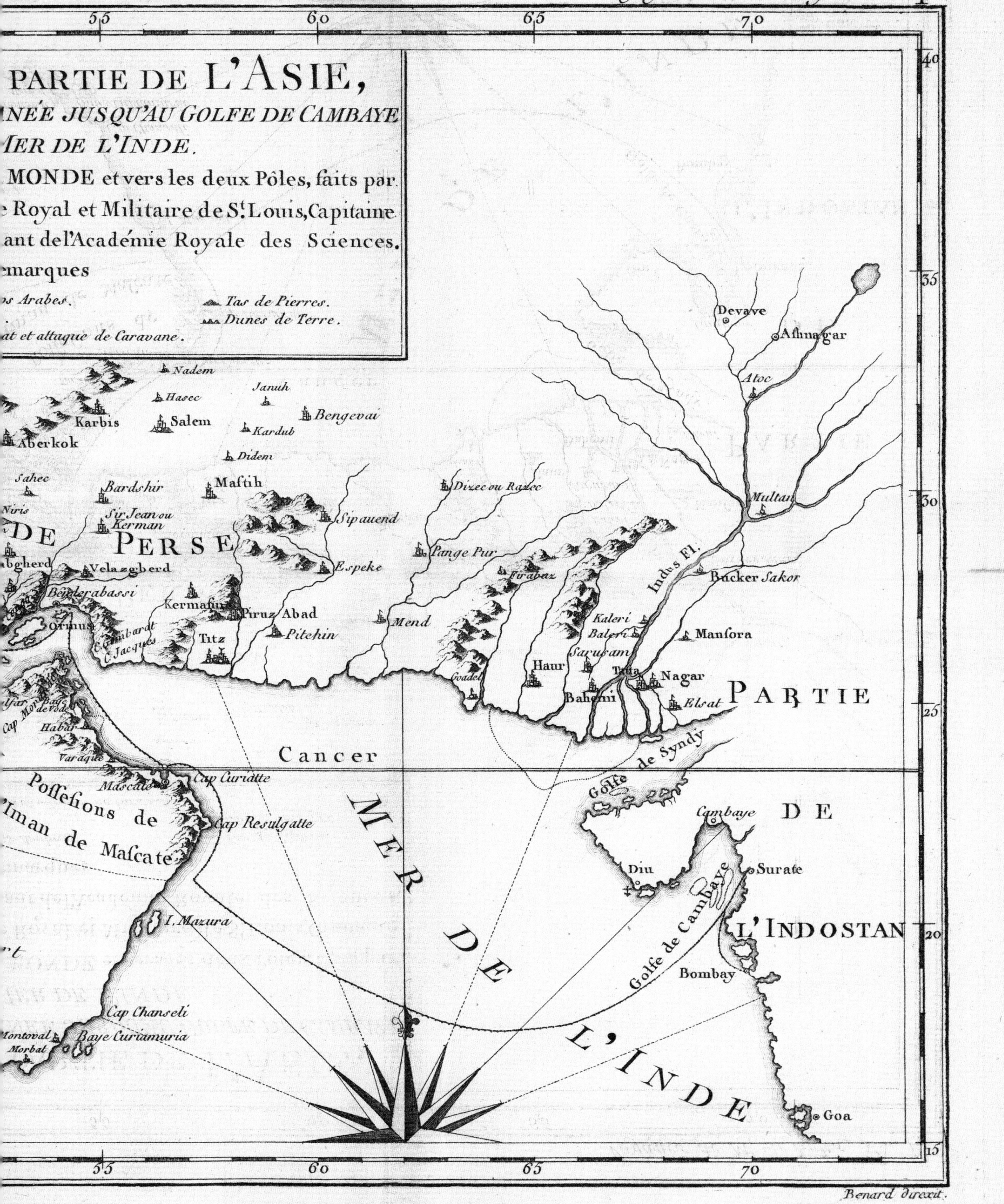


Benard direxit.

Jean-Francois de Galaup

# de Lapérouse

**1741–1788**

KING LOUIS XVI NEEDED A HERO
TO COMPLETE THE DISCOVERY OF THE PACIFIC
FOLLOWING JAMES COOK'S MASSACRE IN HAWAII
AND ENGLAND'S CONSEQUENT ABANDONING
OF PACIFIC EXPLORATION.
JEAN-FRANÇOIS DE GALAUP DE LAPÉROUSE
FULFILLED THE KING'S EXPECTATIONS BEYOND
HIS WILDEST DREAMS. RESPECTED BY HIS MEN
AND HIS ENGLISH ADVERSARIES ALIKE,
LAPÉROUSE WAS A HERO PAR EXCELLENCE—
IMPOSING, TALENTED AND DESTINED FOR TRAGEDY.
FOR IN 2023, THE SEARCH IS STILL ON
TO FIND OUT WHAT HAPPENED TO THE SURVIVORS
AND, WHO KNOWS, FIND THE FOUR HUNDRED
SEALED BARRELS FILLED
WITH TREASURE ...

Plate B, George Wolfgang Knorr, 1766.

# LAPÉROUSE'S VOYAGE AROUND THE WORLD

**1785–1788**

**Lapérouse was forty-four years old when he took command of a great expedition around the world to discover unknown regions. In the event, it turned into an epic tragedy in three acts as his ships were wrecked on the reefs of the Solomon Islands. France, receiving no news, was worried. Even King Louis XVI, on the scaffold, famously asked: "Is there any news of Lapérouse?"**

Jean-François de Galaup de Lapérouse was born in the reign of King Louis XV on August 23, 1741 in Albi, where he grew up with his four sisters on the family estate of Château du Gô. He left the Languedoc at the age of fifteen to enter the Gardes de la Marine—effectively a school for young gentlemen wishing to train as naval officers. He remained there just one year, leaving at the age of sixteen to go and fight against the English at the outbreak of the Seven Years' War in 1757—the spark that set Europe and the whole world on fire. When peace was restored in 1763, he was promoted to lieutenant and far from taking a well-earned rest he asked to go back to sea. So began fourteen years at sea, sailing in the Atlantic, the China Sea, along the Bengal coast and much more besides.

In 1772, Lapérouse went to Mauritius, where he stayed for five years and met Éléonore Broudou, his future wife. Three years later, his defense of the Mahé trading post in the Seychelles saw him decorated and awarded a medal; he likewise distinguished himself in the American War of Independence. In 1782, he led the Hudson Bay Expedition, whose aim was to weaken English positions in the region by destroying two forts: a long and very violent battle from which Lapérouse emerged triumphant. But though he won the battle, he did not win the war; the course of French-American history continued unchanged. The expedition was no less a turning point in his life. Firstly because he distinguished himself as a navigator, strategist and explorer, and proved his humanity, most notably toward his English prisoners. In the process he won an esteem across the Channel that would stand him in good stead in the future. Secondly, it was at this time that he made friends with Paul Fleuriot de Langle and Paul Mérault Monneron, two officers who would soon accompany him on his around-the-world trip. Meanwhile he was promoted to captain and married Éléonore on June 17, 1783, against the advice of his family but with the approval of the Minister of the Navy. This being also the year that marked the conclusion of the War of Independence, the forty-two-year-old Lapérouse headed for home with his bride.

At the end of the following year, the new King of France, Louis XVI, recalled him to Versailles. Passionate about geography and the proud possessor of Cook's collected works, the King now wanted Lapérouse to continue where Cook had left off—reconnect with great discoveries, open up new routes and learn what the English were doing in the Pacific (or Great Ocean as it was called at the time). In short, Lapérouse was being entrusted with a highly ambitious around-the-world expedition requiring detailed preparation and extensive knowledge of previous navigation. To assist him with his preparations, the King enlisted the support of France's historical and geographical societies, the Académie des Sciences, the Académie de Médecine and all the leading astronomers, geologists and botanists of the day, each one making a particular contribution but also leaving instructions ... The Montgolfier brothers even entrusted Lapérouse with one of the very first hot-air balloons.

In June 1785 at Versailles, His Majesty the King personally handed Lapérouse his final instructions: a forty-five-page document setting out in minute detail not only the route, but also the behavior to be adopted when dealing with unknown populations (a must read). On August 1 the whole of Brest celebrated the departure of the frigates *Boussole*, commanded by Lapérouse, and *Astrolabe*, commanded by de Langle. Little did they know that one of the French navy's greatest three-act tragedies had just begun.

On September 29, 1785, having passed by way of the islands of Madeira and Tenerife, the ships crossed the equator. Cape Horn was rounded in early February 1786, and Lapérouse then called at Concepción, Chile, to stock up on supplies and fresh water before embarking on the next leg of the journey: the Pacific crossing, anchoring in the Sandwich Islands [Hawaii] to trade with the natives. Following King Louis XVI's instructions, the ships then continued northwards and began exploring the west coast of America, which was as far as James Cook had gone. On July 2, 1786, the northernmost point was reached: a magnificent bay that Lapérouse named Baie des Français. On July 13, at quarter past seven in the morning, Act I of the tragedy played out, with twenty-one sailors drowned after their longboats were crushed in the breakers. The *Boussole* and the *Astrolabe* then headed south along the American coast to Monterey, where Lapérouse is shocked at the missionaries' behavior. From there the ships proceeded to cross the ocean from east to west, arriving in Macao on January 3, 1787. Lapérouse's descriptions of China and the Chinese illustrate the difficulties of trading between Europe and Asia at a time when neither culture understood the other.

The expedition resumed its course and, after a stopover in Manila, headed north to explore and chart the shores of Korea, Tartary and Siberia, as far as Kamchatka. Featured on these maps were several bays that Lapérouse had named after his officers. But it was contact with the local people that really fascinated him, prompting precise but also intensely moving descriptions. In Petropavlovsk-Kamchatski, Lapérouse received secret instructions from King Louis XVI to make a detour to Sydney to find out what the English were up to, most particularly if it was true that they planned to colonize Australia. With remarkable foresight, Lapérouse instructed his young Russian interpreter Jean-Baptiste Barthélemy de Lesseps to travel overland to Versailles, bringing Lapérouse's journal to the King together with Gaspard Duché de Vancy's drawings, the crew's mail, maps and extensive documentation. It took Lesseps thirteen months to reach France, a journey full of twists and turns that he recounts in his *Journal Historique*, published in 1790.

With the onset of winter, the *Boussole* and the *Astrolabe* headed south, arriving in the Samoan Islands in December 1787. Then came Act II of the tragedy: de Langle's murder on Maouna Island [Tutuila], along with the death of ten other sailors. Lapérouse does all he can to keep his men from exacting a bloody revenge that will serve no purpose whatsoever; and with a heavy heart, he sets sail for Australia. In Botany Bay, south of Sydney, he writes what will be his last letter, and entrusts the English with certain documents to be taken to the King of France. This they do in honor of Lapérouse's exemplary behavior in Hudson Bay, though as we now know they also jumped at the chance to copy his maps! In late February 1788, the frigates sailed north again, reconnoitering and surveying the west coast of New Caledonia and arriving in the Solomon Islands. The scene was now set for Act III of the tragedy: caught in a violent depression in the Santa Cruz archipelago, the ships perished on the Vanikoro reefs ...

Emotions ran high in France as people waited in vain for news of the expedition, so much so that in 1791, the French National Assembly voted to arm an expedition to search for Lapérouse, led by Antoine Bruny d'Entrecasteaux. The expedition failed, the admiral losing his life off the coast of New Guinea after sighting the island of Vanikoro on the horizon, where we now know there must have been survivors of the shipwreck. In 1792, a new expedition was launched by Aristide Aubert du Petit-Thouars. Another failure. Time passed ... The journal of Lapérouse's voyage, written by General Milet-Mureau in the first person based on documents sent by the navigator, was published in 1798 under the title *Voyage Autour du Monde*: a work in four volumes with an atlas. It would be 1826 before the search for Lapérouse's lost expedition resumed. That year Peter Dillon, an English adventurer born in La Martinique, discovered the remains of the frigates in a village; two years later, Jules Dumont d'Urville, having heard of Dillon's find, explored Vanikoro himself and also found traces of the wreck which, together with what he learned from the islanders, confirmed Vanikoro as the location of the tragedy.

The search for the remains of the Lapérouse expedition continued into the twenty-first century. In the 1960s, Admiral Maurice-Raymond de Brossart had discovered a second shipwreck site, nicknamed *La Faille*. But the story didn't end there, as proved by Association Salomon director Alain Conan. An unprecedented twist in the Lapérouse adventure came in 1999 at the end of the fifth excavation campaign in Vanikoro, which revealed traces of a makeshift survivors' camp. Four years later scientists discovered the entire skeleton of one of the scientists. Two years after that they confirmed the locations of the shipwrecks, the *Boussole* in *La Faille*, and the *Astrolabe* in the lagoon. In 2008, new lines of enquiry were launched in search of the *Astrolabe's* survivors and a new expedition is being prepared as we write. The story of Lapérouse's lost expedition is far from over.

GENERAL REFERENCE MAP
For the Voyage of the BOUSSOLE and the ASTROLABE
Led by J.F. de Galaup de Lapérouse
1785–1786–1787–1788
Eastern Longitude, Paris Meridian
Paris Meridian
Northern Latitude
Southern Latitude
ARCTIC OCEAN
Novaya Zemlya
Gulf of Ob
Arctic Circle
SIBERIA or RUSSIAN TARTARY
Lapland
SWEDEN
BRITISH ISLES
Baltic Sea
Petersburg
RUSSIA
ASIA
London
Paris
Brest
FRANCE
EUROPE
Toulon
ITALY
TURKEY
BLACK SEA
CASPIAN SEA
Aral Sea
SPAIN
Madrid
MEDITERRANEAN SEA
Strait of Gibraltar
BARBARY
MOROCCO
Aleppo
Damascus
Jerusalem
Cairo
EGYPTE
Teheran
PERSIA
Isfahan
Basra
Persian Gulf
ARABIA
Muscat
Mecca
Red Sea
NUBIA
Arabian Gulf
INDIA
Surat
Bombay
Calcutta
Bay of Bengal
Madras
Pondicherry
Ceylon
Maldives
Nicobar Is.
SENEGAL
AFRICA
GUINEA
ABYSSINIA
Cape Guardafui
ZANGUEBAR
Seychelles Is.
Mahe Is.
CONGO
Ascension Is.
St. Helena Is.
Madagascar
Isle de France
Bourbon Is.
Rodrigues Is.
INDIAN OCEAN
Hottentots
Natal Bay
Cape of Good Hope
Cape Agulhas
Kerguelen Is.
CHINESE TARTARY
Peking
KOREA
Islands of Japan
Nanking
Shanghai
CHINA
Amoy
Canton
HongKong
Macao
Hainan
Formosa
Cochinchina
Tourane
G. of Siam
Luzon
Manilla
PHILIPPINES
Mindanao
Sumatra
Anambas Is.
Singapore
Borneo
Celebes
Batavia
Surabaya
Java
Sunda Strait
Bali Strait
Timor
Boero
Ceram Is.
New Guinea
New Britain
New Ireland
Salomon Is.
Louisiade Is.
Vanikoro
Ar. du St-Esprit
New Caledonia
Viti Le
July 1787
Sep. 1787
Petropavlosk
Kamchatka
Bering Is.
Tchoka Is.
Kuril Is.
GREAT
Jan. 1787
Dec. 1786
Mar. 1787
June 1788
Feb. 1788
Mariana Is.
Guaham
Carolines Islands
Marshall
Gilbert Is.
Witt's Land
Shark Bay
NEW HOLLAND
Nuyts Land
New South Wales
P. Jackson
Botany Bay
Sydney
Kangaroo Is.
Bass Strait
Van-Diemen's Land
Hobart
Bay of Islands
New Zealand
Tavai-Pounamou
Stewart Is.
Auckland Is.
Macquarie Is.
SOUTHERN
Clarie Coast
Terre Adélie

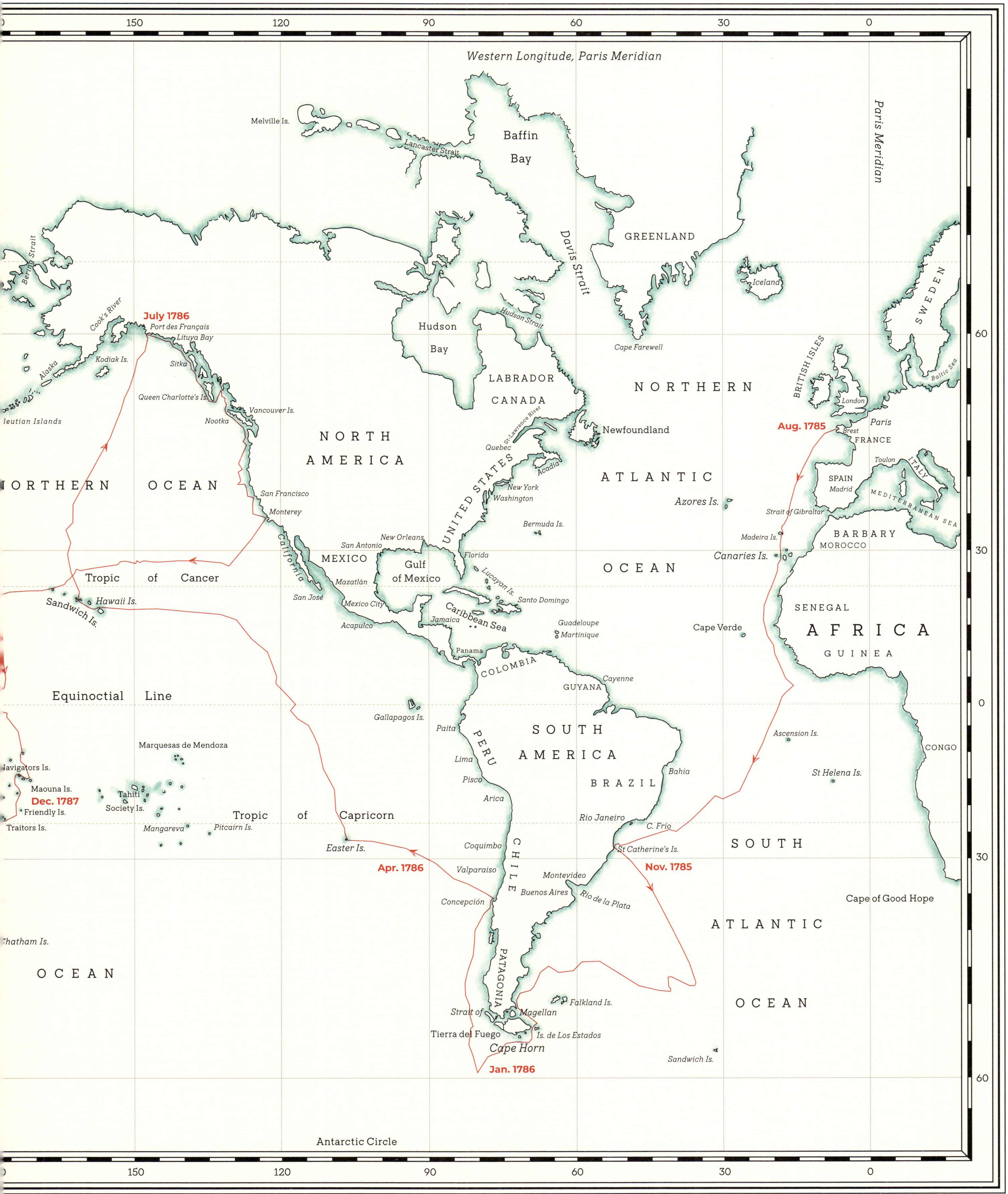
150
120
90
60
30
0
Western Longitude, Paris Meridian
Paris Meridian
Melville Is.
Lancaster Strait
Baffin
Bay
Davis Strait
GREENLAND
Iceland
Bering Strait
Cook's River
July 1786
Port des Français
Lituya Bay
Alaska
Kodiak Is.
Sitka
Queen Charlotte's Is.
Vancouver Is.
leutian Islands
Nootka
Hudson
Bay
Hudson Strait
Cape Farewell
LABRADOR
CANADA
St-Lawrence River
NORTHERN
BRITISH ISLES
SWEDEN
Baltic Sea
London
Paris
Aug. 1785
Brest
FRANCE
Newfoundland
Quebec
NORTH
AMERICA
UNITED STATES
Acadia
ATLANTIC
ORTHERN
OCEAN
San Francisco
Monterey
New York
Washington
Azores Is.
Toulon
ITALY
SPAIN
Madrid
MEDITERRANEAN SEA
Strait of Gibraltar
Bermuda Is.
California
New Orleans
San Antonio
MEXICO
Gulf
of Mexico
Florida
Lucayan Is.
OCEAN
Madeira Is.
BARBARY
MOROCCO
Canaries Is.
Tropic of Cancer
Hawaii Is.
Sandwich Is.
San José
Mazatlàn
Mexico City
Acapulco
Santo Domingo
Caribbean Sea
Jamaica
Guadeloupe
Martinique
Cape Verde
SENEGAL
AFRICA
GUINEA
Panama
COLOMBIA
GUYANA
Cayenne
Equinoctial Line
Gallapagos Is.
Paita
PERU
Lima
Pisco
Arica
SOUTH
AMERICA
BRAZIL
Bahia
Ascension Is.
CONGO
St Helena Is.
Marquesas de Mendoza
avigators Is.
Maouna Is.
Dec. 1787
Friendly Is.
Traitors Is.
Tahiti
Society Is.
Mangareva
Pitcairn Is.
Tropic of Capricorn
Easter Is.
Apr. 1786
Rio Janeiro
C. Frio
St Catherine's Is.
Nov. 1785
SOUTH
Coquimbo
Valparaiso
CHILE
Montevideo
Buenos Aires
Rio de la Plata
Concepción
Cape of Good Hope
ATLANTIC
hatham Is.
OCEAN
PATAGONIA
Falkland Is.
OCEAN
Strait of Magellan
Tierra del Fuego
Is. de Los Estados
Cape Horn
Jan. 1786
Sandwich Is.
Antarctic Circle

## THE SPECIAL INSTRUCTION
## OF KING LOUIS XVI

His Majesty having ordered the outfitting of two frigates in the port of Brest for the purposes of exploration, the *Astrolabe* under the command of captain, Sire de Langle and the *Boussole*, under the command of captain, Sire de Lapérouse, the latter, as commander in chief of these two vessels, will have occasion to acquaint himself with the service that will be his to fulfill on the important expedition whose conduct has been entrusted to him by His Majesty.
The various objects that His Majesty had in mind in commissioning this voyage required that the present instruction be divided into several parts, in order that His Majesty's specific intentions as to the objects for Sire de Lapérouse's attention might be made plainer to him.
The first part will present his itinerary, or navigation plan, in the order of the discoveries to be made or completed; appended to this will be a collection of geographical and historical notes that may be of guidance in the diverse scientific inquiries that he is to undertake.
The second part will deal with subjects relating to politics and commerce.
The third part will set forth activities relating to astronomy, geography, navigation, physics and different branches of natural history, together with the rules applying to the functions of the astronomers, physicists, naturalists, scholars and artists employed in the expedition.
The fourth part will set down the conduct to be adopted by Sire de Lapérouse toward the native peoples of those diverse countries that he will have occasion to discover or reconnoiter.
The fifth and final part will indicate the precautions to be taken by him to ensure the safety of his crews.

VERSAILLES, FRANCE
*(THE INSTRUCTION IN FULL IS GIVEN IN DETAIL IN A FORTY- FIVE-PAGE DOCUMENT)*

### AUGUST 1785
### BREST, FRANCE

*I set sail from the secluded anchorage* of Brest on August 1. The crossing to Madeira was uneventful and we dropped anchor there on the thirteenth day of August. We enjoyed persistently favorable winds, a circumstance that was well-nigh essential to our vessels, which were overloaded at the bow and steered very badly. On fine nights, Monsieur Robert de Lamanon would observe the glowing dots in the sea water, which result in my opinion from the dissolution of marine bodies. Were insects to produce this luminescence, as claimed by several physicists, they would not be so widespread, extending as they do from the pole to the equator, and would be favored by certain climates.

Ere yet we dropped anchor in Madeira, a British merchant by name of Johnston had sent out to my vessel a rowboat laden with fruit. Our arrival had been preceded by several letters of recommendation from London addressed to Mr. Johnston, all of them a source of great astonishment to me, knowing not the persons who had written them. The welcome extended by Mr Johnson was such that we could not have expected a more gracious greeting from our families or closest friends.

### FEBRUARY 1786
### CONCEPCIÓN, CHILE

*This colony, despite its many advantages,* is very far from having progressed in a way befitting its situation. The entire country is under the sway of a prohibitory regime. This kingdom, whose productions were they at their maximum would supply half of Europe, whose wool would suffice for all the manufactories in France and England, whose livestock and cured meats thereof would generate an immense revenue, this kingdom, say I, engages in no commerce whatsoever. Four or five small vessels arrive from Lima every year, carrying sugar, tobacco and a few manufactured goods from Europe, which these luckless inhabitants only acquire second or third hand, and that only subject to payment of the immense excise duties charged on these same goods in Cadiz, Lima and on arrival in Chile. The Chileans meanwhile have but few goods to offer in exchange: wheat, sold so cheap that farmers have no interest in clearing new lands; tallow, hides and a few planks. In consequence, the balance of trade is always to the disadvantage of Chile, whose gold and what few goods it has to barter, are not enough to offset the cost of sugar, grass from Paraguai [Paraguay], tobacco, fabrics, canvases, linen batiste and generally all the different wares necessary to everyday living.

*These monastics where experts at apprising young people of places that priests would have condemned had they known of their existence.*

If Spain does not change her policy; if freedom of trade is not authorized; if the different duties applicable to foreign commodities are not mitigated; and if we lose sight of the fact that a low duty on a major commodity is more profitable to the tax office than an exorbitant duty that annihilates that same commodity, then the kingdom of Chile will never achieve the degree of growth that may be expected from her situation.

Regrettably, this country does produce a small amount of gold, nearly all the rivers being auriferous. Washing gravel and sand is said to earn the local people half a piaster per day, but with food in plentiful supply, there is no motive to work. Deprived of communication with foreigners, they know nothing of our arts nor our luxuries and desire nothing with sufficient force to overcome their inertia. So it is that the land is left fallow. The most active among these people spend a few hours a day washing river sand, so dispensing with the need to learn a trade.

Laziness, far more than credulity and superstition, has peopled this kingdom with monasteries for single women and men who enjoy far more freedom than in any other country. And the misfortune of having nothing to do, of being de facto single without being removed from the world, and of living in seclusion in their cells, has not failed to render them the most wicked subjects in the whole of America. There are no words to describe their effrontery: I have seen some of them remain at the ball until midnight, admittedly set apart from good company and placed among the man servants.

**Pages 86-87:**
*Vue du mouillage des frégates françaises à l'ile de Mowée*
(view of the anchorage of the French frigates, off the island of Mowee).

**Top:**
*Costume des habitants de la Conception*
(residents of Concepción in traditional dress).

**APRIL 1786**
**EASTER ISLAND, CHILE**

*Cook's Bay,* on Easter Island, is the only anchorage on the island sheltered from the southeast and easterly winds that prevail in these parts. At daybreak, I had every thing made ready for our coming ashore. I could look forward to finding friendly people here, having showered with gifts all of those who had come aboard on the day before. However, I was too well versed in travel writing not to know that these Indians were like overgrown children—one look at all our furnishings and they will stop at no thing to grab some item or other.

Four or five hundred Indians were waiting for us on the shore. They were unarmed, some of them dressed in pieces of white or yellow fabric, but most of them naked. Many were tattooed and had painted their faces orange. From their shouts and expressions, they were overjoyed to see us and approached, hands outstretched, to help us ashore.

Soon even the soldiers themselves were exposed to the rapaciousness of these islanders, whose numbers swelled to at least eight hundred, of which no fewer than 150 women. All of them appeared complicit in robbing us, for scarcely had the act been committed than like a flock of frightened birds, they all took flight. But seeing that we made no use of our rifles, they would return within minutes.

The boldest rascals in Europe are less hypocritical than these islanders, whose caresses were entirely feigned and whose physiognomy expressed not a single true emotion. The most suspicious looking one was an Indian to whom we had just given a present and who appeared exceeding eager to serve us in a thousand different ways.

He would forcibly drag young girls of thirteen or fourteen before us, in the hope of receiving payment. The evident repugnance of these young Indian girls was proof enough that the law of the land was being violated in their regard, and no Frenchman stooped so low as to profit from this barbaric invitation. If a few moments were given over to indulging Mother Nature, it was by mutual desire and mutual consent, the women bearing the brunt of the transaction.

*Nobody who has read the latest accounts of voyagers could ever take the South Sea Indians for primitives.*

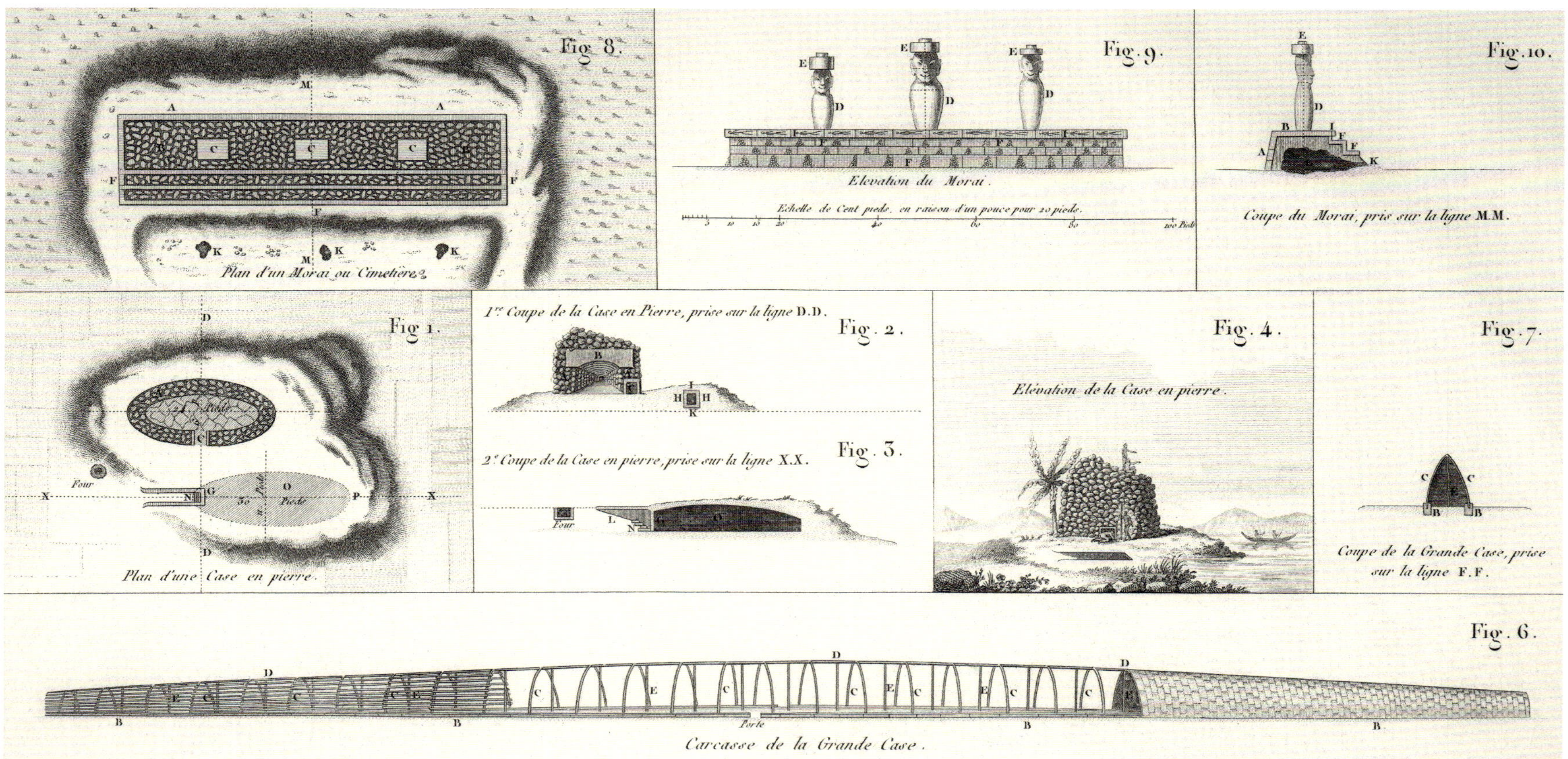


**Top:**
*Détails géométriques des monumens de l'île de Pâque.* (geometric details of Easter Island statues).

**Facing-page:**
*Insulaires et monumens de l'île de Pâque* (islanders and statues, Easter Island).

*It is vain for philosophers to decry this picture. I will concede only that there is no such thing as a society with no virtues whatsoever.*

**JULY 1786**
PORT DES FRANÇAIS [LITUYA BAY] ALASKA

*This port had never been noticed* by any navigator. It is situated thirty-three leagues to the northwest of Punta de Los Remedios, the highest point reached by the Spanish navigators, at a distance of some twenty-four leagues from Nootka and one hundred leagues from Prince William Sound. I gave this place that we had just discovered the name of Port des Français. During our forced sojourn at the entrance to the bay, we had been constantly surrounded by natives in canoes who in exchange for our iron offered us fish, otter and other animal pelts, and various of their accoutrements. To our astonishment they had the air of being very used to such traffic and bargained quite as well as the shrewdest buyers in Europe. Of all our commodities of commerce, the one thing they really coveted was iron. Though we succeeded thereafter in obtaining their acceptance of pewter plates and pots, these articles enjoyed but momentary success, iron excelling all else in precedence. It was a metal not unknown to them; all of them had a dagger hung around their necks. Because we examined these most closely, our interlocutors used hand signs to explain that they only used their daggers against bears and the beasts of the forest. In consequence perhaps, some of the daggers were of red copper but they did not appear to be especially favorite.

The big question on our minds was "whence came these two metals?" It may be supposed that native copper exists in this part of America and is turned into blades and ingots by the Indians. But native iron likely does not exist in Nature, or at least so rarely that no mineralogist has ever seen any.

◄ Facing-page:
*Costume des habitants du port des Français, sur la côte du nord-ouest de l'Amérique* (inhabitants of Lituya Bay, Alaska, in traditional dress).

▼ Below:
*Pirogue du port des Français* (native canoe at Lituya Bay).

▼ Bottom:
*Pirogue, trouvée au port des Français* (canoe found in Lituya Bay).

**JULY 1786**
**PORT DES FRANÇAIS [LITUYA BAY], ALASKA**

*The arts of the Indians* are quite advanced and their civilization, in this respect, shows great progress. But that civility of manners that softens ferocity is still in its infancy. Their way of living excludes all subordination, making them prone to fear and vengeance. Hot-tempered and easily irritated, they are often seen brandishing their daggers and at each other's throats.

I have been traveling the world for thirty years—long enough to testify to the iniquities and cunning of these people whom we are invited to consider good because they live so close to Nature. But Nature is sublime only in a general way; it cares not for the details. It is impossible to delve into forests that have not been thinned by the hand of civilized man, and cross plains strewn with rocks and impassable swamps. It is just as impossible to enjoy any kind of social intercourse with a man of Nature who is barbarous, nasty and cunning. Confirmed in this opinion by sad experience, I did not however feel justified in using those powers entrusted to me to overcome the iniquities of these primitives and teach them that the rights of Man cannot be violated with impunity.

I wanted the wisdom to comprehend them, seeing them always at loggerheads: indifferent fathers, tyrannical husbands, their wives condemned to a life of hard labor. I observed not one thing about these people that might soften the starkness of this picture.

**JULY 13, 1786**
**PORT DES FRANÇAIS [LITUYA BAY], ALASKA**

*On July 4, 1786, this port* welcomed the frigates *Boussole* and *Astrolabe* that had departed Brest on August 1, 1785. Through the good offices of Monsieur de Lapérouse, Commander-in-Chief of the Expedition, and Monsieur the viscount of Langle, Commander of the second frigate, together with the other officers and surgeons, none of the maladies attendant on long sea navigations had affected the crew. Monsieur de Lapérouse congratulated himself and every one else on having traveled from one end of the world to the other, braving every kind of danger and associating with people reputed to be barbarians, without a single man lost or a drop of blood spilt. On July 13 at five o'clock in the morning, our three pinnaces set off to take soundings of the bay as shown on the map. In command was Lieutenant Charles Gabriel Morel d'Escures, a Knight of the Order of Saint Louis, who had received written instructions from Monsieur de Lapérouse ordering him to steer clear of the current. But he found himself caught up in it all the same, and that at the very moment he least expected it. The two La Borde brothers and Flassan in the second pinnace, rushed to the rescue of their shipmates with no thought to their own safety. But alas, it was only to suffer the same fate! The third pinnace was under the orders of Lieutenant Charles Fantin de Boutin who battled valiantly against the breakers and spent several fruitless hours trying to save his friends, only escaping peril himself thanks to the superior construction of his craft, his own enlightened prudence and that of his second, Lieutenant Jérôme Laprise Mouton, and to the actions and prompt obedience of his crewmen, being Master Jean Marie, François Marie L'Hostis, Joseph Le Bas, Corentin Jers and Jean Moncus, mariners all four. The Indians appeared to share in our distress, which touched us to the quick. Wracked by grief but not discouraged, we departed on July 30 to continue our voyage. I gave this place that we had just discovered the name of Port des Français.

**SEPTEMBER 1786**
**MONTEREY, ALTA CALIFORNIA**

*It pains us to say* that we saw men and women in chains and others placed on the block, whipping might also have been heard, this being another punishment that was allowed albeit less harshly administered.

The monks' answers to all our various questions left us in no doubt as to the discipline applying in this kind of religious community—for there is no other name for the system of rules that they have established here. The monks enjoy both spiritual and temporal superiority; they are entrusted with the stewardship of the land. The Indians assemble at the sound of the bell and a monk escorts them to their labors. Corporal

*At the entrance to this port perished 21 brave mariners. Oh ye who ever you may be, weep with us who weep and mourn.*

*There is not a cup of tea drunk in Europe that has not cost the honor of those who purchased it in Canton.*

punishment is inflicted on anyone who fails in their pious obligations. There are seven hours of work per day, two hours of prayer, and four or five hours on Sundays and feast days that are entirely devoted to rest and divine worship. The final point of comparison with religious communities is that from the moment of the neophytes' baptism it is as if they had taken eternal vows. If they escape to return to their parents' home in an independent village, they are summoned three times to return. If they refuse, the missionaries invoke the authority of the governor, who sends soldiers to wrest the runaways from the bosoms of their families and escort them to the missions, where they are sentenced to receive a certain number of lashes.

### JANUARY 1787
### MACAO, CHINA, PORTUGUESE TERRITORY

*Owing to the extreme difficulty* of entering the Chinese Empire, Macao is as separate from China as Europe; and far be it from me to imitate those other voyagers who have spoken of it without knowing it. I shall confine myself to describing the intercourse between the Europeans and the Chinese, the extreme humiliation endured by the former and the paltry protection that may be expected from the Portuguese settlements on the Chinese coast. Lastly, I shall discuss the importance that the town of Macao might possess were Portugal to conduct herself fairly but firmly and with dignity against what is possibly the most unjust and the most oppressive, and at the same time the most dastardly government in the world.

The Chinese engage in trade with the Europeans to the value of fifty millions in revenue, of which two fifths in silver coin and the remainder in English worsted cloth, Batavian and Malaccan pewter (an alloy of lead and tin used most notably to manufacture tea caddies), cotton from Surat and Bengal, opium from Patna, and sandalwood and pepper from the Malabar coast. Other goods from Europe include sundry luxuries such as mirrors of great size, watches from Geneva, corals and fine pearls. In exchange for all these riches, the Chinese have but green and black tea to offer, together with a few chests of ecru silk for European manufactories. There is assuredly no nation in the world that engages in such profitable commerce with foreigners as China, and none that imposes such strict conditions, the while wantonly augmenting vexations and hindrances of all kinds.

I am obliged to report an incident that happened two years ago involving an English gunner who fired a salute by order of his captain, accidentally killing a Chinese fisherman in his sampan who had inadvertently placed himself in the line of fire, the cannon being out of his sight. The governor of Canton demanded that the gunner be brought before him, adding that no harm would come to the man as he, the Governor, was not so unjust as to punish a man for involuntary manslaughter.

◄ Facing-page:
*Vue d'un établissement des habitants du port des Français pour la saison de la pêche* (view of a seasonal fishing camp, Lituya Bay)

▲ Top:
*Vue de Macao en Chine* (view of Macao, China).

► Pages 96-97:
*Naufrage des deux chaloupes, au port des Français* (shipwreck of the two longboats in Lituya Bay).

↑ Bottom:
*Parao, bâteau de passage de Manille* (Parao, a Manila ferry boat).

⟞ Facing-page, top:
*Costumes des habitants de Manille* (dress of the inhabitants of Manila).

⟞ Facing-page, bottom:
*Sarambeau, radeau de pêche de Manille* (*sarambeau*, a Manila fishing raft).

⟞ Pages 100-101:
*Costumes des habitants de la baie de Langle* (dress of the inhabitants of Baie de Langle).

On that condition, the wretch was handed over. Two hours later they hanged him. Honor demanded that the English exact a prompt and blazing revenge on the Chinese, but their merchant ships were not equipped for this purpose and the captains of these ships were too punctilious and honest to undertake resistance activities that would have lost the company forty millions in revenue and sent its vessels home empty. But they undoubtedly denounced the insult and flattered themselves on eventually gaining satisfaction to restore their honor. I would even affirm that all the employees of the different European companies would give a large part of their combined wealth to teach these dastardly Mandarins that there is a limit to injustice and that theirs goes beyond the pale.

**MARCH 1787**
MANILLA, PHILIPPINES

*A few gold-bearing rivers* and an abundance of native spices were undoubtedly what decided the Portuguese to first settle here; but the product fell short of expectations. So it was that covetousness made way for religious fervor and the dispatching of a great number of ecclesiastics of different orders to come and preach the Gospel. Such was their zeal that the Philippines were soon home to between eight hundred and nine hundred newly minted Christians, who were divided into parishes and subjected to the most excessive and maniacal practices. Every fault and every sin is still punishable by flogging; failure to pray and attend mass exacts a price, and punishments are administered at the church doors to women and men alike, by order of the priest.

*Had this zeal been enlightened by a little philosophy, it would have provided the system most apt to ensure the Spanish conquest.*

The effect of this monastic discipline was to enervate the spirit, and further convince an already lazy people rendered slothful by clement weather and abundant food, of the fleeting nature of life and the futility of worldly goods. Added to this is the impossibility of selling the fruits of the Earth for a profit commensurate with effort: the inhabitants having received enough rice, sugar and vegetables to sustain themselves, what remains is pretty much worthless. Under these circumstances, it is not unknown for one pound of sugar to fetch less than one sou and for the rice crop to go unharvested. It is my opinion that not even the most benighted society could imagine a more absurd system of government than that which has obtained in these colonies for the past two centuries. The port of Manilla should be a free port open to all countries but it is only lately that Europeans have gained admission, this being a privilege hitherto

*But they thought only of making converts and never citizens.*

reserved for a few Moors, Armenians and Portuguese from Goa. There is not one single freedom to be enjoyed in Manilla: the inquisitors and the monks supervise the conscience, the *oidor* supervises personal affairs and the governor supervises every activity no matter how innocent. A walk in the island and a conversation are both subject to his jurisdiction. Thus does the most beautiful and the most charming country in the universe rank as the very last place a free man might wish to live.

### JULY 1787
### BAIE DE LANGLE [SAKHALIN ISLAND], TARTARY

*The rest of the day was employed* in visiting the region and its people, who aroused our curiosity and admiration like none we had encountered since leaving France. We knew that these islands were bordered by the most populous, and possibly the oldest, civilizations on Earth; but it seems they have never been conquered by their neighbors because there is nothing here to excite their cupidity. What we did not know and did not expect was that these hunter-fishermen who grow no crops and rear no livestock have milder, more sober manners and perhaps greater breadth of intelligence than any nation in Europe. Assuredly, our European educated classes are vastly more knowledgeable in every way than the twenty-one islanders with whom we communicated in the Baie de Langle. But the islanders were generally better informed than lower-class Europeans, and all of them appeared to have received the same education. There was none of the gawping astonishment of the Baie des Français Indians, just a careful attention to our arts and fabrics, which the Baie de Langle islanders inspected from all sides, talked about amongst themselves and generally sought to discover by what means we had made them.

### SEPTEMBER 1787
### TARTARY

*Our navigation,* from Manilla to l'Île Quelpaert [Jeju Island, South Korea] on the southern coast of Korea, was new only to us, for the Dutch have a long experience of trade with Japan. But I know not whether they steered a course by way of the Strait of Formosa or the east of this island. I am told that before leaving Batavia, the Dutch captains would swear to keep secret the details of their navigation and allow no one to make copies of the manuscript maps with which they were provided. Would not such a precaution suggest that other Europeans might be received in Japan and engage in trade concurrently with the Dutch, or perhaps that the swearing of this oath is merely an ancient custom that they have neglected to reform?

Though the seas of Tartary that we explored constitute the limits of the oldest inhabited continent, they were as unknown to the Europeans as the Strait of Anián or the Archipelago of St Lazarus. And the Jesuits, whose accounts have taught us so much about China, were unable to throw any light on the eastern part of this vast empire. Those who journeyed in Tartary were forbidden to approach the seashores, a precaution that, taken together with the Japanese Emperor's long-standing refusal to allow any ship to navigate to the north of his estates, pointed to this part of Asia's harboring riches that the Japanese and Chinese body politic were at pains to conceal from the Europeans.

*We are approaching that moment when every people will know the breadth of the seas that surround them and of the lands that they inhabit*

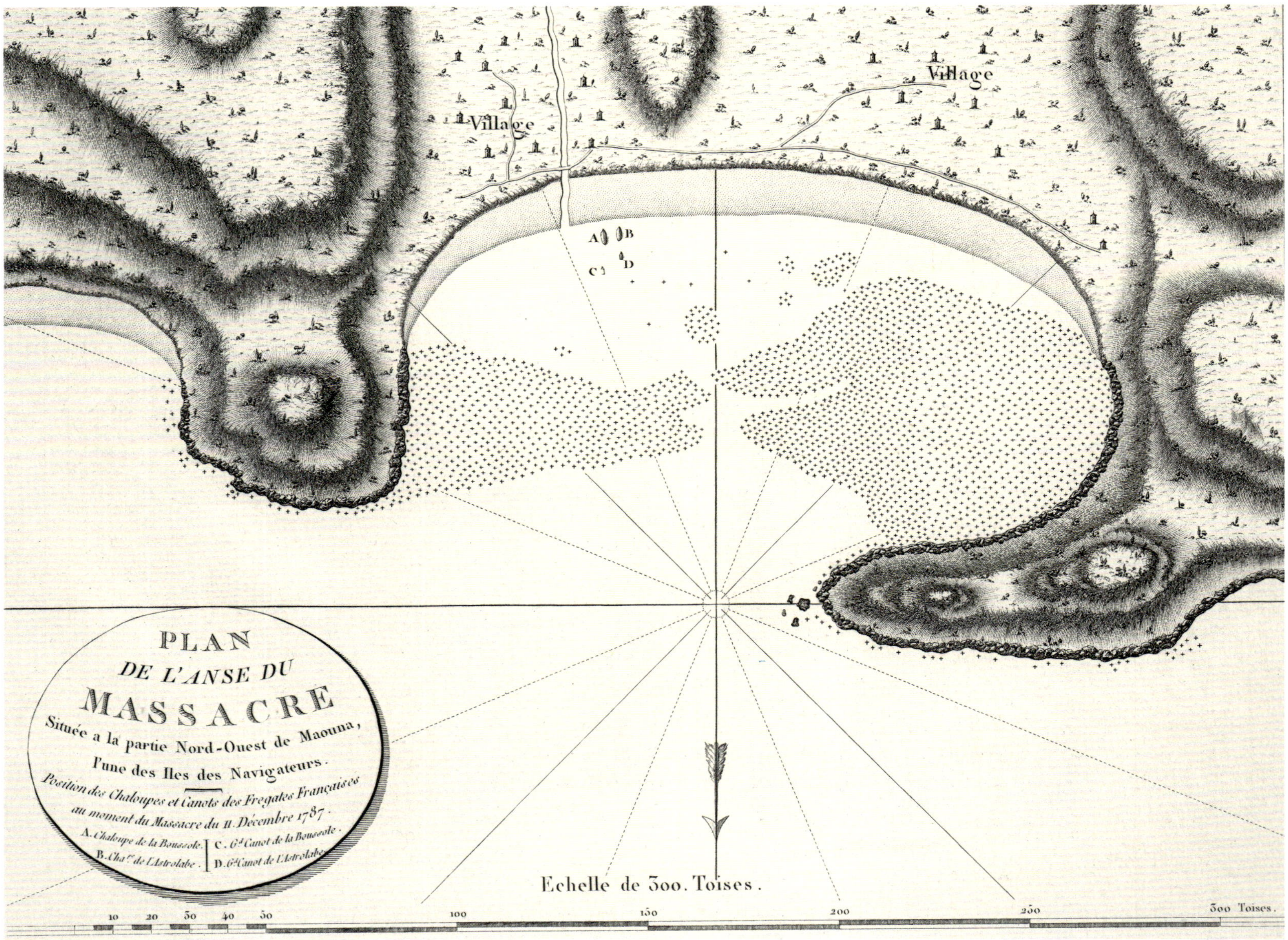

**DECEMBER 1787**
**TUTUILA ISLAND, NAVIGATORS ISLANDS, SAMOA**

Narrative of Monsieur de Vaujuas (account of the massacre)

"On Tuesday 11 December at eleven o'clock in the morning, Monsieur de Lapérouse dispatched his pinnace and longboat, with a detachment of armed soldiers, to undertake an expedition under the command of Monsieur de Langle. At half past twelve the four craft set off to take on water in a cove identified by the latter, the detachment now consisting of sixty-one men including the crews aboard the two pinnaces.

"No sooner were we underway, than we noted with dismay that a large contingent of the canoes along the shore were following us and making for the same cove. We also saw a great many natives from the surrounding villages heading that way, along the rocks that separated the cove from the neighboring bays.

"On our arrival, we were met by some seven to eight hundred savages lined up along the shore, who as a sign of peace threw into the sea several branches of the tree from which they derive their intoxicating liquor. As we came ashore, Monsieur de Langle gave orders that each craft should be guarded by an armed soldier and a mariner, while the longboat crews took on water under the protection of two lines of riflemen on either side extending from the longboats to the watering place.

"By the end of our mission, the natives were still greater in number and growing ever more troublesome. This circumstance decided Monsieur de Langle, who gave orders to re-embark forthwith; but not before, and I believe this was the prime cause of our downfall, he had offered a few glassfuls to the headmen—for so looked to be the natives who had helped to keep back the islanders. Five or six of these individuals were thus honored, stoking such discontent that we were no longer able to contain them. Notwithstanding their displeasure, they allowed us to regain our longboats, albeit followed by a contingent of islanders who took to the sea behind us while the others picked up stones along the strand.

"The longboats having taken the ground some distance from the shore, we were forced to wade out in waist-high water to reach them; in the process, several of our soldiers got water in their weapons. Scarcely had we boarded the longboats, than Monsieur de Langle gave orders to cast off and release the grappling iron. Several of the more robust islanders tried to stop us. The captain, witnessing this resistance together with a growing tumult that occasioned his being hit by stones, fired his rifle into the air to try and intimidate the savages. But far from being frightened they gave the signal to attack. Soon a hail of stones was raining down upon us, thrown with equal velocity and strength as fighting broke out on all sides.

"The first to fall was Monsieur de Langle, despite his having done nothing but good to these ferocious barbarians. No sooner had the hostilities commenced, than he was knocked down and fell into the sea, together with the master carpenter and the master-at-arms by his side. The fury with which the islanders set upon the captain proved the saving of these latter two who both succeeded in regaining the pinnace. Those in the two longboats soon suffered the same fate as our unhappy chief. In less than four minutes they had been overpowered by the islanders, who proceeded to massacre our hapless companions before the horrified eyes of those of us looking on, unable to save them. However, despite the stones and many of us being seriously wounded, we contrived to quit this fateful place and rejoin Monsieur Mouton, the commander of the *Boussole's* pinnace.

"The *Astrolabe*, being the most proximal frigate, arrived to take on the wounded, who were conveyed on board by me and Monsieur Mouton. We then regained the *Boussole* forthwith where I informed the General of the sad news. His surprize was extreme, considering that he had taken every precaution as had Monsieur de Langle in whom he rightly had every confidence. I can only compare his desolation to the grief that I myself felt.

"Words cannot express the feeling aroused aboard the two frigates by this fateful event. Monsieur de Langle enjoyed the trust and friendship of his entire crew and his death threw every one on the *Astrolabe* into despair. Indeed, it took all our powers of persuasion to prevent the mariners from wreaking a bloody revenge on the islanders whom I found drawn up alongside us on my arrival, and who were entirely ignorant of what had come to pass. There can be no finer tribute to this much-lamented man than the general sorrow that reigned aboard ship following his demise."

Signed, Vaujuas (a survivor of the massacre)

◄ **Facing-page, top:**
*Vue des St. Pierre et St. Paul au Kamtschatka* (view of the harbor of St Peter and St Paul in Kamchatka).

◄ **Left, bottom:**
*Plan de l'anse du massacre* (map of the bay where the massacre took place).

► **Pages 104-105:**
*Massacre de MM. de Langle, Lamanon et de dix autres individus des deux équipages* (massacre of Messrs. de Langle and Lamanon and ten members of their respective crews).

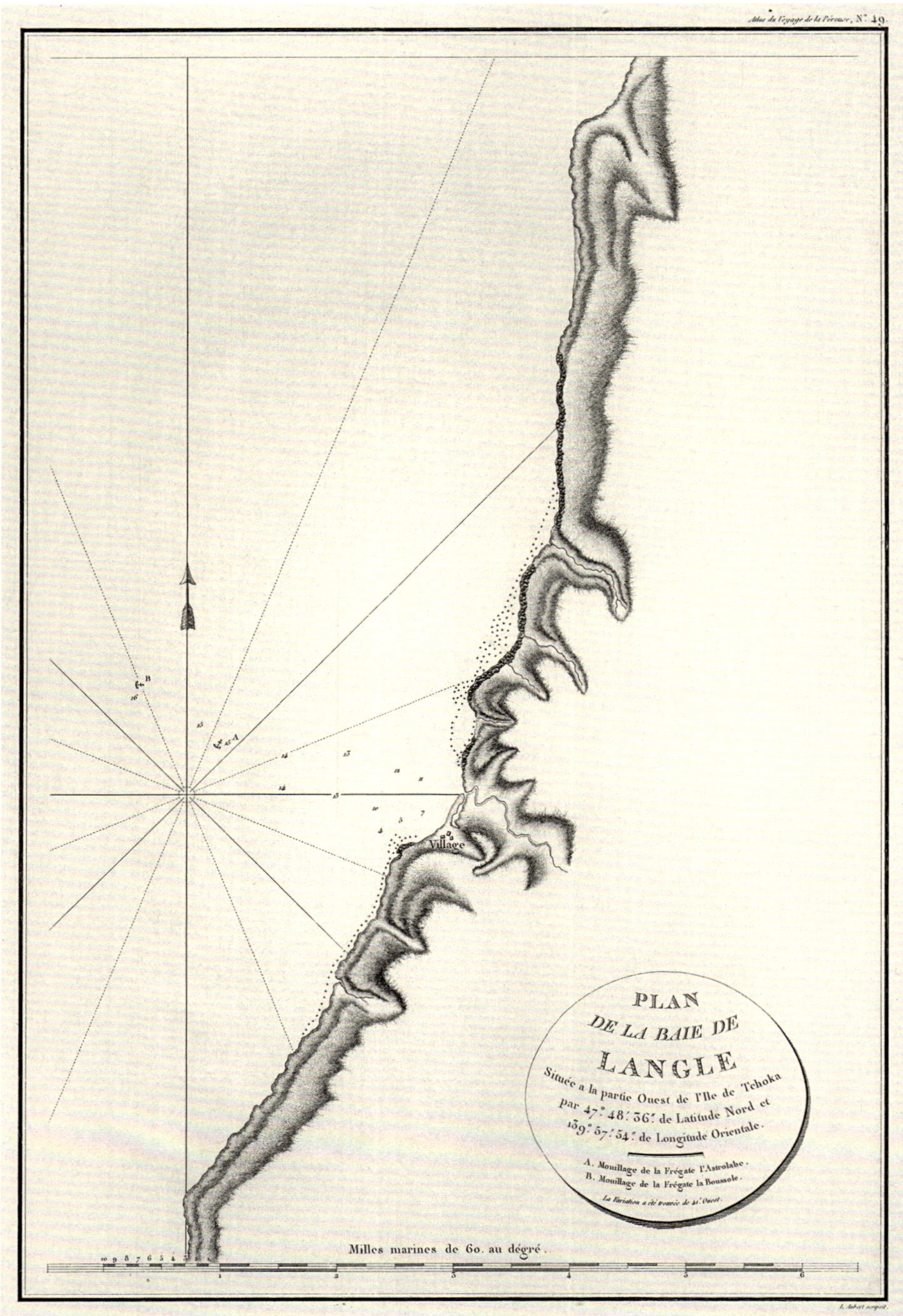


Top:
Detailed map of the Baie de Langle.

Facing-page:
*Inauguration du monument élevé à Lapeyrouse, par le Capitaine d'Urville* (Inauguration of the Lapérouse memorial erected by Captain d'Urville).

## DECEMBER 1787
### TRAITORS ISLAND [NIUATOPUTAPU], TONGA

*The night following our departure* from Traitors Island was appalling. Winds veering from west to east cost us dear, bringing heavy rains that reduced my vision to barely a league. I therefore remained on a beam reach until daybreak, heading south-southwest. The westerly winds continued unabated, as did the heavy rains.

All of those with symptoms of scurvy suffered gravely from the humidity. None of the crew was afflicted by this malady but the officers and especially our servants were beginning to suffer its effects. I attributed the cause to the want of fresh food, a privation most sorely felt by our servants who unlike our mariners had no experience of navigation and were not accustomed to food shortages. Since our departure from Brest, no one on the *Boussole* had died of natural causes. Had we contented ourselves with sailing around the world in the normal manner, we would have returned to Europe without losing a single man.

## JANUARY 1788
### BOTANY BAY, SYDNEY, NEW HOLLAND

*I headed for Botany Bay* under a press of sail, with only three hundred leagues remaining to cover. On the night of the fourteenth, after sunset, I gave the order to drop the mainsail and take a sounding: the Norfolk Island plateau had led me to believe that the basement might extend all the way to New Holland, but this conjecture proved false; so we pursued our route no longer laboring under a mistake, for I had clung dearly to this opinion. The winds from east-southeast to northeast remained steady until New Holland came into sight. We made good speed by day but traveled little by night, our route being charted by no previous navigator.

We became aware of Botany Bay on January 23. Situated at a low elevation, it is hardly perceptible from a distance greater than twelve leagues. The winds turned variable and, like Captain Cook, we encountered currents that carried us daily fifteen minutes to the south of our dead reckoning. In consequence, we spent the following day tacking within sight of Botany Bay. Our ships were unable to battle the winds and the currents at one and the same time.

Europeans are all compatriots this far from home, and we were burning with impatience to gain the anchorage. But the weather was so misty on the morrow that it was impossible to see land, and it was not until nine o'clock on the morning of the twenty-sixth that we finally dropped anchor, abeam of the second bay, one mile off the northern coast. Awaiting our arrival was a sight unseen by us since our departure from Manilla: that of the English fleet at anchor in Botany Bay, their pennants and flags clearly distinguishable.

**FEBRUARY 7, 1788**
BOTANY BAY

## *Lapérouse's last letter*

Thus my dear friend, am I ever to be the bearer of ill tidings, my extreme prudence being eternally confounded by circumstances impossible to foresee but of which I always had, in a way, a dark foreboding. I fear I must reproach myself for yielding, on that unhappy day of December 11, to the importunities or dare I say the extraordinary obstinacy of Monsieur de Langle who would have it that fresh water, lately embarked, was the best anti-scorbutic and that in the absence of such provision his entire crew would succumb to the scurvy ere we reached New Holland. I myself arrived there without a single man falling sick, notwithstanding that our crews had only barrel water to drink. It is my firm conviction that good water, new or old, is equally wholesome.

It was on that fateful day that the sailors manning our longboats, seeing themselves under attack, killed thirty Indians on the shore. Had I not contained the righteous fury of our crews, they might also have massacred those five Indians dispersed among our two frigates and aboard the canoes surrounding our boats. Drawn up alongside us to engage in trade, those canoes together with their occupants who bethought themselves safe, would have been sunk. But it was my belief that such barbaric treatment could neither atone for our misfortune nor console us. Moreover, to inflict harm can only ever be permissible in situations of absolute necessity.

My opinion of these uncivilized people had remained unchanged for many years, a belief that this journey could only reinforce—I have come to know them only too well, and that at my own peril. However, I am one thousand times angrier at the philosophers who so glorify the primitives than at the primitives themselves. Right up to the day they killed him, the luckless Lamanon was always telling me that the natives were worth more than us. And as a strict observer of the orders set forth in my instructions, I have always shewed great restraint toward these people. But I am willing to admit to you here that were I to undertake a new campaign of this kind, I would request different orders. Any navigator who departs from Europe must consider the primitives as enemies; weak enemies certainly whom it would be uncharitable to attack for no reason and barbaric to destroy; but whose proximity the navigator must have the right to avoid where his suspicions are well founded.

I will leave Botany Bay on March 15 then lose no time between now and December when I plan to arrive in the Île de France.

You will find at the end of this journal a map shewing seven of the Navigators Islands. The islanders named ten of them and I believe that the archipelago is not complete without the addition of the islands of Belle-Nation de Quiros, Cocos and Traitors Island; but of this I am not absolutely certain.

Monsieur de Clonard is today in command of the *Astrolabe,* Monsieur de Monti having replaced him aboard the *Boussole*: two officers of the first order. We lost one of still more outstanding merit in Monsieur de Langle, a man of exceptional qualities in whom I discerned no fault other than that of obstinacy, and of being so sure of his beliefs that to say otherwise was to fall out with him. He did not so much obtain as wrest from me the permission that caused his downfall. I should never have yielded had his report of the bay where he perished been accurate. And I will never conceive how such a prudent and enlightened man could have been so grossly mistaken.

As you can see, my dear friend, I remain very much afflicted by this event, which I return to constantly in spite of myself.

AT THE OUTBREAK OF THE FRENCH REVOLUTION,
A MERCHANT MARINE CAPTAIN NAMED ÉTIENNE MARCHAND
PREPARES TO SET SAIL ON THE GREAT OCEAN.
HIS DREAM IS TO MAKE HIS FORTUNE,
AND IN 1790, THE EMBODIMENT OF THE "AMERICAN DREAM"
AS SEEN FROM OLD EUROPE WAS THE FUR TRADE.
MARCHAND'S EXPEDITION FAILED IN TERMS OF TRADE
BUT PROVED A RESOUNDING SUCCESS
IN TERMS OF SCIENTIFIC DISCOVERIES
AND TERRITORIAL CONQUEST.

# Étienne Marchand

**1755–1793**

Plate H VIII, George Wolfgang Knorr, 1766.

# ÉTIENNE MARCHAND'S VOYAGE AROUND THE WORLD

**1790–1792**

**The thirty-five-year-old Marchand sailed his *Solide* to the Marquesas Islands where he claimed his namesake island for France, mapped the northwest coast of America, named many other islands and bays, and filled his holds with otter skins. But his dreams fell apart when he arrived in Canton, faced with the impossibility of selling his furs to the Chinese.**

Is being born in the West Indies an advantage when you want to be a sailor? It certainly had an impact on Étienne Marchand, born in Grenada on July 13, 1755 into a family of five brothers and sisters whose parents came from France to try their luck in the islands. When the English later took possession of the region, the Marchands moved back to France and settled in La Ciotat. Young Étienne studied at the Collège des Pères de l'Oratoire, in nearby Marseille, where he developed a fascination for hydrology and the sea in general that saw him sign up as a volunteer for both the merchant marine and the royal navy. For Étienne Marchand, there could be no better way to learn your trade than on-the-job training. In 1788, on a stopover on St. Helena, he met the man who would change the course of his life: Nathaniel Portlock a former English naval officer turned fur trader who had served under Cook. Having just returned from a long trip around the world aboard the *King George*, Portlock had a hold full of pelts … Marchand wanted to know everything—all the ins and outs of the fur trade and the rich pickings to be had from stopping over in China.

Back in Marseille, Marchand convinced wealthy merchants Jean and David Baux to finance an around-the-world expedition and open up a new trade route between the American West Coast and Canton in China. Having secured funding, he then set about finding a suitable ship: the *Solide*, a potbellied vessel with room for as many pelts as Marchand could buy, but fast with it. She weighed in at 310 tons, measured seventy feet long by nearly twenty-six feet wide (twenty-three by eight meters) and was armed with six cannon; not much but it would have to do. The crew was meanwhile handpicked by Marchand, who wanted only young and experienced men—so no priests, soldiers, draughtsmen, scientists or writers but as many carpenters, gunsmiths, cooks and surgeons as might be needed for the job. He also hired numerous sailors freshly returned victorious from previous expeditions, and officers well versed in the ways of the sea, among them a first-rate second-in-command, Captain Prosper Chanal. Then there was his well-stocked library, not to while away the long evenings on board, but to learn from his predecessors, Cook, Bougainville, William Dampier and Samuel Wallis. Their accounts of their adventures and detailed descriptions of the coasts along which they sailed would provide our man with much food for thought.

In late 1790, a year after the storming of the Bastille but with King Louis XVI still King, the *Solide* left Marseille aiming to round Cape Horn before spring. Leaving Tierra del Fuego, she entered the Pacific Ocean and headed north, just as an epidemic of scurvy hit. With the water on board now unfit to drink, the *Solide* had to make a stopover and headed for the Marquesas Islands, where the crew stocked up on food and water on the Isla Magdalena [Fatu Hiva].

Subsequently, and entirely on his own initiative, Marchand took possession of several islands in the archipelago on behalf of France, without having to make any effort or fight any battles. He spent many weeks living with the islanders and describing their customs.

Back at sea, the *Solide* set sail for Alaska and the promise of the cream of animal pelts. Marchand's strategy was to sail to Sitka, then coast along the American and Canadian shoreline to Vancouver. His strategy paid off, every stopover providing him with an ever-increasing quantity of otter, beaver and bear pelts. On the west coast of Queen Charlotte's Islands (Haida Gwaii), he then took possession of a bay that still bears the name of his first mate: Port Chanal.

In September 1791, the *Solide* sailed west, aiming for the port of Canton—a Pacific crossing with a stopover in Hawaii. But this was where Marchand's luck ran out. On his arrival in Macao, the rules of the game had changed, and ships were now forbidden to sell their skins to the Chinese. The Celestial Empire, newly allied with Russia, wasn't taking any chances … With his holds full of skins and no prospect of a sale, Marchand's only option was to bring his furs back to France. So began the return journey, passing through Indonesia, the Sunda Strait and stopping off at Reunion Island [La Réunion] before finally rounding the Cape of Good Hope.

The *Solide* entered the port of Toulon on August 14, 1792, a dark day for Marchand who learned of the death of his wife Anne, whom he had married in December 1789. Their only son, also named Étienne, would die five years later. France was meanwhile in turmoil. On August 10, the *Sans-Culottes* seized the Tuileries and threw King Louis XVI into prison. Furs being a symbol of the monarchy and nobility, Marchand's luckless cargo was confiscated and sent to warehouses in Lyon, where his furs fell prey to worms and slowly rotted away, until they were reclaimed by the House of Baux once calm had returned.

Marchand himself, on the other hand, was decorated for his achievements. Not only was his circumnavigation of the globe the fastest in recorded history—just twenty months—but it was also of real scientific interest. His observations of the peoples he visited were new and his geographical discoveries were of major importance to France. But he no longer felt at home in this revolutionary France and moved to Reunion Island, where he disappeared a year later, probably in 1793, in very mysterious circumstances. Did he kill himself? Was he murdered? Did he die by accident or did he simply choose to disappear? All we know for sure is that he left behind astronomical gambling debts apparently signed by him in 1793, though this has never been proven.

Charles Pierre Claret de Fleurieu, Minister of the Navy, wrote Marchand's *Voyage Autour du Monde*, published in 1797, based on Captain Chanal's diary and other documents. Marchand's own journal was inherited by his sister Louise and sold by her son to the City of Marseille in June 1843, for the sum of 2,500 francs. In Taiohae, in the Marquesas Islands, a monument erected to the memory of Étienne Marchand was divested of its beautiful copper plaque in 2006. Where is it today? Who knows. But if you have any information, Editions Flammarion would love to hear from you.

Everyday objects of the Tchoutskis, Russia.

Gravé par Adam.

Armes, Vêtemens, Tombeau et Canot.

1 Hache des Tchoutskis, faite de dents de vache-marine.
2 Hache de pierre des sauvages d'Amérique.
3 Instrument des Tchoutskis.

4 Tombeau ......
5 Canot ........
6 Habillement ..
7 Arc ..........
des Ounalaschkans.

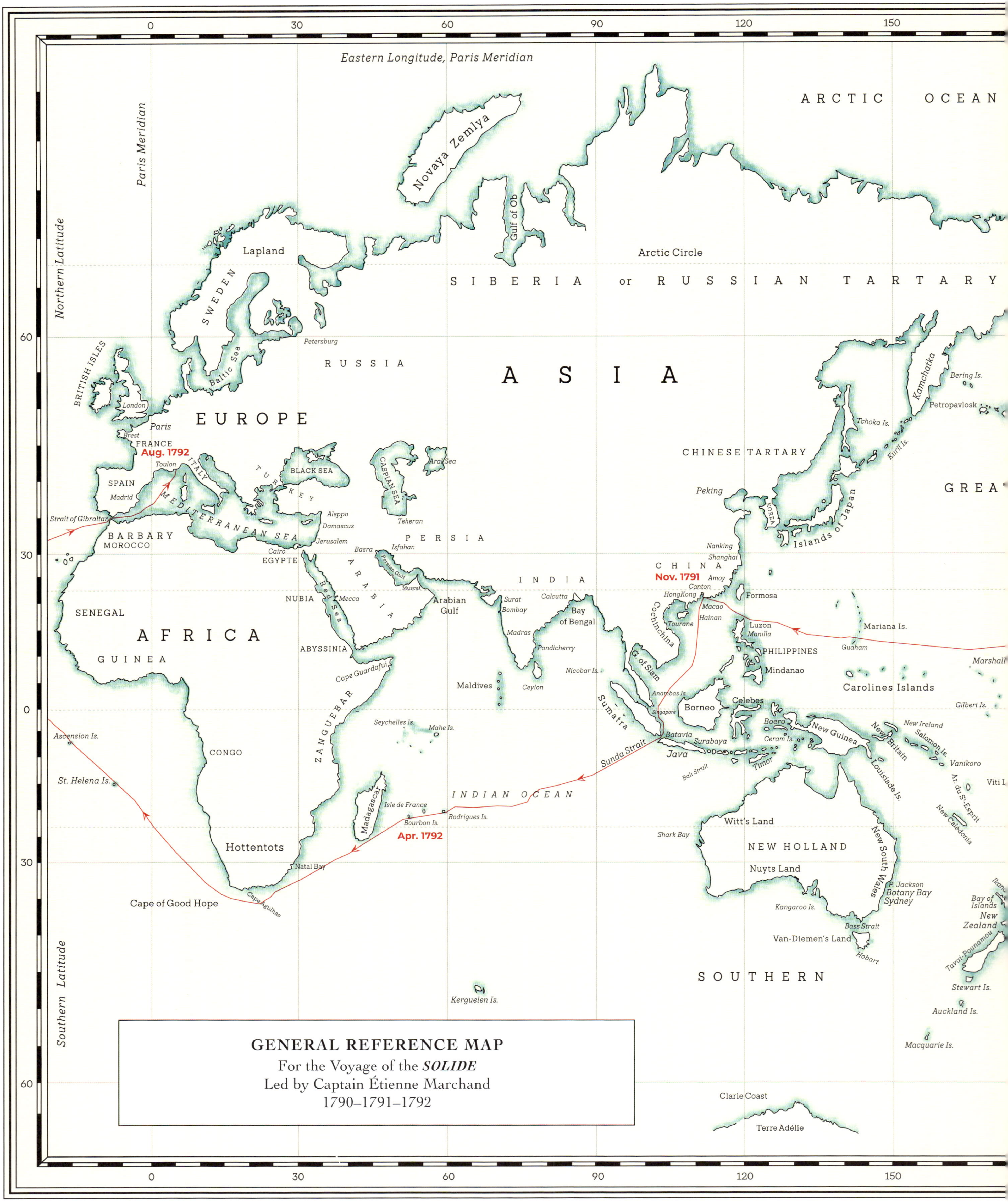
GENERAL REFERENCE MAP
For the Voyage of the *SOLIDE*
Led by Captain Étienne Marchand
1790–1791–1792
Eastern Longitude, Paris Meridian
Paris Meridian
Northern Latitude
Southern Latitude
0
30
60
90
120
150
ARCTIC OCEAN
Novaya Zemlya
Gulf of Ob
Arctic Circle
SIBERIA or RUSSIAN TARTARY
Lapland
SWEDEN
Baltic Sea
Petersburg
RUSSIA
ASIA
BRITISH ISLES
London
Paris
EUROPE
Brest
FRANCE
Aug. 1792
Toulon
ITALY
SPAIN
Madrid
Strait of Gibraltar
MEDITERRANEAN SEA
TURKEY
BLACK SEA
CASPIAN SEA
Aral Sea
Aleppo
Damascus
Jerusalem
Teheran
PERSIA
Isfahan
Basra
Persian Gulf
Muscat
BARBARY
MOROCCO
Cairo
EGYPTE
ARABIA
Red Sea
Mecca
NUBIA
Arabian Gulf
ABYSSINIA
Cape Guardafui
SENEGAL
AFRICA
GUINEA
CONGO
ZANGUEBAR
Seychelles Is.
Mahe Is.
Ascension Is.
St. Helena Is.
Madagascar
Isle de France
Bourbon Is.
Rodrigues Is.
Apr. 1792
INDIAN OCEAN
Hottentots
Natal Bay
Cape of Good Hope
Cape Agulhas
Kerguelen Is.
INDIA
Surat
Bombay
Calcutta
Bay of Bengal
Madras
Pondicherry
Ceylon
Maldives
Nicobar Is.
CHINESE TARTARY
Peking
KOREA
Islands of Japan
Nanking
Shanghai
CHINA
Nov. 1791
Amoy
Canton
HongKong
Macao
Hainan
Formosa
Cochinchina
Tourane
G. of Siam
Luzon
Manilla
PHILIPPINES
Mindanao
Anambas Is.
Singapore
Sumatra
Borneo
Celebes
Batavia
Surabaya
Java
Sunda Strait
Bali Strait
Timor
Boero
Ceram Is.
New Guinea
New Britain
New Ireland
Salomon Is.
Louisiade Is.
Vanikoro
Ar. du St-Esprit
New Caledonia
Viti L
Kamchatka
Bering Is.
Petropavlosk
Tchoka Is.
Kuril Is.
GREA
Mariana Is.
Guaham
Marshall
Carolines Islands
Gilbert Is.
Witt's Land
Shark Bay
NEW HOLLAND
Nuyts Land
New South Wales
P. Jackson
Botany Bay
Sydney
Kangaroo Is.
Bass Strait
Van-Diemen's Land
Hobart
Bay of Islands
New Zealand
Tavai-Pounamou
Stewart Is.
Auckland Is.
Macquarie Is.
SOUTHERN
Clarie Coast
Terre Adélie

Western Longitude, Paris Meridian
150
120
90
60
30
0
Paris Meridian
Melville Is.
Lancaster Strait
Baffin Bay
GREENLAND
Davis Strait
Iceland
Hudson Bay
Hudson Strait
Cape Farewell
SWEDEN
BRITISH ISLES
Baltic Sea
London
Paris
Brest
FRANCE
Dec. 1790
Toulon
ITALY
SPAIN
Madrid
MEDITERRANEAN SEA
Strait of Gibraltar
BARBARY
MOROCCO
Madeira Is.
Canaries Is.
Azores Is.
Bering Strait
Cook's River
Aug. 1791
Port des Français
Lituya Bay
Alaska
Kodiak Is.
Sitka
Queen Charlotte's Is.
Sep. 1791
Vancouver Is.
Nootka
Aleutian Islands
NORTH AMERICA
LABRADOR
CANADA
St Lawrence River
Quebec
Newfoundland
Acadia
NORTHERN
ATLANTIC
OCEAN
UNITED STATES
New York
Washington
Bermuda Is.
NORTHERN OCEAN
San Francisco
Monterey
California
San Antonio
New Orleans
MEXICO
Gulf of Mexico
Florida
Lucayan Is.
Mazatlán
San José
Mexico City
Acapulco
Santo Domingo
Caribbean Sea
Jamaica
Guadeloupe
Martinique
Panama
Tropic of Cancer
Hawaii Is.
Sandwich Is.
SENEGAL
AFRICA
GUINEA
Cape Verde
COLOMBIA
Cayenne
GUYANA
Equinoctial Line
Gallapagos Is.
Paita
PERU
Lima
Pisco
Arica
SOUTH AMERICA
BRAZIL
Bahia
Ascension Is.
CONGO
St Helena Is.
Marquesas de Mendoza
June 1791
Navigators Is.
Maouna Is.
Tahiti
Society Is.
Friendly Is.
Traitors Is.
Mangareva
Pitcairn Is.
Tropic of Capricorn
Easter Is.
Rio Janeiro
C. Frio
St Catherine's Is.
SOUTH
Coquimbo
CHILE
Valparaiso
Montevideo
Buenos Aires
Rio de la Plata
Concepción
Cape of Good Hope
ATLANTIC
Chatham Is.
OCEAN
PATAGONIA
Falkland Is.
Strait of Magellan
Tierra del Fuego
Is. de Los Estados
Cape Horn
Apr. 1791
Sandwich Is.
OCEAN
Antarctic Circle

It appeared to me all the more interesting
to make known the circumstances of the voyage
of Captain Marchand as, quite apart from a discovery
of no small importance in the Great Ocean,
it revealed several new details regarding a part of America
as yet imperfectly known to us, and provided
a great number of astronomical observations that assist
in the perfecting of the arts of navigation and geography;
this being the second voyage around the world
accomplished by the French.
Until now, Bougainville had had no precedent and
no emulator, fortune having denied the unhappy Lapérouse
the completion of his journey.
I have been unable to obtain Captain Marchand's
personal logbook; and I know not into whose
hands have fallen such papers as he might have bequeathed.
But if we may regret the loss of various particulars
contained therein, we may consider ourselves
recompensed by the possession of the logbook kept
by Captain Chanal. His daily journal,
together with that of naval surgeon Robelt, have provided
me with the canvasses on which to paint incidental
details, which I have linked to the principal
subject whose design was furnished
to me by their entries.
But where I have related what the voyager did or saw,
I have adhered scrupulously to his narrative.
And if I do not say things precisely as he said them,
I can confirm that I say exactly the things he wished to say.
Throughout this account, I have sought to express
what the voyagers saw and how they saw it.

CHARLES-PIERRE CLARET DE FLEURIEU
PARIS, 25 APRIL 1797

**DECEMBER 1790**
ATLANTIC OCEAN

*Captain Étienne Marchand* set sail from the port of Marseille on December 14, 1790. He crossed the Straits of Gibraltar and set a course to the Cape Verde islands where he proposed to drop anchor. On the morning of January 4 he sighted the Salvagens Islands. On the morrow, thirty-five leagues thence, he spied Mount Tiede on Tenerife.

Captain Marchand had in mind to drop anchor in the Atlantic Ocean in order to procure water and such victuals as were essential to replace those already consumed, before embarking on the long crossing from the Atlantic Ocean to the Great Ocean and as far as the northwest coast of America, without putting in at any port.

On the morning of January 15, the *Solide* dropped anchor in Praia harbor, which is situated on the southern coast of Santiago, the most substantial of the Cape Verde islands. On the eighteenth day of January, he took to sea anew, heading for the Isla de Los Estados to the east of Tierra del Fuego with which he proposed to acquaint himself before rounding Cape Horn.

**JUNE 14, 1791**
SANTA-CRISTINA AND DOMINICA [TAHUATA ET HIVA OA], MARQUESAS ISLANDS, FRENCH POLYNESIA

*No sooner did the vessel shew herself* at the entrance to the bay than she was surrounded by more than five hundred natives in pirogues. Their ever-increasing number, constantly swelled by those arriving from Dominica Island, might well have aroused alarm had they not appeared of a peaceful disposition. Intent solely on trading, they brought fruits and offered up their womenfolk. They remained in their pirogues until such time as the vessel was safely at anchor. But as the thefts multiplied Captain Marchand felt it prudent to nip such proceedings in the bud and ordered the firing of a one-pounder gun loaded with a charge of gunpowder. Far from intimidating the natives, however, the explosion caused neither fear nor surprise, the islanders believing only that the object was to scare them off and by consequence growing ever more brazen. A four-pounder cannon ball that whistled above their heads, aimed at the rocks on the shore, did however seem to scare them and for a while at least, they were still. But their terror soon passed and ere long they were pelting the ship with coconut shells, breadfruit and sticks, striking certain members of the crew. Others brandished their spears, poking the ship as a sign of defiance; some even had the effrontery to try and wrest the rifle from one of the men charged with defending those craft that had been put to sea. A less prudent, less humane man than Captain Marchand might well have reasoned that his dignity as a European demanded that he make use of his superior weaponry to punish the impudence and avenge the insult; but to his mind these were but children seeking to pick a fight with grown men.

*It seems that every woman is the wife of every man and every man the husband of every woman.*

**JUNE 6, 1791**
LAS ISLAS MARQUESAS DE MENDOZA
[MARQUESAS ISLANDS, FRENCH POLYNESIA]

*From the little that could be ascertained* of the private and domestic life of the natives on the Island of Santa Cristina, it seems unlikely that marriage enters into their way of things. What we do know for sure is that jealousy is as foreign to the men as fidelity is foreign to the women. The latter were offered up indiscriminately and interchangeably for the pleasure and honor of strangers. This practice of offering up daughters and sometimes wives to strangers, a practice in direct contradiction to the principals espoused by European men who live in society, while it admits of no excuse, is not without precedent, not even among the peoples of the Ancient World whose social institutions stand lauded above all others. This may have to do with preconceived opinions, or religious principles or an exaggerated concept of hospitality. It may be considered as a tribute, a pledge of friendship or peace offered to beings with all the markings of superiority. And this usage, intrinsically repellant though it is, must appear even more so to us who would now forswear, as being an outrage against the personal dignity of half the human race, that religious law of a great people of Antiquity that compelled the innocent and timid virgin, before ever she passed into the arms of her loving husband, to submit to the hideous obligation of having her flower brutally plucked by a stranger or a priest.

**JUNE 22, 1791**
ÎLE MARCHAND [UA POU, MARQUESAS ISLANDS, FRENCH POLYNESIA]

*On the morrow at daybreak,* he had the satisfaction of discovering a high land in the northwest toward which he proceeded at full sail to reconnoiter. The officers of the *Solide* elected to name the island the Île Marchand by acclamation.

The southwestern portion of this large island, which continued for half a league, exhibited a few pretty, sandy coves, on whose borders, tucked among banana trees, coconut palms, breadfruit trees and other tall trees, could be seen scattered huts whose occupants were running down to the shore to study the vessel.

It was not possible, in a visit lasting but a few hours, to acquire sufficient knowledge of the country and the nature of its inhabitants to form any firm opinion about the one or the other; all that could be known was what could be deduced at a glance.

Top:
Papeete Bay in 1790.

Pages 118-119:
View of the residence of Pomare IV, Queen of Tahiti, nineteenth century.

Queen Pomare's house Papeite.

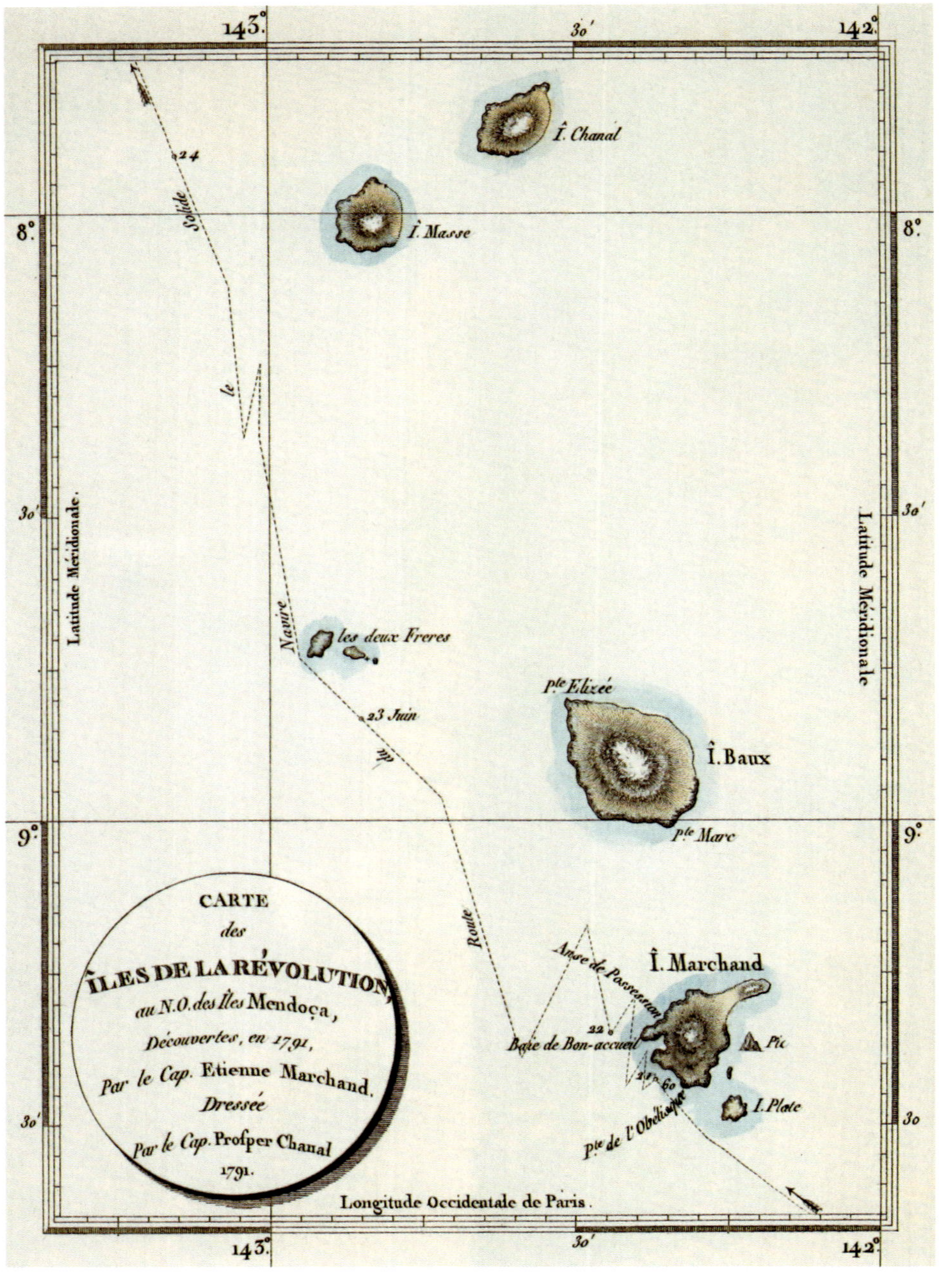


**JUNE 25, 1791**

ÎLE MARCHAND [UA POU, MARQUESAS ISLANDS, FRENCH POLYNESIA]

*Before embarking* on a reconnaissance of the land that could be seen in the northern part, it was decided to land on the northwest coast of Marchand Island, in order to raise a monument attesting to its discovery by the French and thus claiming the island for France. The pinnace was readied to take Captain Marchand thither accompanied by Captain Chanal.

Europeans are convinced that the lands they have just discovered form part of their universal birthright and must now be restored to that dominion having been wrongly alienated by Nature. Fortunate indeed are those primitive tenants of newly discovered lands if the usurper, when claiming the territory for his sovereign, does not bring to bear that devastating weapon of our own European invention that in the space of a century has either slain half of the human race or subjected them to the will of a handful of men. Captain Marchand, taking example from his precursors, believed it incumbent upon him to claim the island he had just discovered in the name of the French nation, which possession, according to received wisdom, gave him the right to claim such other islands as he might discover in the vicinity. This ceremony, which would be but an exercise in futility were it not contrary to the rights of Nature and Mankind, required the use of four nails to fasten to the trunk of a large tree an inscription bearing the name of the vessel, its captain and declaring the island claimed France. There can be no doubt that the natives, who attended with great curiosity to the proceedings of these strange objects of their admiration, had not the least suspicion that these lands wherein lay the bones of their forefathers were being solemnly appropriated by a power that would henceforth be their master, in a hemisphere known neither to them nor to their ancestors.

For greater security, prudence demanded that the inscription be copied on three sheets of paper, which were rolled-up separately then enclosed in three corked and sealed glass bottles. One was placed in the hands of the venerable village headman, the second in the hands of a man of mature age and the third in the hands of a young girl—entrusting this precious deposit to three generations hardly seemed excessive. Out of all the presents given to the inhabitants of a country thus lately joined with France, none was received with more pleasure than these bottles, nor appeared held in greater esteem by the recipients, who little supposed the act of union with a European Empire to be contained therein.

No sooner was this ceremony terminated, than the northwest bay of the Île Marchand was proclaimed the Baie de Possession, its formers owners raising no impediment or objection; their silence was to be taken as their tacit assent. Ever since navigation first revealed to Europeans the existence of parts of the terrestrial globe of which the Ancients were quite unaware, those same Europeans have persuaded themselves that the whole world belongs to them.

**AUGUST 2, 1791**

NORFOLK BAY/BAYA DE GUADALUPA [SITKA], ALASKA

*Trading with the natives continued* in the days that followed. A few pirogues arrived from the coast of Pitt Island, where the vessel dropped anchor, others from the eastern coast or the Grande Terre [Alaska]. All of them were manned by the same men with whom the French had already treated. The fellmongers aboard the *Solide* were kept busy inspecting the hides and skins, beating them to remove dust and vermin, and dressing them to ensure their condition on arrival in China.

Captain Marchand dealt in the following:

- 100 otter pelts of first-rate quality, most of them raw or half finished;
- 250 young, white otter pelts;
- 36 full bear hides and thirteen half hides;

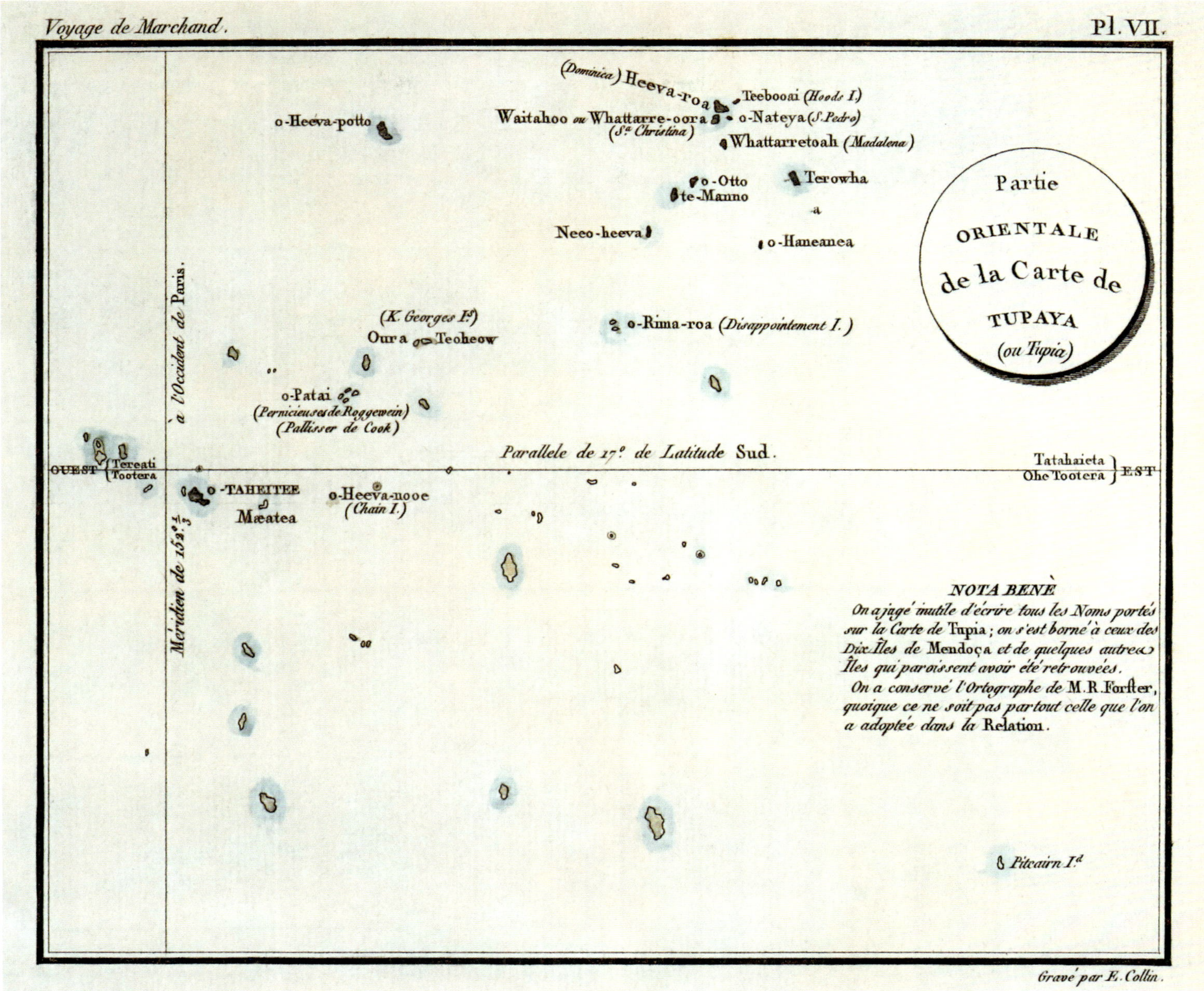


*The Île Marchand shall rank among those too few islands in the Great Ocean whose discovery was not soiled by a great effusion of blood.*

- a considerable quantity of inferior quality otter pelts cut into strips, most of them worn, amounting altogether to some 150 whole pelts;
- 37 common seal skins;
- 60 beaver, mink and other animal pelts;
- a bag containing a few squirrel skins and otter tails;
- a marmot skin rug;
- a marmot and bear skin rug.

### AUGUST 6, 1781
### CLOAK BAY, QUEEN CHARLOTTE ISLANDS [HAÏDA GWAÏI], CANADA

*The natives around Cloak Bay* are not easy to do business with. They pore over every thing with knowing eyes and only ever conclude a transaction after careful deliberation. But they are honest in their dealings. Their slowness in reaching their decision might have something to do with their character but might it not also derive from their having been cheated in the first instance by European traders? Dixon, apparently the first to treat with these people, writes in his *Voyage Round the World* that the islanders would throw their furs into the pinnaces and the vessel of the English without counting them or taking other such prudent measures, and wait patiently to be offered such recompense as the buyer saw fit.

And how did the Europeans respond to this blind trust on the part of the Americans aboriginals? Were the latter not forced to recognize, on more than one occasion, that the goods supplied in exchange for the skins of which they had divested themselves were not of an equivalent level of excellence? Did the Europeans conduct themselves in accordance with the principles of good faith and fair dealing that should form the basis of commerce, and most especially trading, but which is not always the basis of transactions between Europeans?

**Top:** Overall view of Tahiti and her islands, 1790.

**Facing page:** Map of the Marquesas Islands discovered by Marchand.

**Pages 122-123:** Main bay of Huahine Island.

house. Huahine.

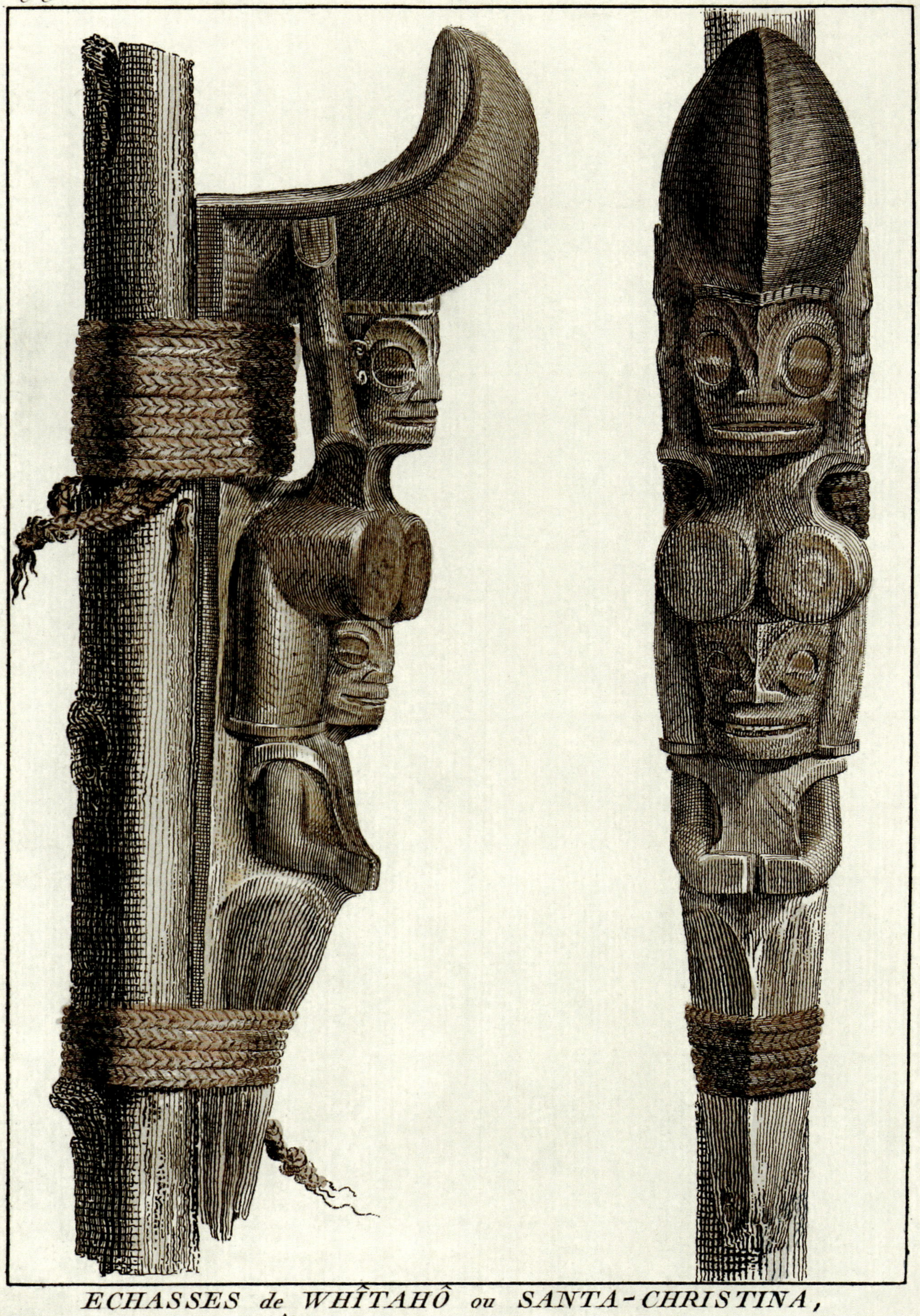

***ECHASSES de WHÎTAHÔ ou SANTA-CHRISTINA,***
*une des Îles* Las Marquesas de Mendoça,
*vues de face et de profil, grandeur de moitié de nature.*

*Dessiné d'après nature par* Genillion *Gravé par* Fortier

**AUGUST 14, 1791**
PORT CHANAL, QUEEN CHARLOTTE ISLANDS [HAÏDA GWAÏI]

### *The deep, narrow canal*

where the new pilot steered his longboat divides into two narrow branches. A short distance from its mouth is a cove, where the ship hove to: this was the location of the dwelling that the dense woods obscured from view. The shouts of the men in the pirogue brought several Americans running, the first of whom threw themselves to the ground in submission gesturing that they would be back ere long. And indeed, they did not keep their visitors waiting. But what a surprise for the French to see these American aboriginals return dressed in English garb! Worsted waistcoat, breeches, round hat—they could have been taken for Thames watermen. But furs had they none; nothing to offer but a few fish. All hopes vanished in an instant. For every one knew that there is nothing remains to be gleaned from fields previously harvested by the English.

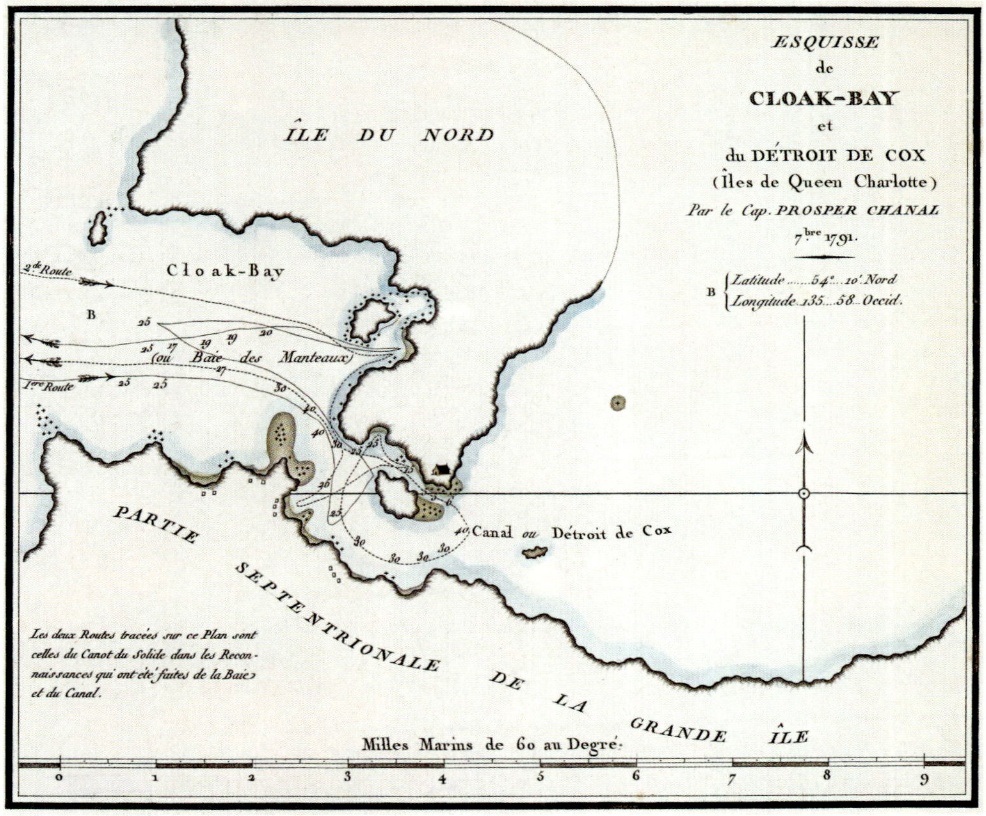


*Have the Europeans nothing to reproach themselves for? Have they never sought to abuse the supposed ignorance of the American aboriginals?*

**AUGUST 18, 1791**
QUEEN CHARLOTTE ISLANDS [HAÏDA GWAÏI]

### *These natives, ever attentive*

to their guests, appeared to concern themselves only with fulfilling their duty as hospitable hosts. We French never saw them either armed or defiant. On the contrary, we were welcomed into the bosom of their families and spent several days on intimate terms with the aboriginals. I am no apologist for primitive peoples. Never have I gone into raptures about Man in his natural state; never have I shared the opinion of certain philosophers who have tormented their imaginations and challenged their eloquence to prove him to be Man par excellence. My question here is not whether Man be good or evil in essence nor what he may have lost or gained as a member of great societies. But let us not be too quick to judge, without knowing them, the peoples of this unhappy land of America which has so much to reproach us for. Were it ever possible for an American aboriginal to write the history of his country, he would portray these men of the Old World, these Europeans who trumpet their civilization and vaunt their philanthropy, he would portray them, say I, as the executioners of his chiefs and his innumerable fellows, bent on returning his entire race to the dust from which it was born.

**SEPTEMBER 1791**
PACIFIC OCEAN

### *The arrival of the Europeans*

stopped Nature in its tracks and condemned America to grow old before it had grown up. That by our actions, a large number of nations on this unhappy continent have disappeared from the face of the Earth that they were to share with us is an affront, equally and simultaneously, to reason and to humanity. The conquerors who devastated the New World have made no effort to heal, if such a thing were possible, the wounds they inflicted on the human race; made no effort to introduce to those peoples who survived where others did not, a social order that, by restoring to them the dignity of Man, might rekindle their love of life and their desire to preserve their species and multiply. It is hardly surprising that the prospect of death inspires no fear in the Peruvian aboriginal! The man is a slave, and death for the slave is merely to shed the intolerable burden of a miserable existence.

◄ **Facing-page:**
Marquesas stilt steps.

▲ **Top:**
Map showing the route of the *Solide* through the Queen Charlotte islands, Canada.

**SEPTEMBER 1791**
**CHINA SEA**

*And having destroyed these men* of the New World, those Europeans proposed to fill the void with a handful of rejects from European societies, men skilled in destruction but with no capacity to create; adventurers for the most part, plucked from the bosom of their mother country, as rotten as a putrid ferment that pollutes the air and seethes with the germs of contagion. If you now combine these impure elements with a few thousand black men bought on the sands of Africa, what you obtain is the species of people that replaced a small proportion of the primitive population reaped by the European scythe in the two Americas. And it is not with new principles of destruction that half of the Earth will ever recover its losses or be regenerated: the masters of the New World are too strong and the American native too weak to admit of his ever being able to shake off the yoke of oppression unaided.

Who can predict what the future will hold for this vast continent in centuries to come? Who can ascertain its destinies in the hands of enterprising European men, greedy for gold and glory, as they become a naturalized population and naturalize their arts and industry in the process? If ever there is to be a revolution in favor of America, it can only ever be at the hands of Europeans who have become Americans. They and they alone can restore America to the weight it merits in the political balance of the globe, just as it has weight in the Earth's physical balance.

**NOVEMBER 1791**
**MACAO, CHINA, PORTUGUESE TERRITORY**

*The news received in Macao* disconcerted the expectations of the owners of the *Solide* regarding their vessel's expedition along the northwest coasts of America; it was plain that the lack of a first success would influence all subsequent operations hinging on a sale that would otherwise have taken place in China. It was discovered that the Chinese government had lately prohibited, on pain of severe punishment, the introduction of all furs in the ports in the south of the empire, and most particularly otter skins. The severity of this prohibition was generally attributed to some condition stipulated in favor of the Russians in a new trade treaty between the Chinese Emperor and the Empress of Russia, a treaty most likely made necessary by various differences between the two powers. But a few people felt that the reason lay rather in the avarice and cupidity of the Mandarins. This combination of adverse circumstances left little chance of bargaining to best advantage. Even should the prohibition come to be lifted while the *Solide* was anchored in Macao, the great concurrence of sellers would necessarily bring down the price of furs to such an extent that the sale would yield a loss, not a profit.

Captain Marchand nevertheless deferred any decision until such time as he had received a response from the Maison Baux, which was established in Canton, to which he had written requesting more precise intelligence. Said response only confirmed every thing he had learned in Macao: namely that the prohibition rendered it impossible to sell his cargo of furs in Canton.

*Were it ever possible for an American to write the history of his country, he would portray these Europeans, and rightly so, ...*

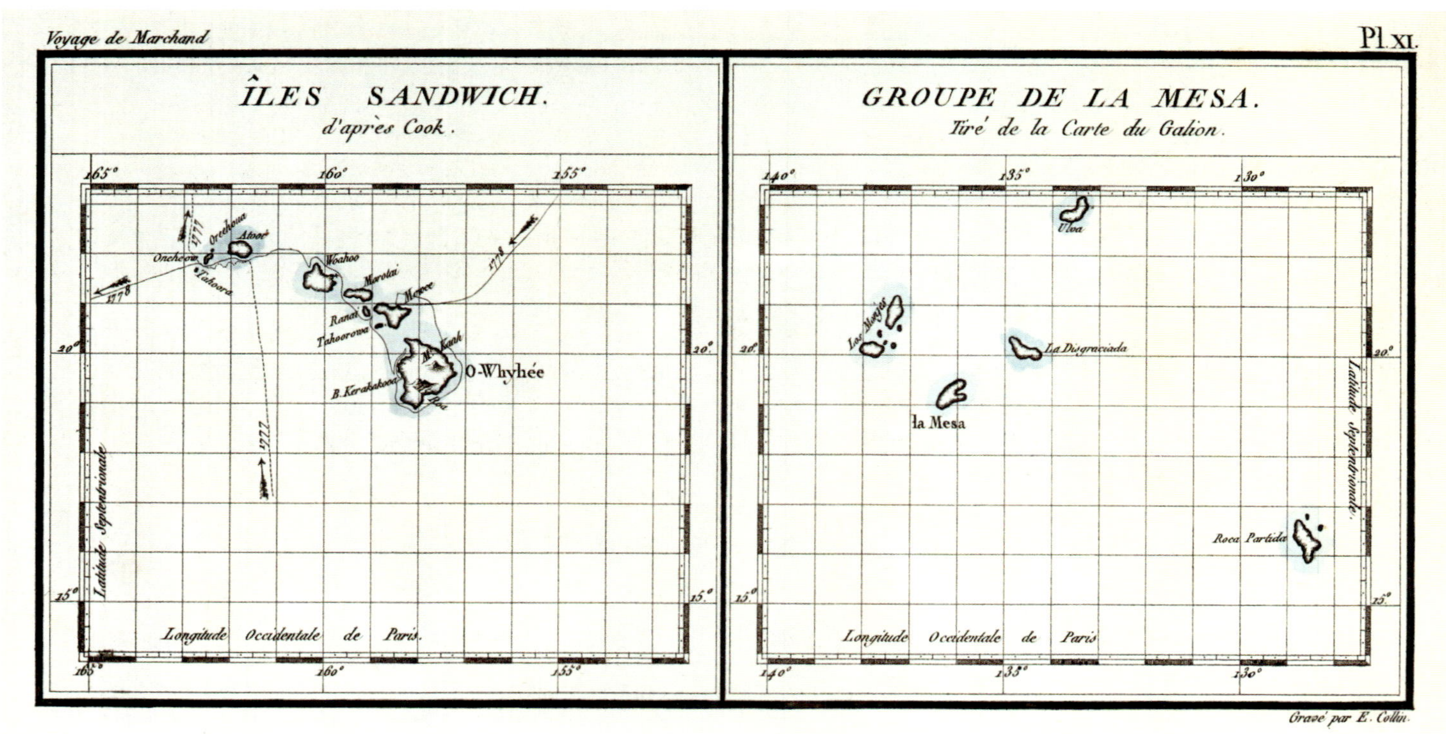

*... as the deadliest scourge ever to have desolated the human race.*

**DECEMBER 1791**

**MACAO, CHINA, PORTUGUESE TERRITORY**

*During his sojourn in Macao,* Captain Marchand frequently had occasion to experience the injustices and harassment of the Chinese government—a source of vexation vehemently decried by any voyager having sojourned in the only port in China that is open to foreigners. That he is not allowed to purchase his own provisions but must address himself to the *comprador* (Chinese buyer or commissionaire) means he must pay twice the normal price for all his supplies.

Most certainly the Chinese empire and its two hundred million inhabitants are cast in a very unfavorable light by the accounts of every navigator since Commodore Anson, each one depicting in ever more hideous detail the rank bad faith of the Chinese government. These same navigators, however, would have us judge China as a whole on the basis of the town of Canton, being the only town of which they are allowed a glimpse without gaining the keys to the city itself. Perhaps, to form an accurate idea of China, we should wait until such time as she grants permission to sensible, unbiased men well versed in its writing and language to remain for some extended time at the court of Pekin, to travel its provinces, live in its countryside and commune freely with Chinese people from all walks of life. But when we consider every thing that is required to permit of this familiarity, we are forced to conclude that we can only ever know China as we know the interior of a convent to whose parlor we may at times be admitted.

◄ **Facing-page:**
Map showing the route of the *Solide* through the Sandwich Islands.

▲ **Top:**
Scene of everyday life in eighteenth century China.

▼ Bottom:
Scene of everyday life
in eighteenth century China.

► Facing-page:
Map showing the route
of the *Solide* between Java
and Borneo.

**AUGUST 4, 1792**
**TOULON, FRANCE**

*At six o'clock in the morning* the *Solide* entered the Straits of Gibraltar together with eight other vessels on the same course. The currents quickly carried the ship to the Mediterranean, and in ten days reached the French coast. At half past five on August 14 the *Solide* dropped anchor in the little harbor of Toulon, bringing its journey around the world to a happy conclusion.

I note that the ship was not what seamen would call a refined sailing vessel. It was solid in name and in fact, built to withstand the rigors of a long crossing and battle the waves in heavy seas; but it did not possess those qualities required for great speed. The time the journey consumed, therefore, is attributable not to swiftness but to ship's always taking the most direct and shortest route between two points, significantly shortening the length of each route traveled.

Over the course of this twenty-month long expedition, notwithstanding the fatigues and privations of a long sea crossing, sailing in all kinds of weather with wildly fluctuating temperatures, the *Solide* lost but one of its fifty-strong crew, and that due to a form of apoplexy. In the normal way of society, there is more than one person in fifty who dies in the course of twenty months, assuming all to be in their third decade.

Captain Marchand's journey is remarkable for the short time he took, circling the globe by way of Cape Horn, and returning by way of China.

*How will Europe ever form an opinion of China, when foreigners are not at liberty to enter and her countrymen are not at liberty to leave?*

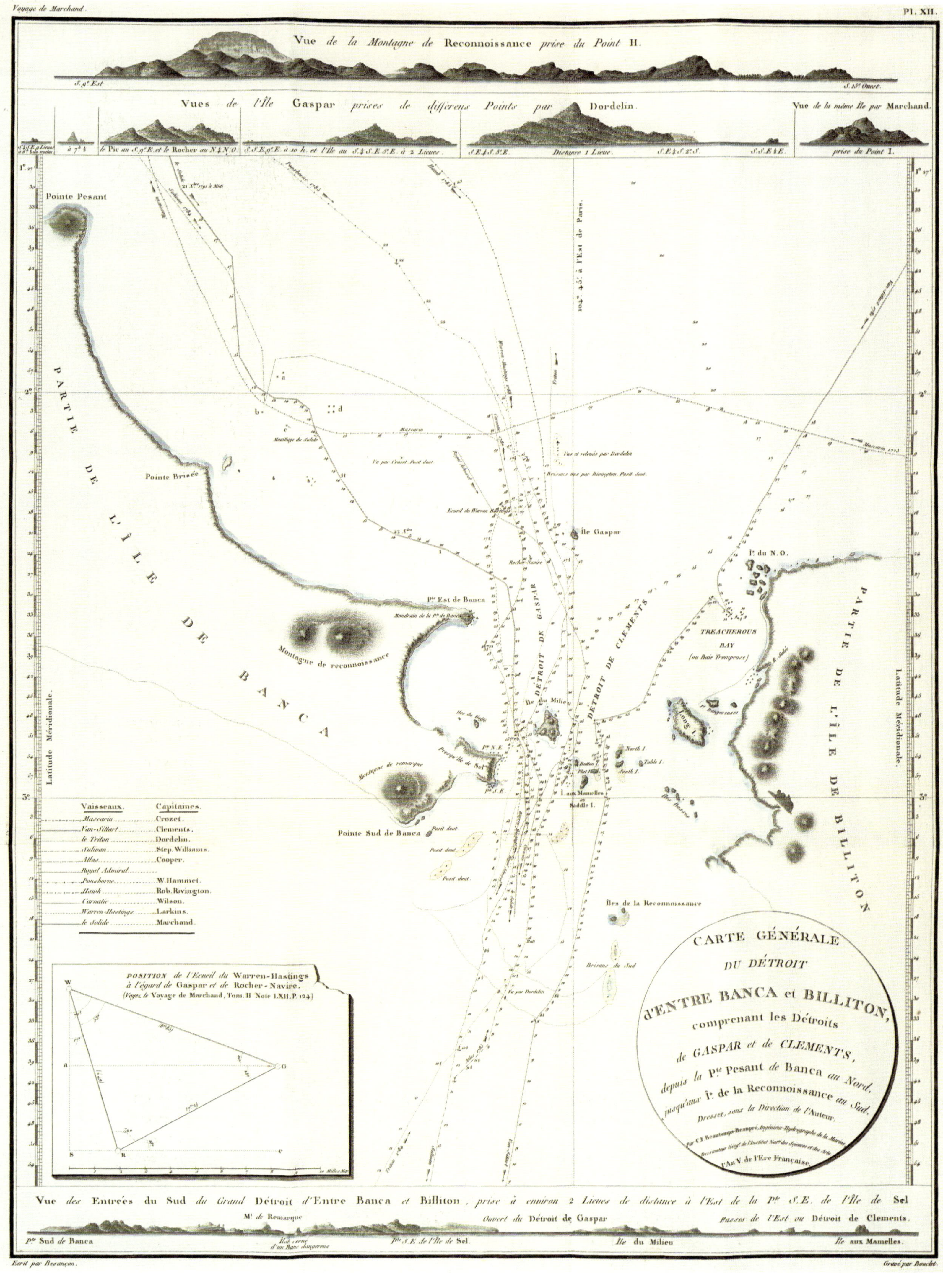

Voyage de Marchand.
Pl. XII.
Vue de la Montagne de Reconnoissance prise du Point H.
S. 9° Est
S. 13° Ouest
Vues de l'Île Gaspar prises de différens Points par Dordelin.
Vue de la même Île par Marchand.
S.S.E. ¼ S.
Distance 1 Lieue
S.S.E. ½ E.
prise du Point I.
Pointe Pesant
PARTIE DE L'ÎLE DE BANCA
Pointe Brisée
Pte Est de Banca
Montagne de reconnoissance
Île Gaspar
DÉTROIT DE GASPAR
DÉTROIT DE CLEMENTS
104° 45' à l'Est de Paris.
Pe du N.O.
TREACHEROUS BAY
(ou Baie Trompeuse)
PARTIE DE L'ÎLE DE BILLITON
Île du Milieu
North I.
Table I.
South I.
Île aux Mamelles
Saddle I.
Pointe Sud de Banca
Îles de la Reconnoissance
Latitude Méridionale.
Vaisseaux. Capitaines.
Mascarin ........ Crozet.
Van-Sittart ........ Clements.
le Triton ........ Dordelin.
Sulivan ........ Step. Williams.
Atlas ........ Cooper.
Royal Admiral ........
Ponsborne ........ W. Hammet.
Hawk ........ Rob. Rivington.
Carnatic ........ Wilson.
Warren-Hastings ........ Larkins.
le Solide ........ Marchand.
POSITION de l'Ecueil du Warren-Hastings à l'égard de Gaspar et de Rocher-Navire.
(Voyez le Voyage de Marchand, Tom. II Note LXII. P. 124.)
CARTE GÉNÉRALE DU DÉTROIT d'ENTRE BANCA et BILLITON, comprenant les Détroits de GASPAR et de CLEMENTS, depuis la Pte Pesant de Banca au Nord, jusqu'aux Îs de la Reconnoissance au Sud.
Dressée, sous la Direction de l'Auteur,
Par C.F. Beautemps-Beaupré, Ingénieur-Hydrographe de la Marine, Dessinateur Géogr. de l'Institut Natl des Sciences et des Arts.
l'An V. de l'Ere Française.
Vue des Entrées du Sud du Grand Détroit d'Entre Banca et Billiton, prise à environ 2 Lieues de distance à l'Est de la Pte S.E. de l'Île de Sel
Passes de l'Est ou Détroit de Clements.
Ouvert du Détroit de Gaspar
Pte Sud de Banca
Pte S.E. de l'Île de Sel.
Île du Milieu
Île aux Mamelles.
Ecrit par Besançon.
Gravé par Bouclet.

# Camille
# de Roquefeuil

**1781–1831**

WARS AND REGIME CHANGE
SINCE THE FRENCH REVOLUTION HAD DIMINISHED
THE RATE OF MARITIME DISCOVERY AND
WEAKENED MARITIME TRADE.
FRANCE'S COFFERS WERE EMPTY.
THE SOLUTION, SAID THE TENACIOUS AND COMBATIVE
CAMILLE DE ROQUEFEUIL, WAS TO SET UP
A VAST, WORLDWIDE BARTER SYSTEM.
AND WITH THAT, HIS HEART FULL OF CONFIDENCE
AND HIS HOLDS FULL OF PRODUCTS
"MADE IN FRANCE", HE EMBARKED ON
AN EXPEDITION JUST AS FRANCE WAS EMBARKING
ON THE RESTORATION. BUT WHILE POLITICAL
UPHEAVAL HAD FADED FROM
FRENCH SHORES FOR THE TIME BEING,
ELSEWHERE REVOLT RUMBLED ON ...

Plate 77, George Wolfgang Knorr, 1766.

# DIARY OF A VOYAGE AROUND THE WORLD

**1816–1819**

**In 1816 Camille de Roquefeuil was the first to relaunch French circumnavigation. At the age of thirty-five he sailed up the American coast and ventured into revolutionary Chile, missionary San Francisco and Russian-dominated Alaska, paving the way for others to follow in his footsteps.**

Camille de Roquefeuil was born in the Tarn region of France, into a family of soldiers and sailors, on January 28, 1781. He was eight when the Bastille was stormed, twelve when King Louis XVI was beheaded, and twenty-three when Napoleon was crowned Emperor of France. Events like these leave their mark on a man. When his family emigrated to Mauritius, Roquefeuil quit the French merchant marine and joined the royal navy. At just twenty years of age, on Fructidor 2, Year IX (August 20, 1801), he saw his first action aboard the *Chiffonne* and the *Flèche* fighting the English in the Seychelles, which earned him his stripes as an officer. Seven year later he was wounded in the right arm fighting the English near Sri Lanka; a year later, in 1809, he lost a finger on his left hand fighting them off Madagascar. For the next eight years, he fought one battle after another, winning honors and promotions galore. Awarded the Légion d'Honneur at an early age, he was made lieutenant and took command of several ships: the *Bellonne*, the *Bayadère* and the *Antigone*. Camille de Roquefeuil was a great sailor and a great Frenchman ...

On June 25, 1816, Roquefeuil lived at 11 Rue du Pont-de-la-Mousque in Bordeaux. He had previously met the merchant, shipowner and slave trader Jean-Étienne Balguerie, known as Balguerie Junior; now he convinced him to finance an around-the-world expedition to source goods from China. France needed to revive its great maritime expeditions, not leave the oceans to the English. Triangular trade was the key to obtaining foreign goods "with no cash outlay". For Camille therefore this expedition served the public good and the King alike.

However, since the purpose of this voyage was commercial, Roquefeuil had to obtain permission from the Minister of the Navy, who refused. In a letter written to his father on the day of his departure, he expressed his astonishment at having been struck off the list of lieutenants eligible for promotion, and stated his conviction that he was being useful to his country and serving the King. Though he raged at the injustice and lack of support, his self-confidence remained intact as did his determination to go ahead with his project. His ship was to be the *Bordelais*, a small two-hundred-ton, three-masted vessel armed with eight cannon and manned by a crew of thirty-four men. He took on board all the current works on circumnavigation, most notably those by George Vancouver, an English explorer who knew the American coast well. Above all, he took along the best chronometer of the time, the Breguet 172, which would enable him to locate his discoveries with precision and correct the maps of his predecessors.

On October 19, 1816, the *Bordelais* set off across the Atlantic, skirting the coast of South America before rounding Cape Horn in early January 1817. Our thirty-five-year-old Roquefeuil meanwhile showed his crew every attention, congratulating himself on having had a metal galley fitted to provide better food for his men and procuring them suitably warm clothing for the journey. On arrival in Chile, his ship was overrun by Spanish soldiers deserting the royal troops, courtesy of a tumultuous war of Independence that forced him to cut short his stopover. Leaving Valparaiso in a hurry, he took refuge in the then more peaceable Peru, where he was able to start trading and loading the *Bordelais* with goods to be bartered in North America. He spent two months exploring Peru on horseback, making encounters that he describes in vivid detail in his journal.

He then headed north up the American coast, reaching the port of San Francisco in August 1817. It was here that Roquefeuil discovered just how powerful were the Spanish and the missions, and just how disgraceful was the decline of the natives in the two Californias. He continued north to the Queen Charlotte Islands (Haida Gwaii) and began trading for skins. With bad weather on the way, he decided to spend the winter in the Marquesas Islands, where he spent two months exploring, mapping and naming new places: Cape Balguerie, Roquefeuil Bay and the Bordelais Strait on the coast of Hiva Hoa. In spring, the ship sailed north again to the Kodiak Islands, then Sitka and Nootka, where Lapérouse suffered his first tragedies. Trade continued, the holds filled with skins, but Roquefeuil came up against the Russian domination of northern Alaska.

By 1819, it was time to head west to China where he arrived in March and, like Marchand, was disappointed by what he found. The price of hides and sandalwood had plummeted, and trading was burdened with disastrous economic conditions, not least because of the taxes imposed by the Mandarins. The *Bordelais* left Macao on April 17 and sailed nonstop towards the Cape of Good Hope. Lacking provisions, he had to beg from passing ships on three occasions. Due to weak and unfavorable winds, it wasn't until November 21 that the *Bordelais* returned to its home port.

Roquefeuil's expedition was not a commercial success but that hardly mattered given the importance of his geographical and ethnological discoveries along the northwest coast of North America—discoveries that saw him decorated but not reinstated by the Ministry of the Navy. Though he wrote an account of his voyage, his failure to have it published meant that his journey was all but forgotten. Resigned to remaining in the merchant navy, he returned to the Indian Ocean as a long-haul captain, handing over his manuscript to his family before leaving France. It was they who published his story in 1823.

In October 1825, Roquefeuil married Émilie Lavergne-Dennemont in La Réunion, with whom he had a son, also a soldier. He then became captain of the port of Saint-Paul, but contracted dysentery and died on November 7, 1831 at the age of fifty-one, without ever seeing his beloved France again.

Map of northwest coast of America, by Camille de Roquefeuil.

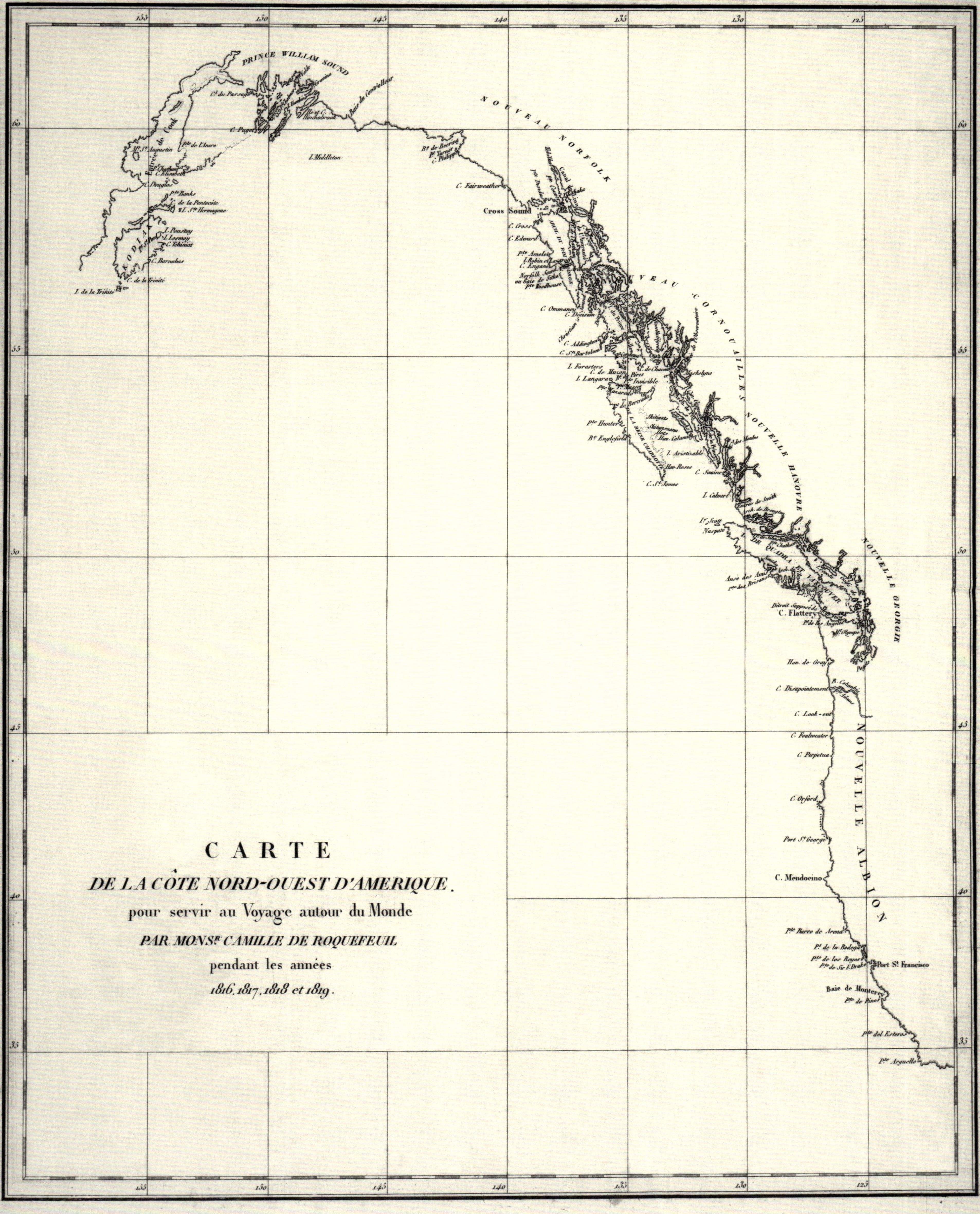
CARTE
DE LA CÔTE NORD-OUEST D'AMERIQUE.
pour servir au Voyage autour du Monde
PAR MONSR. CAMILLE DE ROQUEFEUIL
pendant les années
1816, 1817, 1818 et 1819.
PRINCE WILLIAM SOUND
Baie du Contrôlleur
NOUVEAU NORFOLK
NOUVEAU CORNOUAILLES
NOUVELLE HANOVRE
NOUVELLE GEORGIE
NOUVELLE ALBION
I. Middleton
C. Fairweather
Cross Sound
C. Ommaney
C. Addington
I. Langara
Pte. Hunter
I. Aristizable
I. Calvert
C. Flattery
Hav. de Gray
C. Disapointement
C. Look-out
C. Foulweater
C. Perpetua
C. Orford
Port St. George
C. Mendocino
Pte. Barro de Arena
Pt. de la Bodega
Port St. Francisco
Baie de Monterey
Pte. de Pinos
Pte. del Esteros
Pte. Arguello
I. de la Trinité
C. de la Trinité
C. Barnabas
C. Douglas
C. St. James
C. Edward
C. Gross
Mt. St. Augustin
155
150
145
140
135
130
125
60
55
50
45
40
35

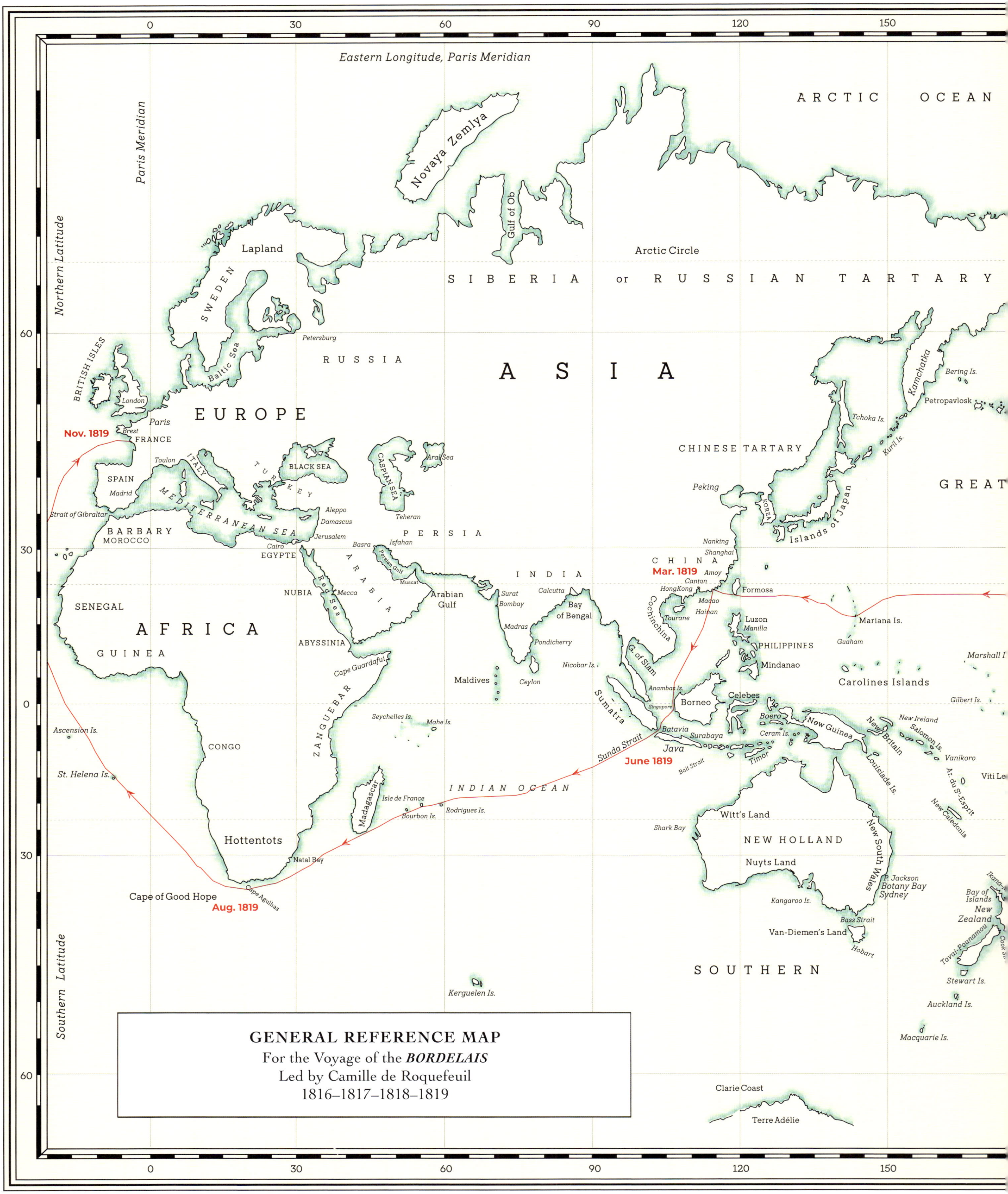


GENERAL REFERENCE MAP
For the Voyage of the ***BORDELAIS***
Led by Camille de Roquefeuil
1816–1817–1818–1819

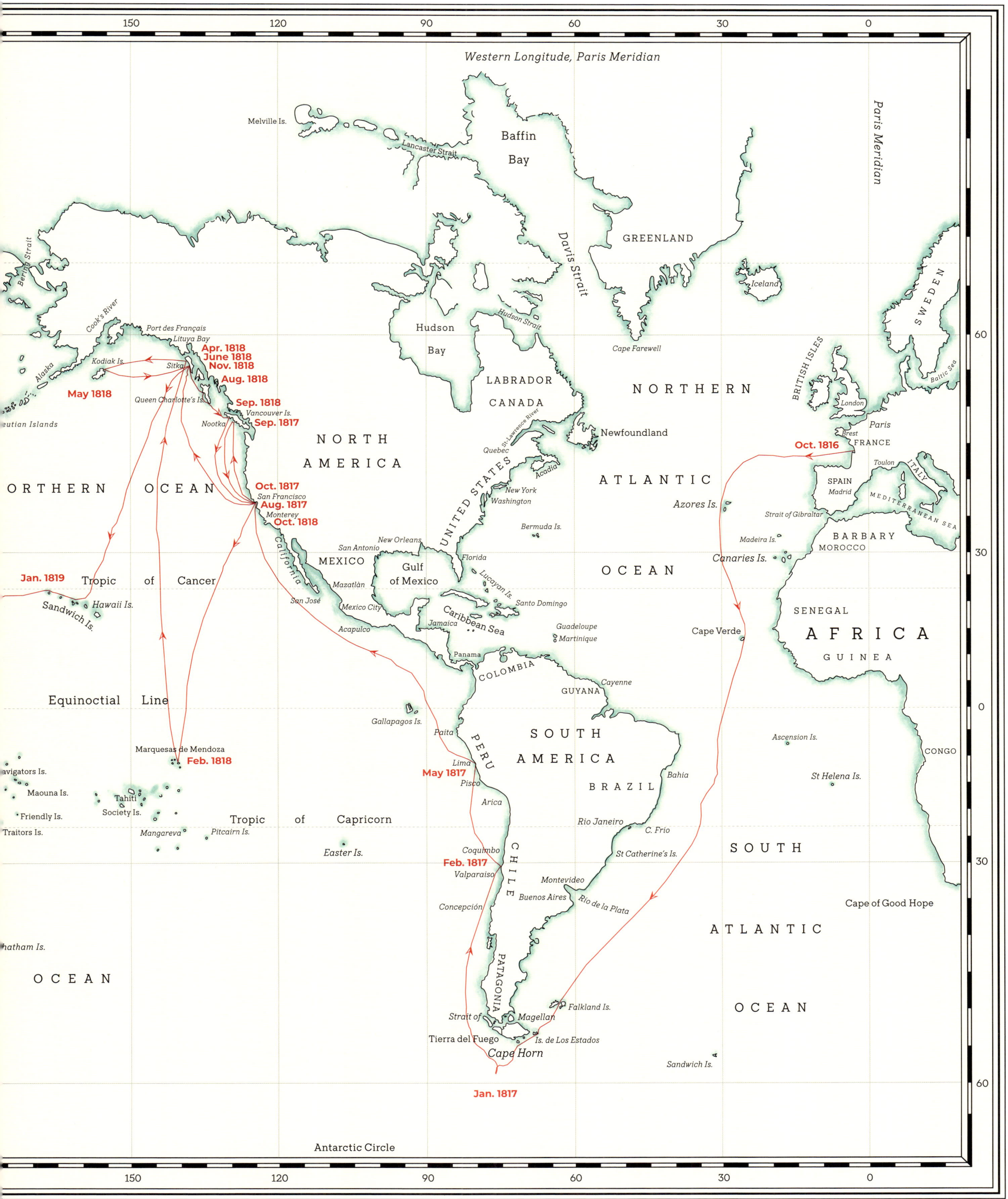

Western Longitude, Paris Meridian
Paris Meridian
Melville Is.
Lancaster Strait
Baffin Bay
Davis Strait
GREENLAND
Iceland
Hudson Strait
Hudson Bay
Cape Farewell
LABRADOR
CANADA
NORTHERN
BRITISH ISLES
SWEDEN
Baltic Sea
London
Paris
Brest
FRANCE
Oct. 1816
Toulon
ITALY
SPAIN
Madrid
MEDITERRANEAN SEA
Strait of Gibraltar
BARBARY
MOROCCO
Madeira Is.
Canaries Is.
Azores Is.
ATLANTIC
OCEAN
Newfoundland
St-Lawrence River
Quebec
Acadia
New York
Washington
UNITED STATES
Bermuda Is.
NORTH AMERICA
Bering Strait
Cook's River
Port des Français
Lituya Bay
Apr. 1818
June 1818
Nov. 1818
Aug. 1818
Sitka
Kodiak Is.
Alaska
May 1818
Queen Charlotte's Is.
Sep. 1818
Vancouver Is.
Nootka
Sep. 1817
Aleutian Islands
NORTHERN OCEAN
Oct. 1817
San Francisco
Aug. 1817
Monterey
Oct. 1818
California
New Orleans
San Antonio
MEXICO
Gulf of Mexico
Florida
Lucayan Is.
Mazatlàn
San José
Mexico City
Acapulco
Jamaica
Caribbean Sea
Santo Domingo
Guadeloupe
Martinique
Panama
COLOMBIA
Jan. 1819
Tropic of Cancer
Hawaii Is.
Sandwich Is.
Cape Verde
SENEGAL
AFRICA
GUINEA
Cayenne
GUYANA
Equinoctial Line
Gallapagos Is.
Paita
PERU
SOUTH AMERICA
Ascension Is.
CONGO
Marquesas de Mendoza
Feb. 1818
Navigators Is.
Maouna Is.
Tahiti
Society Is.
Friendly Is.
Traitors Is.
Mangareva
Pitcairn Is.
Tropic of Capricorn
Easter Is.
Lima
May 1817
Pisco
Arica
BRAZIL
Bahia
St Helena Is.
Rio Janeiro
C. Frio
SOUTH
Coquimbo
Feb. 1817
Valparaiso
CHILE
St Catherine's Is.
Montevideo
Buenos Aires
Rio de la Plata
Concepción
Cape of Good Hope
ATLANTIC
Chatham Is.
OCEAN
PATAGONIA
Falkland Is.
OCEAN
Strait of Magellan
Tierra del Fuego
Is. de Los Estados
Cape Horn
Sandwich Is.
Jan. 1817
Antarctic Circle
150
120
90
60
30
0

Having been all but annihilated by the Revolution,
by wars and the errors that persisted thereafter,
French maritime commerce by the time
of the Second Restoration found itself confined
within those narrow limits in which it was held
by an implacable rivalry with England.
Its resources were further depleted by the enormous
amount of cash that was periodically withdrawn
from our coffers in execution of the latest treaties.
In such circumstances, France had good reason
to go in search of new openings for its national industries:
to try, by means of our own products,
to revive and nurture our former relations
with a country whose products had until lately
only been procurable by Europe at the cost
of a considerable portion of those precious metals
supplied to us by the New World.
The goal of the expedition was to furnish China
with valued objects that would be redeemed in Chinese
merchandize for consumption in France,
thus supplying our own markets through the useful
employ of the fruits of own soil and our own industries
but with no cash spent whatsoever.

CAMILLE DE ROQUEFEUIL,
BORDEAUX, 1821

**OCTOBER 1816**
BORDEAUX, FRANCE

*The* Bordelais *departed Bordeaux* on October 11, 1816. Fitting-out was undertaken in the preceding summer and left nothing to chance. The ship, a three-master specially chosen for the expedition, was furnished with every item required for a campaign lasting two years: a pinnace-cum-longboat as big as the vessel could carry; two whalers including one in parts; and a pinnace at the stern. It was armed with two four-pounder and six eight-pounder cannon, plus a sufficiency of weapons of every sort for the crew and its vessels. The greatest care was taken in the choice of provisions and the preparation of food; and no less important, the ship's navigational instruments included an excellent Breguet chronometer, Number 172.

**JANUARY 1817**
TIERRA DEL FUEGO

*I determined on a night-time* crossing of the Straits Lemaire, the weather being cloudy but not threatening. The Straits Lemaire are situated between Tierra del Fuego and the Isla de Los Estados. The Dutch had granted but one company the exclusive privilege of sailing to the Indies by way of the Straits of Magellan; to avoid contravening that privilege, Lemaire set sail for Brazil with two ships fitted out in Cape Horn, hoping to find a way through what are now the Straits Lemaire. His hopes were crowned with success, marked with the discovery of the straits that have immortalized his name ever since.

The weather was dark and misty with large hailstones and heavy rain: there was a big sea running but with regular swells, and the temperature never dropped below four degrees. The wind then veered northwest and grew stronger. On January 14 we encountered a whaler on an east-southeast course. It hoisted an English flag and we hoisted our own. At night, the wind dropped. In less than an hour we caught thirteen albatrosses on a line, some of them very big, one with a wingspan of ten feet.

**MAY 1817**
LIMA, PERU

*The self-respect and ambition* of the Peruvians is offended by the preference shewn by the Government to Europeans, notwithstanding that Creoles are often entrusted with important offices. Wealthy, well-traveled men with a degree of knowledge uncommon in most of their compatriots believe themselves qualified to hold the very highest office. They therefore resent not being called upon to administer their own country and will use every means at their disposal to accelerate progress towards an independence that will bring them out of their obscurity. It is to be feared, however, lest such a gift prove fatal to a country where the rights and duties of citizens

*Vue des Etablissements de Schelikoff dans l'Ile de Kadiak.*

1 *Eglise portative*
2 *Tente servant d'Observatoire*
3 *Galiotes*

*Extrémité Septentrionale de l'Ile d'Yanaga.*

*Ile d'Atcha.*

*The man who travels will feel his ideas multiply and expand; his memory will become enriched with new thoughts and images.*

are largely ignored; where superstition and apathy discourage patriotism and impede the emergence of public spirit; and where speechifying and the lust for power invariably go hand in hand with the inability to direct the public spirit for the public good. The circumstances of the time would most likely lead to the adoption of a republican government, though the customs and manners be monarchical. Considering the insufficiency of the means deployed by Spain to quell the uprising in the south, which asserts its independence more forcefully every day, there is every likelihood of the Peruvian's wishes being soon fulfilled.

**AUGUST 1817**
**SAN FRANCISCO, ALTA CALIFORNIA**

*On the fifth day of the month* at two o'clock, the mist finally began to lift and the coast of California came into view. Though land was barely two leagues distant, our view was hampered by atmospheric conditions and we ran aground. The fog soon clearing, we recognized the coast which extends in the west from Point Reyes to the port of San Francisco in the southwest. The point was hidden by the mist but we could distinguish the entrance to Port Drake.

Top: View of Kodiak Islands, Alaska.

Pages 140-141: Views of the American coastlines as seen from the ship.

The sea was beautifully calm and the weather was superb. Three big whales played for a while around our vessel. We found the entrance to the port and shortly afterward recognized the fort at the southeastern tip of the entrance. We hoisted our colors, supported by a single cannon shot. The fort did likewise, hoisting the Spanish flag. The ship was soon borne in on the waves, coming alongside somewhat to the south. At six o'clock we passed under the fort and were hailed by an officer. We answered that our ship was lately arrived from Lima. The view depicted by George Vancouver of the coast of San Francisco is accurate in every detail.

**AUGUST 1817**
**SAN FRANCISCO, ALTA CALIFORNIA**

*We dined at the mission,* together with Don Louis, at the invitation of Father Ramon who had officiated. The meal was neatly served, consisting of a small number of simple, wholesome dishes prepared in the Spanish manner. The bread, meat, vegetables and fruits were produced at the mission and were of good quality. The bread however did not seem to me to be properly kneaded and the wine, which was from Santa Barbara, was vinegary for want of correct keeping.

Conversation mainly revolved around the alarming decline in the indigenous populations being witnessed by missions in Old and New California. It was agreed that indigenous people were all but extinct in the former, where the number of missions had fallen from seven to just two; one person went so far as to suggest that all that remained there were *presidios*. It was also admitted that in New California, ever more fertile and more populous, there remained but one single mission where the birth rate equaled the death rate.

**SEPTEMBER 1817**
**FRIENDLY COVE, IN NOOTKA BAY, BRITISH COLUMBIA, CANADA**

*It was only with the aid of pinnaces* that we were got to Friendly Cove, where we dropped anchor at half past six in ten fathoms of water, over a bed of mud and shells. We immediately laid our kedge anchor in a fork mooring aft. At night the officers stood watch as at sea, with a pilot and four men on guard fore and aft. The scramble nets were hung, the cannon loaded and a quantity of weapons were arranged on the deck.

In the morning we laid out various objects that might best serve us in trading with the natives, who came in great numbers and remained a very long time in their canoes alongside the ship. This late in the season, the villages were almost entirely deserted, big chief Macouina and the greater part of the Indians having already removed to Tachès, the chief's winter residence. The natives appeared to make rounds at night.

During our short sojourn in Nootka, the Indians gave us no cause for concern and their peaceful behavior rendered all our security measures unnecessary. With the greatest of ease, we took on water, wood and every provision offered by the locality. Sad to say furs are far less abundant than in former times having been traded to the point of exhaustion, such that we were only able to obtain twenty pelts.

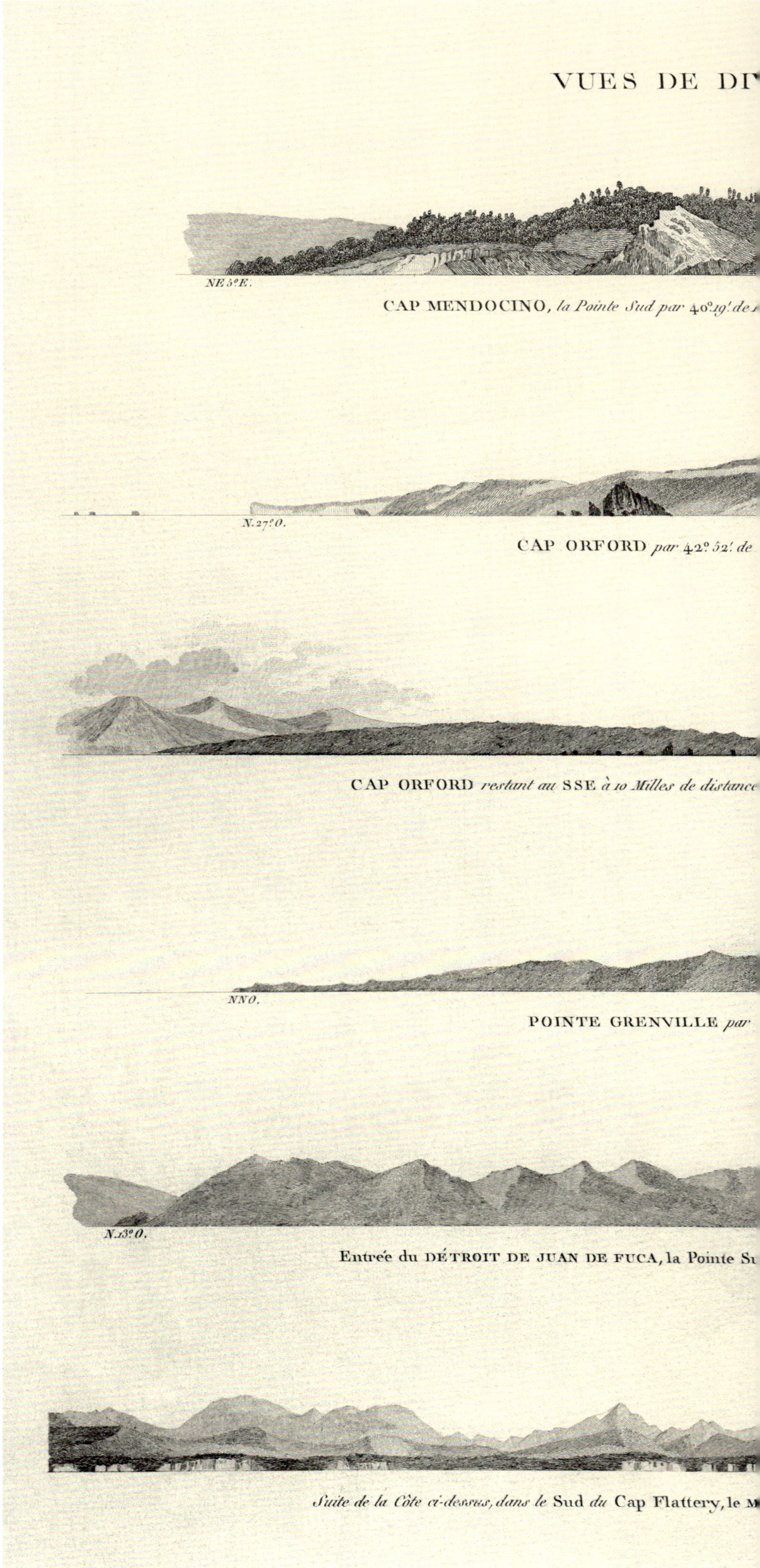

ES PARTIES DE LA CÔTE NORD-OUEST D'AMÉRIQUE.

t 235°. 53'. de Long. Orient. restant au S.E. du Compas, à la distance de 5 Lieues. la terre la plus proche à l'E$\frac{1}{4}$NE à 10 Milles de distance. Variation 15°. NE.

et 235°. 35'. de Long. Orient. restant au N. 27°. O. à la distance de 5 Lieues, la terre la proche au N. 64°. E. à 3 Milles et $\frac{1}{2}$ de dist.ce Variation 16°. NE.

CAP GREGORY par 43°. 23'. de Lat. Nord et 235°. 50'. de Long. Orient. restant au NNE à 5 Milles de distance.

Lat. Nord et 235°. 58'. de Long. Orient. restant à l'E$\frac{1}{4}$ SE à la distance de 7 Milles, la terre la plus proche au NE$\frac{1}{4}$N à 4 Milles de dist.ce Variation 18°. NE.

P FLATTERY par 48°. 23'. de Lat. N. et 235°. 38'. de Long. Orient. restant au N. 5°. E. à 6 Lieues et $\frac{1}{2}$ de distance. la Pointe Nord de l'Entrée au N. 13°. O. Variation 18°. NE.

YMPE par 47°. 50'. de Lat. Nord et 236°. 34'. de Long. Orient. restant à l'E à 15 Lieues et $\frac{1}{2}$ de distance, la terre la plus proche au NE$\frac{1}{4}$ E. à 5 Milles de dist.ce Variation 18°. NE.

**FEBRUARY 1818**
**MARQUESAS ISLANDS**

*— The Marquesas Islands are a good port* of call for ships that, having rounded Cape Horn, might be required by the nature of their expedition to travel to certain parts of Australasia, and for whalers that operate in the great Southern Ocean.

The Island of Nuku Hiva [Île Marchand] produces the best sandalwood in all the archipelago. The American seafarer, Captain Rogers, was the first to engage in the trading of sandalwood, which he chanced to discover from the fragrance given off by a fire into which the Indians had thrown a few sticks of this precious wood. In 1810, he obtained more than 260 tons of sandalwood in exchange for objects with an initial value of less than one thosand piasters, among them axes and other tools, liquor and a few whale teeth that he happened to have on board, one of which was then worth three or four tons in itself. He sold his cargo in China for twenty piasters a pickle and returned for a second shipload in exchange for goods worth three thousand piasters. This time he took on ivory that he fashioned aboard ship into the shape of whale teeth, of which he had only been able to obtain a small quantity. This deception netted him a goodly profit but the natives soon got wise to his tricks and would not now allow themselves to be hoodwinked in this fashion. Axes and other tools remain much in demand, iron less so. Kerchiefs and blue and white canvas are mainly popular with the women, who also have that predilection for mirrors typical of their sex. Ornamental feathers, particularly red feathers, are another favorite.

That said, all of these items take second place to weapons and gunpowder, which remain the basis of trade even though their value compared to sandalwood has significantly diminished. Until quite recently, a rifle was worth one stere of sandalwood.

The whites on these islands are not to be trusted, most of them being naval deserters with all the vices of civilization but none of its attributes in terms of education. Despite their small number, they contribute not a little to the loss of those qualities that continued to distinguish the Indians at the end of the past century, or so we are told by the navigators of those times.

*The resources of this little island have been well-nigh exhausted by the exporting of nearly eight thousand tons of sandalwood.*

**SEPTEMBER 1818**
**QUEEN CHARLOTTE ISLANDS [HAÏDA GWAÏI], CANADA**

*There is every reason to believe* that the Nootka Indians have practiced anthropophagy and I would be tempted to believe that this abominable custom persists among many of them, even if they do not admit it. They offered various parts of the human body, including some cooked pieces, to Captain Cook's people and to the Spanish aboard the *San-Carlos* and several other ships. Little realizing how abhorrent were such practices to civilized people, they admitted to the Spaniards of the settlement, when first they made contact with them, that a few of their most revered chiefs and warriors continued to indulge this execrable taste when they were preparing to march on the enemy.

*No sooner did they capture their enemy than they slit his throat and skinned him while his companions celebrated their deliverance.*

**OCTOBER 1818**
**QUEEN CHARLOTTE ISLANDS [HAÏDA GWAÏI], CANADA**

*Polygamy is an established custom* among the *tahis* (chieftains) and the nobles, who consider it a sign of wealth and prestige. Indeed, girls are only given up for marriage in exchange for gifts to their parents of furs, canoes, European-style garments, rifles, &c; thus do they represent a source of wealth for their families, provided they have the advantage of being well-favored. The poor *mistchimis* (slaves) dispose of a mere fraction of the fruits of their labors and can rarely afford such expense. The best placed among them may have a woman bestowed upon them by the *tahi* as a reward for their services, but celibacy for most is their lot. Notwithstanding this purchasing of brides, so to speak, women are treated with great kindness by their husbands, who demand of them only the performance of such household chores and other tasks as are suited to their sex. According to various accounts worthy of belief, women exert a marked supremacy over the stronger sex. Those in a few of the northern tribes have been observed to mistreat men most cruelly. Even assuming, as I do, that the aforesaid facts relate to the mistreatment of lower-ranking men by high-ranking women, this particularity would still not suffice to justify an act of authority that is the more disgusting for being in direct opposition to the laws of Nature. It is almost always women who steer the war canoes. The destruction of the first Russian settlement in Sitka was brought about by women in the neighboring tribes who took offence at the disdain shown by the governor's mistress. Not the least of her affronts, this Creole woman from Kodiack had spat in the decorative lip plate worn by a chief's wife.

**NOVEMBER 1818**
**AT SEA, WEST COAST OF NORTH AMERICA**

*The Russians, whose foothold* on the American continent has been expanding since their conquest of Kodiack in the time of George Vancouver, presently occupy the whole of the crescent-shaped coast line from Cap Cross to the tip of Alaska. They have outposts at Cook River, Prince

◄ ▼ Facing-page and bottom:
Traditional burial ground, Marquesas Islands.

*The women said that if the men did not have the courage to fight the Russians, they would do it themselves and launch an attack on their fort*

▲ Top:
Coastal village, Marquesas Islands.

► Facing-page:
Chief of the Marquesas Islands, and his dwellings.

► Pages 146-147:
*Intérieur d'un jardin chinois à Canton*
(a Chinese garden in Canton).

William Sound and all the best spots for hunting and trade, over which they enjoy an exclusive prerogative. It is some eighteen years since they founded a settlement in Sitka which, in 1808, was destroyed in a surprise attack by natives. Governor Alexander Baranoff, having driven out the hostile tribes in this part of King George Island, established the present settlement on the eastern side of the bay, a few miles to the southeast of its predecessor. He named it Novo-Arkhangelsk (New Archangel) and designated it as the seat of the General Government of all Russian possessions in this part of the world.

A decree issued by the Russian Emperor, prompted by the need to protect Russian trading posts by warding off so-called "interlopers" (foreign competitors), contains the following provisions:

Art. I. The Russians alone may frequent the northwest coast of America and that of Siberia, together with the Aleutian Islands, Kurdish lands &c. Foreigners are banned from engaging in navigation of any kind in the area extending from the Bering Strait to latitude fifty-one degrees west and latitude forty-five degrees forty minutes east.

Art. II. No foreigner may approach these coasts to within less than one hundred miles of Italy on pain of having his cargo confiscated.

One is dismayed to see this powerful monarch of the North, a man rightly renowned for his fairness and moderation, sanction an act of usurpation worthy of the ordonnances of sixteenth century kings Ferdinand and Philip, to similar ends but with less regard for the law. Excepting in the eyes of the natives, the Russians enjoy full rights over the lands to the north and west of Cross Sound, by virtue of their having been first to discover and settle these regions and notwithstanding the Acts of Possession performed by the Spanish and the English subsequent to the arrival of the Russians.

## MARCH 1819
### CANTON [GUANGZHOU], CHINA

*The delays we incurred* as much in America as in the Sandwich Islands had unfortunate consequences for our operations in China, where we arrived in late fall. In those days, the impediments to trade were rendered more troublesome still by the unthinkable number of American traders who, having preceded us, had brought down the price of imports and exhausted the supply of local goods thereby pushing up prices. Our copper fetched but nineteen piasters a picuel (136 pounds). Sandalwood from the Sandwich Islands, hitherto never priced at

3. Chef de Tao-Wati.

4. Vallon à Nouka-hiva.

less than thirteen piasters, only fetched nine and Marquesan sandalwood just six. Otter pelts had retained their value having arrived in very limited quantity, fetching us fully thirty piasters despite nearly half of them being Californian pelts of inferior quality. A few furs from the Northern Straits reached fifty-four piasters. Raw silk, tortoiseshell and rhubarb were exhausted and green tea was very rare; so-called "gunpowder", much in demand in France, was only procurable at the rate of eighty piasters a picuel . Sugar was the only reasonably priced product, at seven and a half piasters a picuel and constituted a goodly proportion of our small cargo despite its being of a variety as yet little known in France, and my only being authorized by letters waiting for me in Macao to take on a few quintals as samples.

**APRIL 1819**
**CANTON** [GUANGZHOU], **CHINA**

*While the circumstances* in which we find ourselves preclude the possibility of seeing trade with China reap benefits comparable to those enjoyed by the present masters of India, we may at least aspire to pursue said relationship without losing any money. There may even come a day when we see our fabrics compete with those of England. In the meantime, trade in furs on the northwest coast of America and trade to the South Sea provide us with the wherewithal to meet demand in our own markets. The favor enjoyed by the fruits of our soil and our industries in southern America promises to bring even greater rewards, such that there is every hope of forming extensive and advantageous trade relationships with those wealthy countries situated in the Great Ocean (Pacific). But in the weakened and tremulous state of French commerce following a quarter of a century beset by disasters, rekindling the spirit of enterprise may require encouragement. The appearance of a few well-armed ships of His Majesty's fleet would be most advantageous in terms of enforcing the respect due to the French flag, and evidencing the protection accorded by His Majesty to His subjects' trading activities.

*England has succeeded in turning its commerce with China into a mine of opportunities, so further and enriching its Asian possessions.*

Bottom:
Canton main street in 1820.

Facing-page:
*Arrivée à Bordeaux* (arrival in Bordeaux).

Pages 150-151:
*Vue du détroit de la Sonde* (view of the Sunda Strait).

*By what right do the Russians arrogate to themselves sovereignty over the seaboard up to the 51st North and forbid its access to other nations?*

**NOVEMBER 1819**
BORDEAUX, FRANCE

*The* Bordelais*, having left* Macao harbor on April 17, was buffeted by monsoon winds as it sailed up the China Seas, such that it was not until May 30 that it made its exit by way of the Karimata Strait, proceeding through the Sunda Straits on June 7. A leak discovered at the bow required the ship to put in at the island of Mauritius, where it remained at anchor from July 1 to July 17. On the twentieth day of the month, it entered the harbor of Saint-Denis to collect parcels being dispatched by the Governor of Bourbon Island. The ship rounded the Cape of Good Hope on August 13. Its crossing to the Azores was then delayed by weak and nearly always contrary sea breezes, and by a sequence of often fierce northeast winds to the east of these islands. In consequence, the *Bordelais* was obliged to beg provisions from three different ships and did not land on the shore of Oléron Island until November 19. Three days later it returned to the Gironde, bringing to a conclusion a voyage around the world lasting thirty-seven months and two days, of which twenty-two months and six days under sail.

The crew still numbered twenty-three but of the thirty-four men who had joined the ship in Bordeaux, only seventeen returned, one man having died, five having been put ashore for reasons of health and eleven having deserted.

In considering the general utility of the voyage undertaken by the *Bordelais*, it should be noted that by virtue of its operations on the coasts of the Great Ocean, it was able to secure its return by means of a cargo derived from our own soil and industry; whereas had it loaded its cargo in Macao, as is usual in dealings with China, the cost to France would have been several hundred thousand francs in cash. These considerations will engage the attention of a wise government that fully appreciates every factor affecting the national interest and is disposed to look favorably on a new career that, in commercial terms, has the advantage of substituting the use of goods for that of cash and, in naval terms, promises to provide invaluable schooling for subjects of the highest excellence working in the service of His Majesty's fleet.

I should count my efforts well spent were they to result in the achievement of one or the other of these important objects. I deeply regret having been unable to obtain good enough results to compensate for the sacrifices made by that courageous and patriotic merchant, Balguerie Junior. But whatever the consequences of my journey and notwithstanding its being undertaken aboard a privately-owned vessel, I dare hope that its very nature will prevent its being considered unconnected with the service of the nation.

# Louis Claude de Saulces de Freycinet

**1779–1842**

ONE NIGHT IN SEPTEMBER 1817, IN THE PORT OF TOULON, THREE SHADOWY FIGURES SLIP ABOARD THE *URANIE*. AMONG THEM IS A WOMAN, FORBIDDEN ON BOARD SHIP, HENCE THEIR CAUTIOUS ARRIVAL. HER NAME IS ROSE AND SHE IS THE WIFE OF LOUIS DE FREYCINET, CAPTAIN OF THE *URANIE*. IT IS THROUGH THE PAGES OF HER DIARY THAT TWENTIETH CENTURY READERS WOULD EVENTUALLY COME TO DISCOVER THE JOURNEY OF THE CORVETTE COMMISSIONED BY KING LOUIS XVIII TO STUDY THE SOUTHERN HEMISPHERE. THE TIME HAD COME FOR FRANCE TO REVIVE HER PRESTIGE AND LAUNCH NEW SCIENTIFIC EXPLORATIONS.

Plate 57, George Wolfgang Knorr, 1766.

# VOYAGE OF THE *URANIE*

**1817–1820**

**It was a difficult voyage punctuated by illness, desertion and a violent storm that sent the ship crashing into a reef off the Falkland Islands; but the scientific findings of the Freycinet expedition were considerable, thanks to experiments carried out in the Atlantic and Pacific.**

Louis de Freycinet was born on August 7, 1779, in Montélimar. In 1793, at the height of the French Revolution, he and his brother Henri joined the French Navy as cadets, Louis having just turned fourteen. So it was that he cut his teeth as a sailor amid revolution and battles with the English, developing his knowledge of the sea, gaining experience and becoming inured to hardship. In 1800 the two brothers embarked on an expedition led by Captain Nicolas Baudin, bound for Oceania and in the company of a certain Hyacinthe de Bougainville, Louis aboard the *Naturaliste* and Henri aboard the *Géographe*. This was the first expedition launched by the government since Entrecasteaux' mission in search of Lapérouse, and it was dogged by tragedy. Over the course of a forty-two-month tour of duty, many of the crew members were lost to malnutrition and scurvy, Baudin among them. Baudin died in Mauritius and Louis Freycinet, aged just twenty-five, took over the helm.

On his return home in March 1804, France was in the throes of the Napoleonic Wars. In the absence of far-flung explorations, Freycinet accepted a few commands, took charge of the Dépôt des Cartes de la Marine, and above all, began to write. In particular, he completed the historical part of the travel account by François Péron, a naturalist on the Baudin expedition who had died before he could finish. In the process, Louis Freycinet grew ever more impatient to embark on a new major scientific expedition.

In 1811 he was promoted to frigate captain and on June 6, 1814, he married Rose Marie Pinon. Two years later, he suggested to King Louis XVIII, who had acceded to the throne in 1815, that he should undertake another scientific circumnavigation. The King agreed and Freycinet, now thirty-seven, set to work finding naval officers who were competent in the subjects to be studied: astronomy, natural history, botany and ethnography. He knew from his experience with the Baudin expedition just how important it was to have the right men for the job. The *Ciotat*, a 350 ton corvette, was renamed *Uranie* after the muse of astronomy, and refitted to carry twenty cannon and a crew of one 126 men.

The expedition cast off from Toulon on September 17, 1817, with ship's ensign Louis-Isidore Duperrey, draughtsman Jacques Arago and Rose, Freycinet's twenty-three-year-old wife who had embarked in secret. Women had been forbidden aboard royal ships ever since Jeanne Barret, and this ship was no exception. Her contribution to the trip, however, would be exceptional thanks to the record she kept throughout the voyage, in the form of letters to a friend describing her adventures and dramas aboard the *Uranie*. After a difficult start to the voyage, due to very strong winds, the death of one officer and the departure of another in Gibraltar, the *Uranie* arrived in Rio de Janeiro on December 6, where Freycinet made his first scientific observations.

He set sail again two months later, new desertions forcing him to take on reinforcements before reaching Mauritius in July 1818. But Freycinet was impatient to pursue his research and headed for Shark Bay in Western Australia, where he set up his first observation camp in September.

The *Uranie* then headed north to Timor and the Portuguese trading post in Kupang. Despite a royal welcome from the governor, Freycinet and his crew struggled to obtain all the provisions they needed. So they coasted from island to island in search of food, trading glassware for chickens, the while plagued by dysentery that saw the *Uranie* take refuge in Dili in December 1818. The situation there was much better, and Freycinet was finally able to stock up with provisions, refusing in passing the governor's offer of four young slaves ...

With her crew in better shape, the *Uranie* set sail for the Moluccas and New Guinea, and the Caroline and Sandwich Islands. Scientific observations continued, data were gathered and magnificent drawings were added to the ship's coffers. Following a year at sea in the heart of the Pacific Ocean, Freycinet already knew that his mission would be a success. Finally, in November 1819, the *Uranie* entered Port Jackson. It was back to civilization: twelve months of island living gave way to non-stop entertainment including balls, sumptuous dinners and outings on horseback. In short, Governor Lachlan Macquarie did his utmost to show his French guests the power of the yet nascent British Australia.

Several days after the *Uranie* set sail again, the crew discovered ten stowaways in the holds: ten convicts who would continue the adventure with the sailors. On January 21, 1820, the first icebergs were sighted. Next came Cape Horn, which was rounded at the beginning of February thanks to favorable but overly strong winds that split the mainsail in two. To make the repairs, Freycinet decided to head for the Falkland Islands and it was there that tragedy struck. On February 14, in fine weather, at five forty-five in the afternoon, the *Uranie* hit an uncharted underwater rock and her holds filled with water. Despite numerous efforts, the ship listed twenty degrees and ran aground at half past one in the morning. By daybreak the situation was desperate, not least because Freycinet himself was in the grips of a raging fever. Fortunately however, the scientific documents and drawings were saved. It was not until March 20 that a sloop belonging to a whaler joined them and allowed them to board the seal-hunting ship that would take them and their cargo back to Brazil, where Freycinet found and bought a suitable ship for their return, a vessel that he renamed the *Physicienne*.

Having left Toulon in September 1817 aboard the *Uranie*, it was aboard the *Physicienne* that Freycinet and his crew returned to Le Havre on November 13, 1820, her holds filled with scientific treasures. He himself was no less summoned before the council of war for having lost a boat and taken his wife with him. But given his immense contribution to science, and his already acclaimed success, he was acquitted and even promoted. In the years that followed, Freycinet's health remained fragile, and he devoted his time to writing up his adventures. Only partially completed for lack of funding, the work was nevertheless published in 1827. He died in Saulce-sur-Rhône in 1842 at the age of sixty-three, ten years after the death of Rose, whose own account was not published until 1927.

Ile Timor, vue de notre premier débarquement à l'établissement portugais de Dillé (Timor Island, view of our first landing at the Portuguese settlement of Dille).

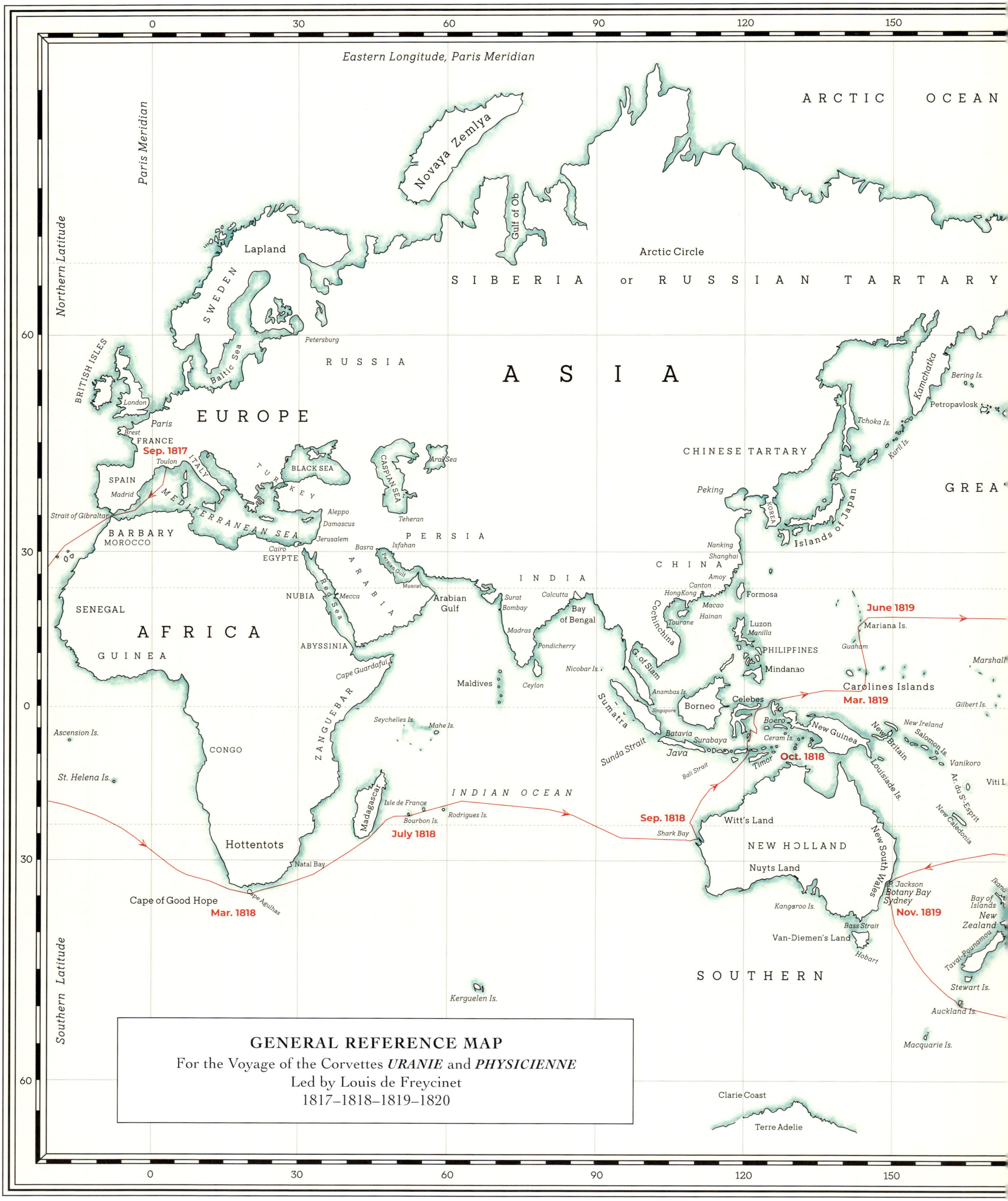


GENERAL REFERENCE MAP
For the Voyage of the Corvettes ***URANIE*** and ***PHYSICIENNE***
Led by Louis de Freycinet
1817–1818–1819–1820

150
120
90
60
30
0
Western Longitude, Paris Meridian
Paris Meridian
Melville Is.
Lancaster Strait
Baffin
Bay
Davis Strait
GREENLAND
Iceland
SWEDEN
Baltic Sea
Bering Strait
Cook's River
Port des Français
Lituya Bay
Alaska
Kodiak Is.
Sitka
Hudson Strait
Hudson
Bay
Cape Farewell
BRITISH ISLES
London
Queen Charlotte's Is.
LABRADOR
CANADA
NORTHERN
leutian Islands
Nootka
Vancouver Is.
St-Lawrence River
Newfoundland
Nov. 1820
Paris
Brest
FRANCE
NORTH
AMERICA
Quebec
Acadia
ORTHERN
OCEAN
UNITED STATES
New York
Washington
ATLANTIC
Toulon
ITALY
SPAIN
Madrid
MEDITERRANEAN SEA
San Francisco
Monterey
Azores Is.
Strait of Gibraltar
Bermuda Is.
Madeira Is.
BARBARY
MOROCCO
California
New Orleans
San Antonio
MEXICO
Gulf
of Mexico
Florida
OCEAN
Canaries Is.
Tropic
of
Cancer
Mazatlàn
Lucayan Is.
Hawaii Is.
Sandwich Is.
San José
Mexico City
Santo Domingo
SENEGAL
Aug. 1819
Acapulco
Jamaica
Caribbean Sea
Guadeloupe
Martinique
Cape Verde
AFRICA
Panama
GUINEA
COLOMBIA
Cayenne
GUYANA
Equinoctial
Line
Gallapagos Is.
Paita
PERU
SOUTH
AMERICA
Oct. 1819
Marquesas de Mendoza
Ascension Is.
CONGO
Lima
lavigators Is.
Pisco
Bahia
St Helena Is.
Maouna Is.
BRAZIL
Tahiti
Society Is.
Arica
June 1820
Friendly Is.
Tropic
of
Capricorn
Rio Janeiro
Traitors Is.
Mangareva
Pitcairn Is.
C. Frio
Easter Is.
Coquimbo
CHILE
St Catherine's Is.
SOUTH
Valparaiso
Montevideo
Buenos Aires
May 1820
Rio de la Plata
Concepción
Cape of Good Hope
ATLANTIC
Chatham Is.
OCEAN
PATAGONIA
Mar. 1820
Falkland Is.
OCEAN
Strait of
Magellan
Tierra del Fuego
Is. de Los Estados
Cape Horn
Feb. 1820
Sandwich Is.
Antarctic Circle

## LOUIS DE SAULCES DE FREYCINET

"Our expedition is, I believe, the first maritime journey that did not have hydrography as its focus. The purpose of the mission was essentially threefold: to determine the shape of the Earth in the southern hemisphere; to observe magnetic and meteorological phenomena; and to study the three kingdoms of Nature. It was also to conduct research into the manners, usages and languages of indigenous peoples. The instruments destined for use in our experiments came from the workshops of our best artists and were submitted, before our departure, for verifications to validate their accuracy. We embarked with a great quantity of maps, together with a selection of excellent books on natural history, astronomy, &c. A medal was struck on the occasion of the voyage, destined for distribution en route to such people as might seem to show some interest in our mission. Being restricted to a single vessel, I gave preference to men of equal merit who were both seamen and workmen. So it was that out of 120 men, at least 50 could practice a useful trade such as carpentry, rope-making, sail-making, smithing, &c. The practice of using tanks for water storage had only lately been introduced to England and had yet to come to France. Several such tanks were constructed for the *Uranie*, a most salutary innovation to which we owe the inestimable benefit of having kept our drinking water sweet and free from putridity throughout the course of our voyage. Quite apart from these water containers, I also had embarked a retort capable of distilling large amounts of seawater. Practical and economical, the apparatus could provide us with 7 measures of water for just one measure of coal, amply satisfying our daily needs. The ship weighed anchor on September 17, 1817 at half past seven in the morning; no sooner was this done than the *Uranie* raised its sails and steered a course out of the bay."

LOUIS DE FREYCINET,
*VOYAGE AUTOUR DU MONDE*, 1827

The following extracts are taken from the journal of de Rose de Freycinet.

**SEPTEMBER 1817**
**TOULON, FRANCE**

*I will begin this logbook* at the very moment when, midnight having struck, I left the house where I lived in Toulon and betook myself alone to the residence of a lady friend of mine. There I spent a most troubled night ruminating on the somewhat harzardous enterprize that I was about to undertake.

Every one in Toulon believed that I had left by the post that departed from the town at midnight, headed for a relative of my husband's in Marseille.

Accompanied by Louis and one of his friends, we betook ourselves to the port where we were to embark. The moon hid itself in clouds as if to prevent my being recognized by the other people present. But as we were leaving the harbor we had to wait to give the watchword, at which a light was cast and I knew not where to hide myself. When at last we came alongside the *Uranie* I climbed aboard as nimbly as possible despite my shaking like a leaf, only to find myself surrounded by officers who enquired as to my identity. The friend accompanying us assured them that I was his son, who is indeed about the same size as me.

**DECEMBER 1817**
**RIO DE JANEIRO, BRAZIL**

*On December 5 we rounded Cape Frio* and on the sixth we entered the splendid bay of Rio de Janeiro. Before ever we had dropped anchor a canoe came alongside. In it was an officer of the King's Household with orders to ensure that this was indeed the long awaited *Uranie*. He gave us to understand that the King would welcome the French in every way possible, procuring for them such provisions as were necessary for their journey.

Louis himself went to see the admiral and attended on those persons for whom he had letters of recommendation. I remained aboard alone, the harbor being very safe and the weather magnificent. It was on that day that Louis saw Countess de Roquefeuille, a French emigree residing in Brazil where she receives an allowance from the King. Though she yearns to see France again, she dare not return to France lest the King should discontinue her allowance and leave her bereft of all means of support.

One day we set off at five o'clock in the morning, the women in a large calash drawn by eight mules and the men on horseback or in a cabriolet. Rain had churned up the ground rendering the paths impassable and forcing us to leave the highway and follow a track leading to the forest. Mules and saddle horses were waiting for us but several of the ladies preferred to walk and I followed their example. The mud was ankle deep and we nearly broke our necks twenty times! But funniest by far was a small personage about whom I must say two words.

She was the wife of the Dutch Ambassador, and what an astonishing sight she was in her hand-embroidered muslin dress, lilac-colored spencer and profusion of flourishes and flounces. Though apprised of the fact that we should be walking in an ever-wet forest, she arrived wearing silken slippers. To see her slithering in the mud was truly comical. Presently, her slippers no longer staying on her feet and her embroidered dress knee-deep in mud, they put her on a horse. To the regrets that we expressed at seeing her in this condition, she replied that it was all the same to her as she would have other dresses in Paris. Fortunately for me, I possessed no such finery; I could never have solaced my wretchedness over losing so much money on futilities through my own fault (she assured us that her dress alone had cost six hundred francs, to say nothing of the rest of her toilette.)

We were amply rewarded for these minor inconveniences by the charming prospect afforded by our walk.

JULY 1818

## BOURBON ISLAND [LA RÉUNION]

*We headed for Bourbon Island,* where we were to stock up with provisions from the King's storehouses. The wind blowing quite strong but against us, the crossing proved long and unpleasant. We carried several passengers, two of whom were particularly recommended to our attention. One was the commanding officer of the Bourbon artillery and the other was an employee of I know not what government department. Both were seasick and our cabin pretty much became a hospital. Even my servants were under the weather. I spent the entire duration of the crossing sending for cups of broth, coffee, &c.

Being forced to tack and sailing against strong currents in a heavy sea, we made little headway. We finally dropped anchor in the port of Saint-Denis at five o'clock on the evening of the nineteenth. On the morrow a grand dinner was held at the official quarters of the intendant, to which were invited all the town

Bottom:
Rose debarking at the Dille settlement, Timor.

Pages 162-163:
*Nouvelle-Hollande, baie des chiens marins Camp de L'Uranie, sur la presqu'île de Péron* (camp site of the Uranie on the Peron Peninsula, Shark Bay, New Holland).

*I spent the whole of the morrow writing my letters of adieu and at nightfall, toward half past eleven, I donned men's clothing.*

authorities and the staff officers of the *Uranie*. However, the latter got so very wet disembarking that nearly all of them had to make their excuses.

The dinner was a sumptuous affair. The table was large and well-dressed but fifteen out of the original twenty-five guests were not sufficient to do it justice and the table seemed deserted. One thing shocked me, not least because I had received no notice, and that was the presence of a lady who has very much the air of a pantry servant and who fulfills the functions of housekeeper. We were at table when she and her daughter entered by the door of the pantry and came to join the assembled company. That the general, when he is alone, should allow his housekeeper to eat with him in the manner of priests, admits of no reproof; but that she should be admitted at table on the occasion of a formal dinner is beyond ridiculous. The baroness whispered a few words on the subject in my ear after the meal and I could tell from her manner that we were of like mind.

**SEPTEMBER 1818**

SHARK BAY, NEW HOLLAND, AUSTRALIA

*No sooner had we arrived* than we saw several personages at the top of the hills that border the sea. Having first taken them for seamen who had gone hunting, we realized on approaching that they were in fact natives, all of them naked and armed with spears and lances. Louis was for marching out to meet them, but knowing not the numbers he would be confronting, he decided instead to call for reinforcements from the camp about a league away then proceed with them to greet the aboriginals in a friendly manner.

On return to the place where we had disembarked, we found our breakfast waiting and after stretching out a sail to shelter us from the sun, we made a good meal, not only with those provisions that came from the boat but also with excellent oysters that we found on the rocks, more delicious than any I had ever eaten in Paris, seated most comfortably at a table.

*I remember what you said in one of your letters from Toulon, ...*

Top:
Inside a house in Coupang, Timor.

Facing-page, top:
Countryside around Coupang.

Facing-page, bottom:
Scene from everyday life at the Coupang bazaar.

**OCTOBER 1818**

TIMOR, DUTCH EAST INDIES, INDONESIA

*We found the palace* to be most charmingly decorated with foliage and flowers in anticipation of our visit. Music played as we crossed the palace courtyard, a large well planted esplanade surrounded by gardens. The entire troop was under arms. The ladies were already gathered at the governor's residence and came out to welcome me at the foot of the staircase. They were all richly attired in the Portuguese manner, pretty much as was the fashion in France forty or fifty years since, their hair cut in a straight fringe over the forehead and gathered at the back in a loose bun that was fastened to the head with gold pins.

Judging from my own toilette, however, that theirs was somewhat outdated, they seemed embarrassed by their apparel and most especially their shoes. These were what I believe were called mules, and were made of colored silk embroidered with sequins. Thus did they hasten to say that great articles of fashion had been ordered from Macas and were awaited with such impatience by these good ladies that the sight of the *Uranie* had sent their pulses racing in the mistaken belief that this was the much longed-for vessel. They reserved special praise for my short, curly hair and all the benefits it offered in hot countries, their own hair being very long, very beautiful and, from what I hear, very dear to them. Their husbands and the governor, however, invited them to cut their own locks short, and one of the ladies appearing prepared to make that sacrifice, the governor asked for a pair of scissors and made as if to do the job himself. In the event, she plainly had no such inclination.

Crouched beside each of these women was a slave holding a kerchief and a betel bag, nearly all of them pretty, richly appareled as is their custom and festooned in golden chains and jewelry.

*... that you 'would hide behind Louis' coat tails at first sight of the natives.'*

**DECEMBER 1818**
**RAWAK ISLAND, PAPUA NEW GUINEA**

*Louis says that while the indigenous* Papuans in the vastly distant, southern part of the island may well be anthropophagus, those hereabouts are on the contrary gentle and spiritual in nature. We see some every day at present, and there is certainly nothing formidable about them.

A Papuan canoe very like the first one we saw came alongside directly we dropped anchor in Rawak. The natives offered us tortoise eggs and tortoises but the price was too high so we bought nothing fearing lest we should get off on the wrong foot. Their chief speaks a mixture of Malay and Papuan. When he meets the chief of the other canoe, they greet each other by placing their hand on each other's head, heart and shoulder then rubbing noses.

Louis resolved to drop anchor between the small islands of Rawak and Waigiou; the former indeed is where we pitched the tent for the observatory. Keep in mind, however, that we are located precisely below the equator, under a leaden sun, and that these isles are humid in the extreme and covered in dense forests.

*You would quake with fear were I to tell you that these natives are Papuans, or inhabitants of New Guinea, with a reputation for anthropophagy.*

**MARCH 1819**
**CAROLINE ISLANDS**

*On the twelfth of this month* we reached the Caroline Islands. We did not stop however, hoping that a few days more would take us to the Mariana Islands where there is a Spanish settlement, unlike on the Caroline Islands where there are only natives—but such very amiable natives!

No sooner had our corvette been noticed, than we were surrounded by a vast number of small and large canoes that were remarkable for the manner of their manufacture and their elegance. Trading commenced forthwith providing us with fresh fish, coconuts, sarongs, &c. Nearly all these natives were naked, as is the wont of natives, but the Carolinians have blue tattoos on their arms and legs. The chiefs wear a sarong that they draw down over their heads by means of an opening that gives this garment the form of a chasuble. Their earlobes are slit and hang very low and some of them wear big yellow flowers and stems of foliage in the slit.

They are remarkably adroit in the handling of their canoes, these being admirably crafted and of the most graceful shape, with a red varnish finish and black border like the vases that come from China.

**Top:**
Portrait of a Pahoro (detail from Île Rawak plate).

**Above:**
*Île Guam: Principal personnage des danses de Montézuma* (Guam island: main character in the Dances of Montezuma).

Top:
*Île Guam. Travail d'agriculture* (farming on Guam Island).

Pages 168-169:
Everyday objects from the Mariana Islands and the Guebe Islands, brought back by Louis de Freycinet.

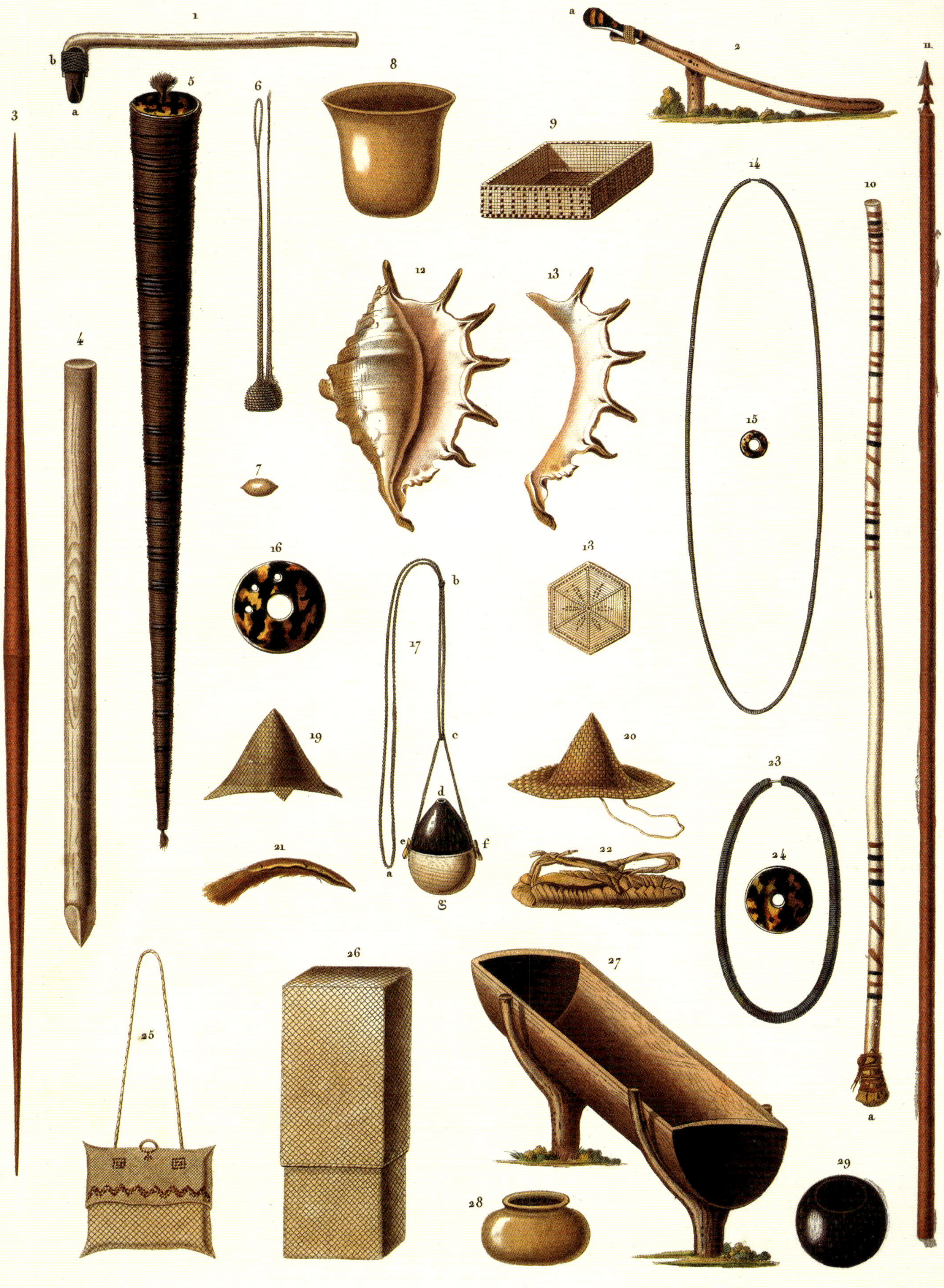


Bévalet *pinx.t* Coutant *sculp.t*

**ÎLES MARIANNES: DIVERS OBJETS À L'USAGE DES ANCIENS HABITANS.**

*(Voyez l'explication des planches.)*

Bévalet *pinx*t Coutant *sculp*t

**ÎLE GUÉBÉ. CHAPEAUX, ARMES ET USTENSILES DES HABITANS.**

*(Voyez l'explication des planches.)*

Top:
*Îles Sandwich. Un officier du roi en grand costume* (an officer of the king in full ceremonial dress, Sandwich Islands).

Facing-page:
Plate from a work by Rose de Freycinet.

Pages 172-173:
*Îles Sandwich: baptême du premier ministre du roi à bord de L'Uranie* (Sandwich Islands: baptism of the King's First Minister aboard the *Uranie*).

**AUGUST 1819**
**SANDWICH ISLANDS** [HAWAII]

*Chief Pouï is a man* of some five and forty years, tall and of powerful constitution, with a noble bearing for a native. He wears a single piece of homespun cloth draped about his body, leaving his arms and chest and shoulders bare. On his head he wears a straw hat with a very tall crown and a wide brim.

He came bearing gifts for the captain in the form of a pig, coconuts, onions and bananas, in exchange for which he received as much and more. Afterward, he readily accepted to dine with us, requesting only that he be permitted to fetch his wife who had staid in the canoe. Oumaye, Pouï's wife, wore a sarong wrapped around her waist like a skirt, and another that she sometimes draped over her shoulders but more usually let fall to leave her arms free for eating. I should rather say "devour", for had I not seen the natives of Rawak I should have been aghast at the voracity of those I saw here. They guzzled every thing set before them at such alarming speed that we might have thought them starved and on the run. What is more, no sooner had they finished than for the sake of peace, they must be given the glass, plate, bottle and even the napkin they had just used; and far from appearing satisfied, Pouï only became more fretful still.

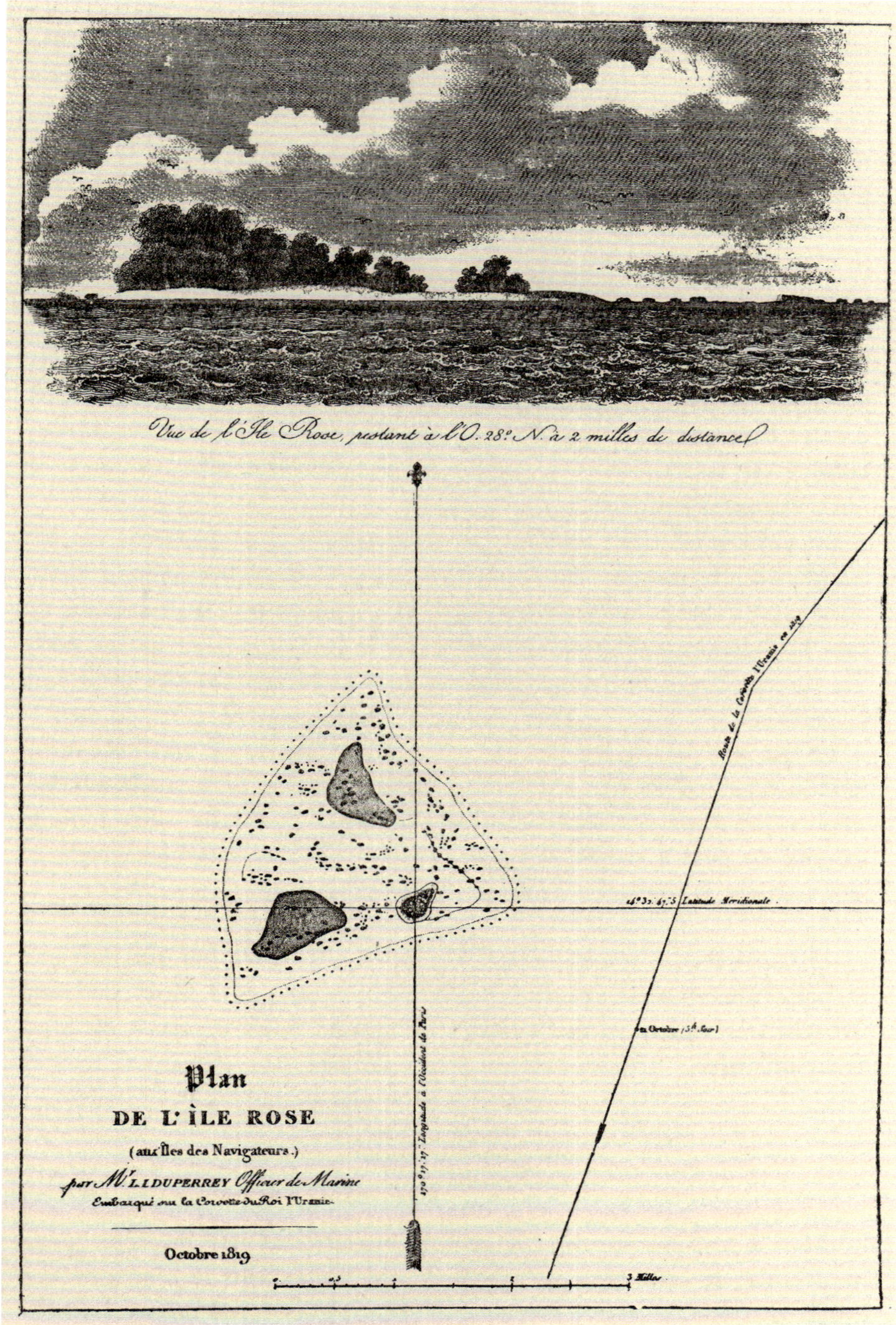


**Top:**
Map showing route of the *Uranie*, and discovery of the Ile Rose.

**Facing-page:**
Camp site on the Falkland Islands following the shipwreck of the *Uranie*.

**OCTOBER 1819**
**ÎLE ROSE [ROSE ISLAND] NAVIGATORS ISLANDS, SAMOA**

*I shudder to think* what you would say were you only to see our present location—were the *Uranie* suddenly to appear before your eyes as you follow our progress on the world map. Even I tremble to see how far science has brought us, but my exclamations meet with nothing but praise for this peaceful sea on which I pursue my long journey. To my mind the Pacific Sea is like those people of good character who always appear to share one's opinion—pleasing at first, but tedious over time for want of showing a little spirit.

In those rare moments of leisure when Louis is not working himself or supervising others at work, our favorite pastime is to make plans. Retiring to the country often seems to us a return to the promised land. Planting, planning, watering—there is no folly that does not cross our minds.

Thus the deed is done. My name is now attached to a little spot on the planet. So little indeed that envious people might begrudge its being called an island at all and reduce it to an isle. Such as it is, encountered at night, it could have proved fatal to our vessel. Instead, it is now clearly marked on our expedition maps, warning every one to give it a wide berth and preventing any one, I hope, from coming to grief around Rose Island.

**FEBRUARY 1820**
**CAPE HORN, CHILE**

*My husband being ever concerned* for the safety of our vessel, his sole preoccupation was to preserve us from the hurricane in the offing. Suddenly, the words "cut the cable" resounded through the ship and he ordered the sails to be raised. The *Uranie* was dragging its anchor and being pushed forcibly by the wind toward the coast. In no time at all we found ourselves at the exit to the bay and headed for the strait.

I really believed our hour had come. I entrusted my soul to God and recoiled at the prospect of the ghastly death that was about to befall us. I prayed that my head might be smashed against some thing hard, sparing me an agonizing wait poised between life and death. It was then that we noticed that what we took for land was none other than a mighty wave rearing up on the horizon.

We endured two days and nights of this horrific tempest, being pushed ever northward by a wind from the north that determined Louis on dropping anchor in the nearby Falkland Islands, instead of returning to Bahia Buen Suceso and wasting time in the process.

On the fourteenth the weather having improved, we headed for the Bay of Fundy where we planned to drop anchor. But the vessel struck a rock just as we were about to round the very last point that hides the entrance, and to judge from the water rushing into the hold, a piece of that rock had pierced the hull and been drawn off by the ship's wake.

The place where we planned to hove to was still very far away and we were surrounded by craggy shores and rocky cliffs—the *Uranie* would have been lost. At that point, Louis ordered all hands to the pumps and resolved to make toward a sandy beach to save what the *Uranie* contained—instruments and the expedition findings—if not the *Uranie* herself. To see two years of work swallowed up by the waves was too heartbreaking to contemplate! Meanwhile the water was still rising, the wind had dropped to next to nothing and the men were running out of strength, having had nothing to eat for days on end.

Overwhelmed by the horror of our situation but powerless to offer any succor, I retreated to my cabin consumed by dark thoughts. What was to become of us should the ship sink? What fate would be ours, stranded on a desert island bereft of all expedients? The wind was violent in the extreme and about to smash us against the rocks. The memory will remain with me forever. The last sail had ripped and the ship was out of control.

**FEBRUARY 1820**
**FALKLAND ISLANDS**

*My first night ashore* was not such that might compensate for those last nights aboard. Our tent was not pitched and as it had rained all night, we were soaked in our bed. When at last it was daybreak we arranged our canvas house a little better. But rain or no rain, the coolness of the night combined with the dew contrived to keep our sheets permanently damp, our being only protected from the elements by canvas. Louis is having a deck installed on the longboat to send it to Montevideo, but he remains indisposed and keeps his bed. He lacks strength to an extraordinary degree and for two months now has been beset by a great fatigue.

At low tide we were able to extract a great many objects from our wretched *Uranie*. Among these was a vast quantity of hardtack recovered from the hold that will prove a welcome addition to the crew's soups, the men now eating nothing but meat. They even refuse to drink broth when there is nothing in it, on the grounds that it is devoid of nourishment.

*The* Uranie *has discovered a small uncharted island to the east of Navigators Islands, to which the captain has given the name of Rose Island.*

I slept well on a makeshift bed that I cobbled together from a plank, two chairs and a long cushion by way of a mattress. It is at times like these that my kimono is such a blessing: the thickness of the cotton keeps me wonderfully warm, to such an extent indeed, that sleeping fully dressed under my kimono I never feel a moment's cold.

**MARCH 1820**
**FALKLAND ISLANDS**

*As we were walking,* we heard a great shouting coming from the camp. Every one climbed to the top of the sandy hillock that overlooks the sea but we could see nothing despite our eyes being fixed on the entrance to the bay. Presently Louis spotted a cutter tacking to enter the harbor. He straight away ordered his pinnace, such as it was, to be made ready, and the firing of two cannon shots together with the raising of a white flag at the highest point on the island. The vessel was a sloop belonging to a whaler at anchor twenty leagues hence on the western coast of the Falklands. Its handler appeared little disposed to do as Louis wished, namely to return forthwith to his ship with one of our officers charged with asking the master to come and fetch us—us and our effects—and conduct us to Rio de Janeiro.

Though Louis assured him that we would make good any losses, the man replied that he still had eight days fishing remaining and durst not take it upon himself to return empty-handed to his ship beforehand. This did not at all fit with Louis' plans, hence his inviting the man into his tent and offering him a shot of rum, the while showing him a passport from the USA ordering every ship of whatsoever description to offer the *Uranie* succor and assistance on its voyage. The man read and re-read this document, plainly more affected by what he saw there than anything he might have heard before.

**MAY 1820**
**ABOARD THE *MERCURY*, ATLANTIC OCEAN**

*The latest proposal from the captain* of the *Mercury* is that Louis should buy his boat at the price agreed, plus two thousand piasters more, to take us to Rio de Janeiro. After consulting his officers, Louis is persuaded of the advantage of such a transaction and at midnight a bill of sale is drawn up and translated into English for the sake of the *Mercury's* captain. It is henceforth agreed that the ship is no longer the latter's but will remain under his command until Montevideo, where we will stay just three days to allow him to disembark his merchandize and passengers. Thereafter Louis will take possession of the vessel as it stands, complete with its cannon, spare masts and general sea worthiness. He will call it the *Physicienne*.

**JUNE 1820**
**RIO DE JANEIRO, BRAZIL**

*The sea being slightly calmer* and the wind continuing favorable and quite strong, we sighted land early this morning. We had hoped to make it to shore by nightfall but the wind dropped and we were forced to spend the night with our sails hanging limp.

It grieves me to see my poor Louis kept from his bed night after night owing to our ship approaching land. On the other hand, this little delay is propitious to my finishing a dress of black silk that I bought in Montevideo in the apprehension that the French Embassy in Rio might still be in mourning for the Duc de Berry, and that I might be required to pay court to Madame Ambassador.

*It is now two months since the* Uranie *came aground on the sands of the Falkland Islands never to rise again.*

There is an infinity of vessels here, including several English vessels and other warships. We passed before them decidedly shamefaced at the unwarlike aspect of our *Physicienne* whose condition, quite apart from the damage sustained during the crossing, is vastly inadequate. So little does it resemble a ship of the line that the mercantile agencies here in Rio sent for confirmation as to our identity—were we the *Cécile* from Le Havre or the *Mutuelle* from Marseille? As if this were not humiliating enough, to add insult to injury they dispatched excise men to inspect our ship. Imagine their discomfiture when Louis gave them to understand, loud and clear, that he would have nothing to do with them as his was a warship!

**NOVEMBER 1820**
LE HAVRE, FRANCE

*Land was looming through the fog* and the sea was high. The lead line was pretty much our only guide; if it was occasionally possible to distinguish the coasts of England and Normandy, this intelligence was too vague and uncertain as to be of any great benefit.

We hailed a coaster coming toward us, only to be told that gaining Le Havre was impossible but that the pilot would conduct us to Cherbourg. We navigated in his wake and dropped anchor within just a few hours in a French harbor. Other pilots arrived, who spoke our language and all but addressed us by name.

I set foot on land. On my native land. My heart is beating fit to burst and my blood is boiling in my veins. I need rest but rest weighs heavy on my spirit.

**Facing-page:**
*Vue d'une distillerie sur l'île Guam* (view of a distillery on Guam Island).

**Top:**
*Vue de l'île Pisang: Corocores de l'île Guébé* (view of Corocora Geba from Pisang Island).

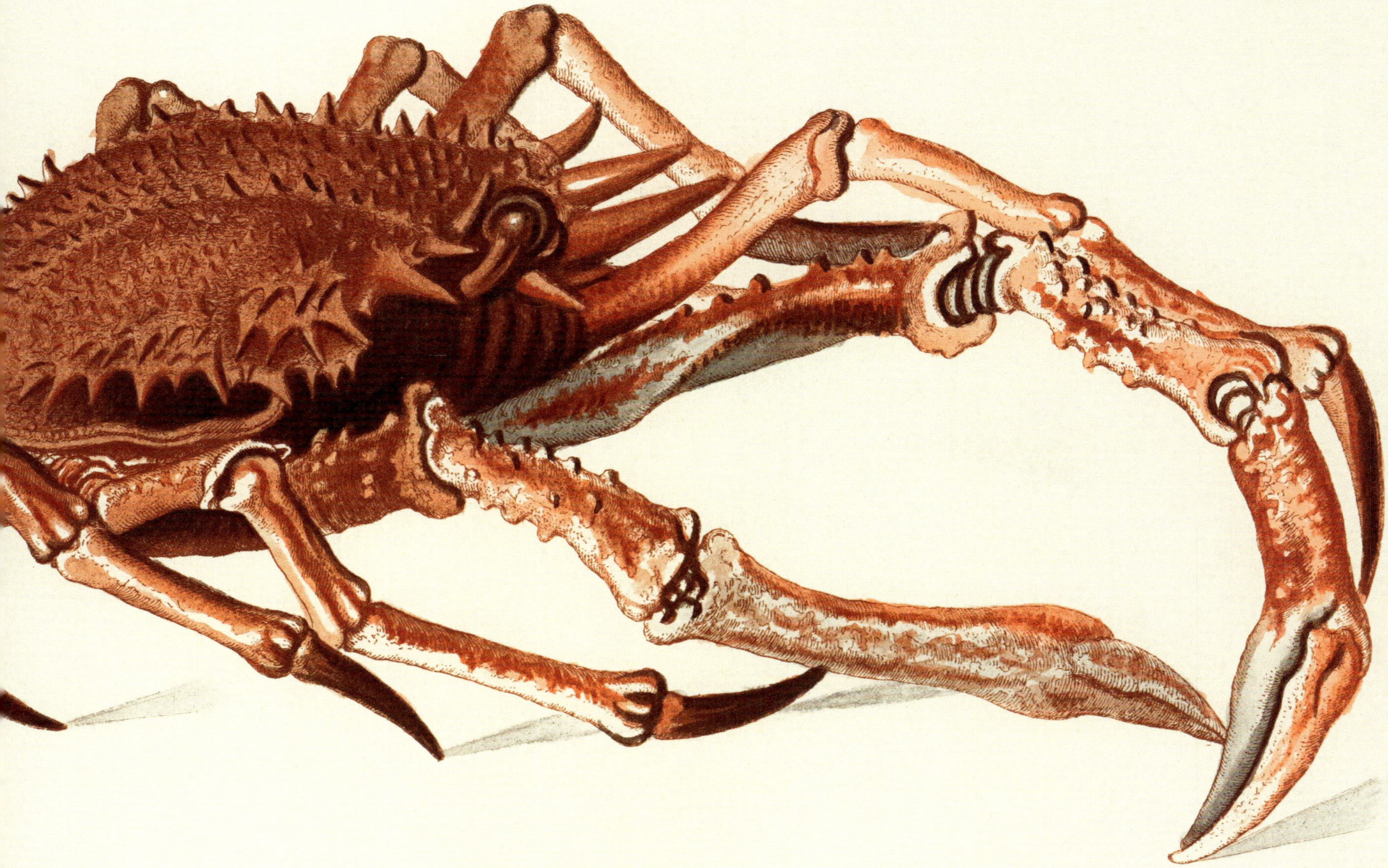

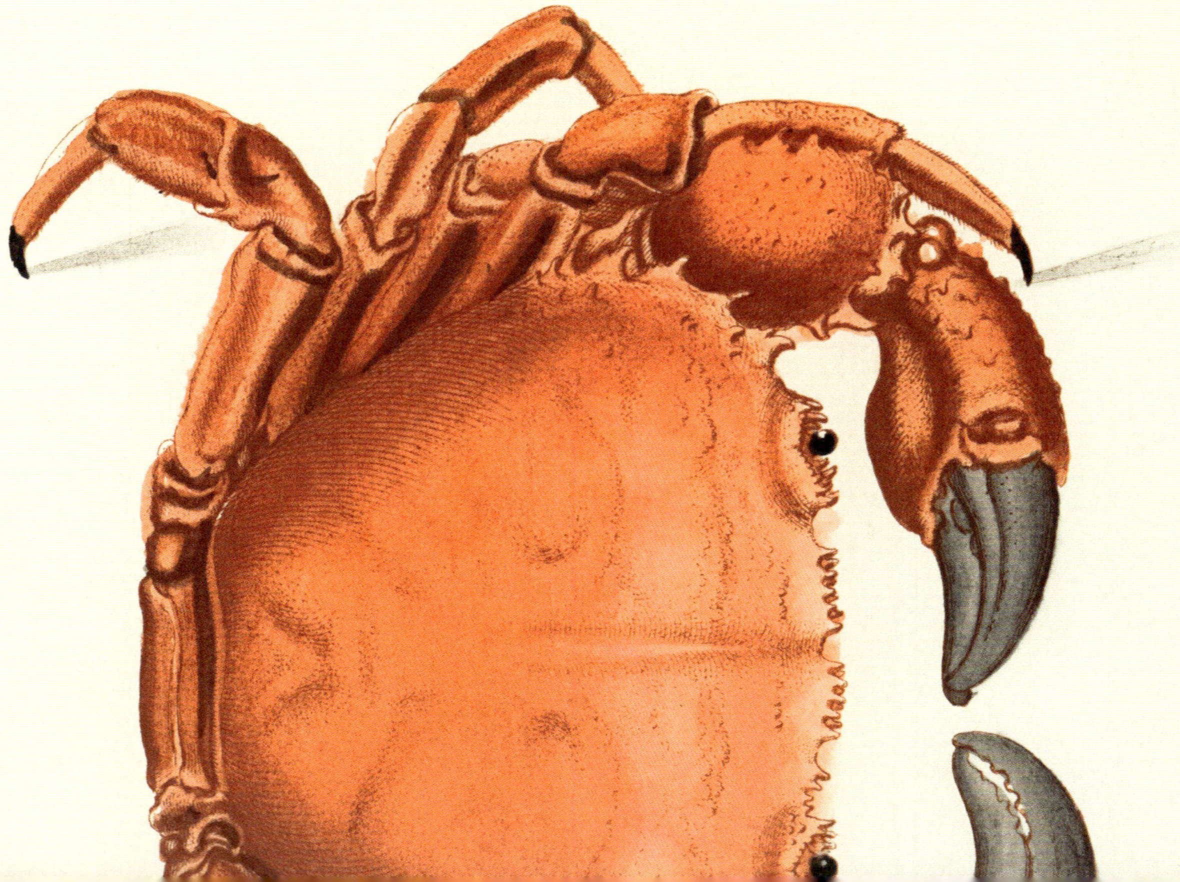

# Louis-Isidore Duperrey

**1786–1865**

HOW DID LOUIS-ISIDORE DUPERREY
AND HIS MEN MANAGE SUCH AN IMPRESSIVE
CIRCUMNAVIGATION OF THE GLOBE?
WITH NO LOSS OF LIFE OR DAMAGE,
THE EXPEDITION WAS NOT MERELY A FEAT OF DARING
BUT ALSO A MAJOR SCIENTIFIC SUCCESS.
THE HOLDS OF THE *COQUILLE* WOULD RETURN
FILLED WITH UNKNOWN SPECIMENS OF PLANTS,
ANIMALS AND ROCKS, ALONG WITH MORE
THAN A HUNDRED DRAWINGS, SKETCHES AND MAPS.
BUT ALREADY AT EVERY PORT OF CALL,
DUPERREY AND HIS OFFICERS COULD SEE
THE WORRYING CONSEQUENCES
OF EVANGELIZATION AND THE BEGINNINGS
OF COLONIZATION …

Plate 8, George Wolfgang Knorr, 1766.

# VOYAGE OF THE *COQUILLE*, BY LOUIS-ISIDORE DUPERREY

**1822–1825**

**Duperrey was thirty-six when he took command of the *Coquille*. And as he crisscrossed Oceania in pursuit of discovery, his observations of Pacific peoples and their maritime practices, together with his findings in hydrography, cartography, zoology, botany and entomology, would win him the admiration of the Académie des Sciences.**

Louis-Isidore Duperrey came into this world on October 21, 1786, in Paris. An enthusiast for Science and Nature since childhood, he was not predestined for a career in the navy but enlisted in 1803, at the age of sixteen, out of admiration for Napoleon Bonaparte and to serve his country. As an ensign, he took part in the Empire's campaigns, notably the 1809 attack on English fire ships. But when Napoleon fell, Duperrey went down with him. His reinstatement came in 1816; France needed good sailors and Duperrey was too good an asset to pass up.

Following Freycinet's success, Duperrey and his cadet, Jules-Sébastien César Dumont d'Urville, suggested to King Louis XVIII that they continue his work, and made him an offer he could not refuse: they would waive the usual indemnities, depend on their own instruments and make do with a single vessel, the *Coquille*, a sturdy, well-armed merchant ship reclassified as a corvette for the occasion. Proposal accepted! The voyage was planned down to the last detail. Duperrey was thinking big, and wanted to cover all the sciences: hydrography, of course, but also zoology, ethnology, botany, geography, meteorology and astronomy. To that end, he recruited the best officers and scientists, notably pharmacist and botanist, René Primevère Lesson, and master draughtsman Jules-Louis Lejeune. To ward off the dreaded scourge of scurvy, he engaged Nicolas Appert, inventor of a new food storage technique based on heat sterilization (appertization) to can the food for his expedition. It was a brand-new technique, and it worked: Duperrey lost not a single sailor to scurvy.

The *Coquille* departed Toulon on March 11, 1822 with just sixty-five men on board, a year and a half's worth of provisions, and holds filled with the various objects and gifts needed for trade. Heading south after a stopover and quarantine in Tenerife, the vessel reached Santa Catarina. Brazil had declared its independence a few days earlier, and the island was still in turmoil: they could barely stock up on provisions before rounding Cape Horn. In the Falkland Islands, Duperrey and Dumont d'Urville made a melancholy stopover in the bay where the *Uranie* was wrecked. The remains of the ship were still visible, half-buried in the sand. In Chile, with the War of Independence still raging, the situation was unstable, General Ramón Freire having just made a triumphant entry into Concepción at the head of his insurgent army. But what mainly interested Duperrey and his team of scientists were the customs of the tribes living inland and, as at every stopover, they built an observatory, took measurements and collected all manner of plant, mineral and animal samples.

The *Coquille* then sailed up the American coast to Paita, Peru, before crossing the Pacific Ocean to the Tuamotus, Duperrey claiming possession of Reao Island along the way, which he named Clermont-Tonnerre after the Minister of the Navy. Tahiti on the other hand was a big disappointment. Far from resembling the smiling island described by Bougainville, the place portrayed by Duperrey and Lesson had been laid waste by the Protestant missionaries, especially the English ones, and showed no signs of its supposed two-month-old independence.

Disillusioned, Duperrey continued eastward to the Friendly Islands then headed south to Australia until a westerly storm accompanied by violent rain forced him to turn back and make for the Solomon Islands, bypassing Australia to the north. For the scientists, this detour was a godsend. They continued collecting at every port of call in the Savage Islands, Bougainville Island, New Ireland and New Guinea, filling the holds with samples of every description: animals, minerals, multicolored shells and more besides. When Duperrey and his men did finally enter Sydney Harbor, a forest of ninety masts rose up before their eyes: Sydney was one of the most visited ports of the time due to its strategic position on the Australian continent.

Back at sea, Duperrey made a stopover in New Zealand, where the local people quite won him over, before heading north to the Caroline Islands, reconnoitering several islands along the way. Then it was southwards by way of Indonesia to Bourbon Island [La Réunion] followed by the Cape of Good Hope. In St. Helena, he was speechless at the decrepit condition of the place where Napoleon had spent his last years. How could France have let this happen?

On March 24, 1825, the *Coquille* entered Marseille harbor with its holds overflowing and its crew still intact. The consensus in Paris was that the expedition had been a great success, both scientifically and in terms of health. Duperrey's collections were of the highest quality, Lejeune's illustrations were strikingly realistic, and the maps, fifty-four in all, were precise to the nearest detail and complete with magnetic parallels and meridians—a first in the history of cartography! Though Duperrey's own account of the voyage, published in 1828, was incomplete, the wealth of information contained in the botanical, zoological, physical and hydrographical sections is no less a testament to the uniqueness of his achievements.

A full account of the voyage would be published by Lesson more than ten years after his return, in 1839. Duperrey became President of the Académie des Sciences, was honored by King Charles X, and passed away in Paris on August 25, 1865.

Everyday objects brought back by Louis-Isidore Duperrey from the Society Islands.

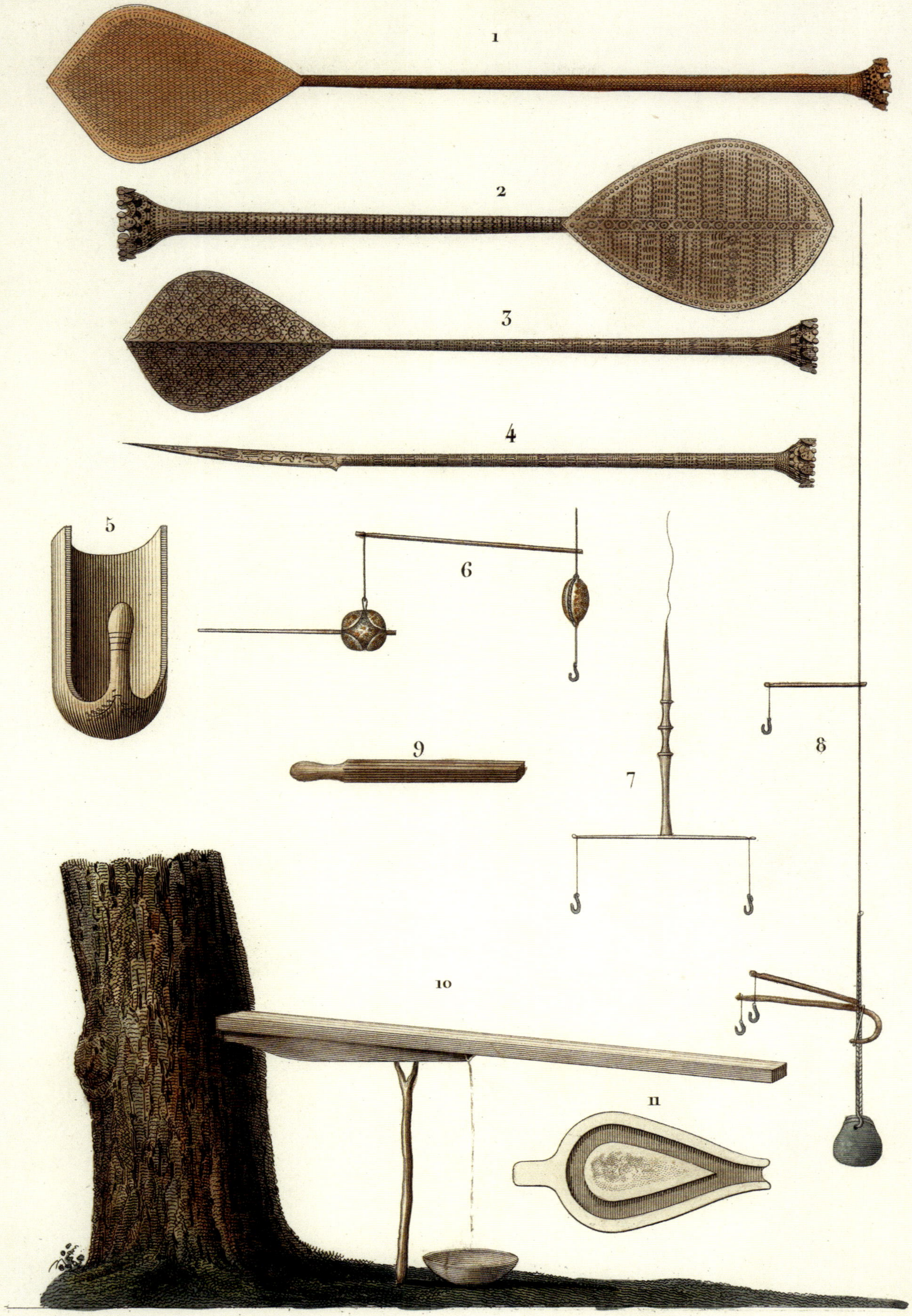


## ILES DE LA SOCIÉTÉ .

1.2.3.4. Pagaies vues de face et de profil.

5. Escope à main .

6.7.8. Instruments de pêche .

9 Battoir à étoffes

10, 11. Appareil pour extraire l'huile de Coco .

GENERAL REFERENCE MAP
For the Voyage of the Corvette *COQUILLE*
Led by Louis Isidore Duperrey
1822–1823–1824–1825

Western Longitude, Paris Meridian
150
120
90
60
30
0
Paris Meridian
Melville Is.
Lancaster Strait
Baffin
Bay
GREENLAND
Davis Strait
Iceland
Bering Strait
Cook's River
Port des Français
Lituya Bay
Kodiak Is.
Sitka
Alaska
Hudson
Bay
Hudson Strait
Cape Farewell
SWEDEN
Baltic Sea
BRITISH ISLES
London
LABRADOR
CANADA
NORTHERN
Queen Charlotte's Is.
Vancouver Is.
eutian Islands
Nootka
NORTH
AMERICA
St-Lawrence River
Quebec
Newfoundland
Brest
Paris
FRANCE
Aug. 1822
Toulon
ITALY
UNITED STATES
Acadia
ATLANTIC
SPAIN
Madrid
MEDITERRANEAN SEA
ORTHERN
OCEAN
San Francisco
Monterey
New York
Washington
Azores Is.
Strait of Gibraltar
Bermuda Is.
New Orleans
San Antonio
California
MEXICO
Gulf
of Mexico
Florida
Madeira Is.
BARBARY
MOROCCO
Canaries Is.
OCEAN
Tropic of Cancer
Mazatlàn
Lucayan Is.
Sandwich Is.
Hawaii Is.
San José
Mexico City
Santo Domingo
Caribbean Sea
Jamaica
Acapulco
Guadeloupe
Martinique
Cape Verde
SENEGAL
AFRICA
GUINEA
Panama
COLOMBIA
Cayenne
GUYANA
Equinoctial Line
Gallapagos Is.
Paita
PERU
SOUTH
AMERICA
Ascension Is.
CONGO
Marquesas de Mendoza
Mar. 1823
Lima
avigators Is.
May 1823
Pisco
BRAZIL
Bahia
St Helena Is.
Maouna Is.
Tahiti
Society Is.
Arica
Friendly Is.
Traitors Is.
Mangareva
Pitcairn Is.
Tropic of Capricorn
Rio Janeiro
C. Frio
Oct. 1822
St Catherine's Is.
SOUTH
Easter Is.
Coquimbo
CHILE
Valparaiso
Montevideo
Janv. 1823
Buenos Aires
Rio de la Plata
Concepción
Cape of Good Hope
lar. 1824
ATLANTIC
hatham Is.
OCEAN
PATAGONIA
Nov. 1822
Falkland Is.
OCEAN
Strait of
Magellan
Tierra del Fuego
Is. de Los Estados
Cape Horn
Sandwich Is.
Dec. 1822
Antarctic Circle

The satiety that comes of riches would one day yield
to the glory that follows from virtue.
Gone are the days when navigators rounded daunting
headlands believing themselves to have reached the ends
of the Earth; when they descended upon remote lands
purely to sew discord, seize treasure and garner the laurels
of their too often bloody victories.
Today a more honorable sentiment directs the course
of their travels; and trust, moderation and enlightenment
take root in whatever place on Earth they may venture.
Prejudice is put to flight by reason, and the broadening
of social connections would seek but to unite
all the peoples of the Earth as a single family.
Posterity shall remember France as among those powers
that have contributed the most to the exploration
of the Globe and to ensuring that the precious benefits
of her brilliant expeditions are fully enjoyed
by the sciences, the arts and geography.
With constancy and disinterest, she has pursued
her scientific research notwithstanding the disasters
that have befallen her most celebrated navigators:
the unfortunate Lapérouse, still mourned by the mother
country that he was never to see again;
d'Entrecasteaux, so drained by fatigue that he collapsed
at the end of his journey; Baudin, destined to the same
fate after seeing almost all his companions claimed
by contagion; and most lately, Monsieur de Freycinet,
for was he not left to perish on the rocks
of the Falkland Islands where fate in its fury had hurled
the ship under his command?
Painful memories indeed, but though too many
for comfort they have not dulled France's zeal for doing
good and conducting research around the world
that will enrich the knowledge of every nation.
Hence her applauding the launch of this expedition
whose details we are about to relate.

LOUIS-ISIDORE DUPERREY
JULY 1822

**AUGUST 1822**
TOULON, FRANCE

*It was a fine morning* on August 11 when the corvette *Coquille* weighed anchor and set forth from Toulon harbor. This was the second time we were departing this port to travel about the Globe, driven by a burning desire to make certain discoveries useful to science. Yet, overlooked by the tomb atop this headland, our feelings of glory could not but be tinged with a sense of oblivion. The summits illuminated by the dawning rays of the sun, the gently swaying sea and the soft splash of the waves breaking against the bow combined in a vast moving tableau that kindled a thousand memories. Our eyes fixed on the shores as they fell from view in our speeding wake, silence befell us and we were struck by that incontrovertible truth that love of motherland is never stronger than as one leaves her behind.

**OCTOBER 1822**
SANTA CATARINA ISLAND, BRAZIL

*Wishing to provision* the ship with fresh food, we had chosen the Bay of Santa Catarina on the counsel of sea captains Johann Adam von Krusentern and Otto von Kotzebue, whose writings had led us to expect that we would find an abundance of every thing. In the event, we could not have been more disappointed; for all we were able to obtain, and that with great difficulty, was beef, poultry and a few fruits. It must be said that our two Russian captains put into the bay at a time of unalloyed peace, unlike the circumstances that prevailed on our arrival when the whole world was consumed with fear and the dread of a future war. Every body here was ready to flee at the first sign of a Portuguese squadron. So terror-struck were the usually peaceable local inhabitants that on the night we dropped anchor, rumor had it that a Portuguese division was about to enter the bay. The inhabitants were no less under the persuasion that enemy troops were about to descend on their land and re-colonize them, which according to them was as good as enslaving them.

*The scythe of revolution had already reaped a grim harvest in this charming land, and was now being honed anew in the name of liberty.*

**NOVEMBER 1822**
FALKLAND ISLANDS

*It is but a few days* since we were moored on Brazil's charming shores, contemplating dense forests and wandering through groves studded with all manner of trees laden with delicious fruits and fragrant flowers. Fluttering about us were brightly colored birds whose dazzling plumage was but equaled by a melodious singing that nourished our imagination. Now we find ourselves exiled in a country where the vistas are blocked only by denuded mountains and our feet tread only on vast plains as monotonously alike as the pampas of America. In our ears, meanwhile, the incessant howling of the wind is punctuated only by the shrieking and croaking of birds and amphibians.

The bad weather continued almost uninterrupted throughout our sojourn in the Bay of Fundy but we pursued our scientific research all the same. Having spent three whole months in this desolate land following the shipwreck of the *Uranie*, we were naturally interested in commencing in the place that had served as our refuge and thither did we direct our footsteps.

**JANUARY 1823**
BAY OF CONCEPCIÓN, CHILE

*No sooner had morning broken on* the shores of the bay, than we beheld the sad spectacle of a coastline that extended desolate and solitary as far as the eye could see, with a few dilapidated canoes and ragged fishermen visible in places. The houses in Talcahuano appeared in a state of degradation that defies description, and in its silent streets could be seen wandering the dregs of a population reduced to the most extreme indigence. Here and there along the coast stood what may rightly be called huts, on the thresholds of which could be distinguished women helping one another to clean and comb their hair, their heads exposed to the rays of the sun. To judge from their peculiar attire and the manner of their acquitting themselves of this part of their toilette, the native state had long ago taken over from the Spanish civilization imported from Europe.

And yet, every thing about this country spoke of a fertile land, from its verdant hills to its clusters of charming woodlands and from its gardens to its orchards that, albeit in tatters, were bedecked in a thousand wildflowers. Beneath a magnificent sky unfurled a landscape where wealth and poverty were illuminated with equal eclat.

Top: *Habitants du Chili* (Chileans).

Pages 188-189: *Temples et maisons du village de Caïeli, île Boutou* (temples and houses in the village of Caeïli, Boutou Island).

**FEBRUARY 1823**
LIMA, PERU

*My heart was racing* as we approached Lima, the city generally regarded as the capital of South America, long the source of all the gold and silver that gushed from Peru, and the seat of a government that laid its foundations in the bloody debris of the peaceful empire of the Incas. The renown of this city spread o'er land and sea and resounded in Europe but oh! how many of those reputations that are impressive from afar come tumbling down under close scrutiny!

While the *Coquille* was hove-to, Lima was in the grip of civil war. The political parties were vying for possession of the city, the people were harassed, molested and taking care to hide their valuables. The convents, though protected by religious exclusiveness, were divesting themselves of their altar statues of solid gold and silver. The city, in a word, was but a shadow of its former self, its erstwhile splendor greatly dimmed since the time of the Castilian viceroys.

Working women now gave free rein to their passions, openly bathing with men, teasing them with unambiguous gestures and proving by all their actions that modesty is a virtue that has yet to round Cape Horn.

**MARCH 1823**
ÎLE DE CLERMONT-TONNERRE [REAO]

*On the twenty-second day* as we sailed in fair weather through a sea with unencumbered views, Monsieur d'Urville thought he heard the muffled roar of waves crashing on rocks—we hove to. At sunrise, we found ourselves within sight of a low-lying island half a league hence, hardly visible above the waves. Its sloping shores were composed of white sands that glistened afar in the sunlight, and its verdant center was clad with coconut palms. The inhabitants revealed their presence by lighting a big fire but seeing that we made no answer to their signal they straightway sent out a small outrigger canoe manned by three islanders that we could not entreat to come alongside the *Coquille*. These were, however, the very first islanders that we had ever set eyes upon, the first men of this race of Oceanians whose features we were so eager to study. The natives were stark-naked except for a strip of cloth around their loins. Their bodies glistened with coconut oil and they were unarmed and brought no objects to barter. They merely repeated, with much laughing and gesticulating, the word "friendship" that was ever on our lips.

*Even the clergy were swept up in this torrent of abuse—reduced to such a state of moral degradation ...*

INDIGÈNES DU PÉROU.

**APRIL 1823**
TAHITI

*It grieves me to say* that the physiognomy of the indigenous peoples of Oceania grows daily more spoiled through contact with Europeans. It is certain that the indolent and effeminate lifestyle of the Tahitians, or what we disparagingly call their libertinage, was a long way from being as dissolute as in the towns in our own corrupt civilization. They practiced prostitution as part of a cult of courtesy and its impact has been greatly exaggerated. Sacrificing for love was a kind of ritual to which they would only submit in very particular circumstances. In the end their penchant for love making was but the consequence of a human economy whose every other need is supplied; of a fertile land that guarantees a life of ease and freedom from want, populated by peaceful inhabitants content simply to exist and greatly disinclined to wage disastrous wars. I have pondered at length on every thing that has been written on the natives of the Polynesian islands and I must confess that all too often these people have not been understood by the writers, or seen only through the prism of our own vainglorious civilization.

Bottom:
Peruvian couple in 1820.

Facing-page, top:
Natives of the Society Islands.

Facing-page, bottom:
*Vue de l'île de Bora Bora* (view of Bora Bora Island).

Pages 192-193:
*Vue de la partie méridionale de l'île Maupiti et du village d'Atipiti* (view of the southern part of Maupiti Island with the village of Atipiti in the foreground).

*The Christian religion* brought to Tahiti by protestant ministers has modified a few local usages without bringing spiritual enlightenment. As an example, the Tahitians are assiduous in their attendance at the temples built in several districts, but purely because their chiefs have been won over by the missionaries' many gifts and made non-attendance punishable by chastisement. Games, dancing, and distractions of any kind are forbidden on Sundays, the islanders' only consolation being their right to join in the singing of mediocre hymns translated into bad Tahitian. Their voices being soft and supple and their language melodious and rich in vowels, the Tahitians have acquired quite a taste for these hymns, which have replaced their ancient chants and provide them with a form of recreation. But these European missionaries sent out to the South Sea to preach the gospel are in fact former craftsmen with bigoted opinions, little if any education and a mind full of pettifogging dogma but not one whit of true religious doctrine.

The Tahitians, for their part, are now well versed in deception, helped by their associating with the convicts who escape from Port Jackson and infest all the South Sea islands. That these islanders are fully capable of pernicious thought is proof enough of the ideas planted in their minds by Europeans. The distinction that once characterized these peoples is now hidden by the veneer of dishonesty laid upon them by the rigid fervor of Protestant priests. Most of the natives forcefully rejected the Christian religion long ago, and though they tremble at the name of God, they may hardly be called religious.

*... by wealth and power that far from hiding their wickedness they flaunted it for all to see.*

**JUNE 1823**
**PORT-PRASLIN, NEW IRELAND** [PAPUA NEW GUINEA]

*In the interests* of my natural history collections, I explored the environs of the waterfall in Bougainville, then proceeded to climb the flank of the mountain on which it is situated and tarry a while following its sinuosities. As I left the dense woods to regain the sea, I encountered a few natives who had built a kind of *ajoupa* with palm tree leaves. Two fires were burning to roast *laka* fruits, which have something of the shape but especially the flavor of chestnuts. *Laka* fruits strew the ground and are the main source of sustenance for the natives in their sorties through the woods. At the behest of my hosts, I sat down among them and shared their frugal repast, biting into the chestnuts as they came off the fire, with an appetite that rendered even their firm texture a delicacy. Flowing alongside us was a pure stream where we quenched our thirsts. On completion of this pastoral repast, I gave myself over to the pleasure of hunting butterflies and terrestrial mollusques, this pursuit being made singularly more fruitful by the efforts of two young natives who climbed the trees like monkeys and whose keen eyesight could spot even the smallest insect tucked among the leaves. I made the most of my companions' benevolence to enrich my notes with the profusion of names that they give to the various products of Nature that surrounded us.

**AUGUST 1823**
**RAWAK ISLAND, NEW IRELAND** [PAPUA NEW GUINEA]

*So charming did I find this isolated island* that I would spend a few hours of every day in the company of the Papuans, whose friendship I cultivated with gifts in very much the same way as I might have done in a Paris salon; for whether wearing a tailcoat or a piece of cloth about the loins, man is everywhere self-regarding. My rewarding of the natives' complaisance with beads of colored glass, which are tantamount to diamonds in their culture, would often find me invited to share in their cooking, the natives tolerating me as a friend of the tribe in the belief, no doubt, that they were doing me a great honor. So it was that one morning, scullion-like, I attended on the preparation of a feast for the gala day ahead. Appetite and the singularity of the occasion, or perhaps that self-interest of the voyager who was to relate his adventures on return to France, led me to take quite some pleasure in this native banquet. Wild tortoises proliferate in these parts, finding plentiful nourishment along the coast where they are speared by the natives using a makeshift harpoon made of long bamboo sticks tipped with a metal fork. I had already eaten my fill when one of the older natives tendered me some fish that had been cooking to one side. To judge from the conscious smile on his lips, he was plainly a connoisseur of good food and his fish, wrapped in tender leaves and cooked in the embers, was most assuredly delicate of flavor. However, nothing could have induced me to eat raw white maggots squeezed between bamboo sticks, and I was careful to make my exit well before the time came to indulge in those barely cooked lizards and other reptiles to which my dark friends appeared so partial.

*After all, in Europe we eat frogs, and in some places the people are even said to eat snakes; ...*

*... in America, people feast on giant iguanas. The Papuans taste in food is therefore the indisputable mark of people* comme il faut.

**OCTOBER 1823**
AMBON, DUTCH EAST INDIES, INDONESIA

*Seen from the harbor, Ambon,* celebrated capital of the Maluku Islands, presented itself at the time as a verdant landscape beneath that now familiar warm sky. The mountains were covered with a carpet of fresh green grass and the ravines were dark and densely wooded, their shadows contrasting with the chiaroscuro of the escarpments. We saw no other European ships in the bay. The Dutch are extremely wary of admitting foreigners to their remote eastern possessions and confine themselves to provisioning only such ships as are forced to drop anchor in the bay, the latter being forbidden from trafficking unless they have obtained a license to that effect in Batavia by acquitting themselves of the duties payable on their cargo.

The welcome extended to the expedition in Ambon is a memory that will remain with me forever. French being pretty much the lingua franca, we felt ourselves transported back to our homeland and swept up in the arms of our families and loved ones, the privileged recipients of a lavish hospitality the like of which we could not hope to enjoy in our own colonies or even France itself.

**OCTOBER 1823**
SERAM ISLAND, DUTCH EAST INDIES, INDONESIA

*It is a widespread* and long-established custom among the young people here to go hunting for men and return to their village with a freshly severed human head that glorifies them in the eyes of their tribe and wins them the right to marry. It is easy to see how such a barbaric practice might engender hatred against the perpetrators. The skulls of neighboring tribesmen, bleached by time, are made into glorious decorations that betoken the right of the old to the esteem of the young and steel the latter to follow the example of their elders. The upshot of these perpetual hostilities between the different tribal peoples is a savagery of the most brutal kind. Armed to the teeth, no tribesman ever ventures forth alone without having taken every precaution to anticipate the ambushes laid for him by others and laid by him for others. Once an enemy's fate is sealed, his pursuer can spend many years waiting to surprise him. Sleeping in the trees, crawling through grass, steering a course between Scylla and Charybdis—every means at the disposal of cunning is employed to profit

◄ Facing-page:
*Tombeau près du Doreri, Nouvelle-Guinée* (mausoleum near Doreri Bay, New Guinea).

▲ Top:
*Fête religieuse des habitants du Caïeli, île Bourou* (Caïeli villagers celebrating a religious festival, Bourou Island).

*One day, Australia will dictate her laws to Oceania and rival in strength Great Britain and the USA, the latter having shewn her the way to independence*

from whatsoever opportunity might occur. This is the Corsican vendetta in all its glory. The customs of the Seram Alifurus are in no way different from those of the Borneo Ida'an people.

**JANUARY 1824**
**SYDNEY, NEW HOLLAND, AUSTRALIA**

*What struck us most forcibly* about the settlements established in New South Wales was the English genius for colonization: England is assuredly deserving of the very highest praise for its careful attention to detail in installing the machinery of government necessary to the functioning of a civilization implanted on the borders of a region hitherto overrun by wretches. Consequently, during my two months' sojourn I redoubled my efforts to obtain such documents as might raise awareness of this process of agricultural colonization in my own country.

In summary, I think that the colonization of New South Wales, which was established at the cost of great sacrifice, is organized in such a way as to allow her parting company with England no sooner than she has reached the limits of riches and prosperity: a distant time no doubt but hastened by the hatred felt for their mother country by the emancipated convicts and their children, an aversion that keeps these people born in England quite separate.

**FEBRUARY 1824**
**SYDNEY, NEW HOLLAND**

*Being both Frenchmen and voyagers,* we wished to pay a tribute of regret to the sadly famous Lapérouse by visiting the spot where he wrote the last dispatches ever to reach Europe. It was here, in the camp that he established at the northern tip of Botany Bay, that he planted a vegetable garden for the benefit of his disease-ridden crew. Known to the English as the French Garden and preserved by them out of respect for its founder, the garden is now partly overgrown but its loamy, heathland soils continue to produce a few vegetables for the soldiers stationed in one corner of the bay. One of them led us to the place where Lapérouse's shipmates are said to be buried: Friar Père Le Receveur, De Langle and de Lamanon, who died here following their fatal wounding by the natives of the Navigators' Islands. We were shown the mound under which Père le Receveur is inhumed.

*The whole village came rushing out to greet us, congregating in our wake with the same enthusiasm as the crowds that mass in the streets of Paris.*

We spent several hours carving a new epitaph into the bark of the tree that stood over this sorrowful episode. The day will come when a more durable monument will be erected to the shipwreck of this great expedition that was dedicated to the discovery of knowledge in the service of progress and civilization.

**JUNE 1824**
**CAROLINE ISLANDS**

— *We were still quite far from the coast* when our little pinnace, steered by one of our domestics, could go no farther. We straightway jumped into the water and came ashore before a large hut where more than one hundred natives were crouched at their repast. They pressed us to sit down with them, whereupon each in turn came to satisfy his curiosity. One sought to discover if the color of our skin were not the effect of white paint; and all of them stared long and hard as we removed our hats, shoes and jackets, our good hosts no doubt thinking that these effects formed part of our bodies. One of their number rushed to bring us coconuts and breadfruit. Having rewarded the natives' attentions with a few trinkets that made them happy, we gave them to understand that we were in want of guides to accompany us to the big village on the other side of the island. They perfectly understood our sign language, three of them detaching themselves from the group and walking before us. One seized my vasculum and carried it carefully all the way to the village, talking to me all the while despite my being unable to understand one word of the no doubt pretty things that he had to say. In every hut we were received with the same assiduous hospitality and innate thoughtfulness. Our surprizing appearance struck fear into the hearts of the women and young girls, but our guides and our behavior were enough to reassure them and build a foundation of confidence.

◄ Facing-page:
Native house in Doreri Bay, New Guinea.

▲ Top:
*Naturels de l'île Rotouma* (Rotuma Island natives).

**DECEMBER 1824**
**SAINT HELENA ISLAND**

*Having sojourned* from July 26 to August 9 in Papua New Guinea, we set sail for Surabaya where we landed on August 28. We then steered a course for the island of Mauritius, arriving on October 2 and staying but a few days before making sail for Saint Helena that shortly came into view.

On visiting the house so ennobled by great misfortune, we found it beset by an abject disorder far worse than mere neglect. The place had been transformed into a grain sieve for the winnowing of wheat for the colony workers confined to this part of the island. Not one thing about this dwelling attested to its having served as the refuge of an emperor banished by kings! Only the barriers that obstructed even unto his daily walk were still standing roundabout, their labyrinthine presence hampering the passage of visitors to this very day.

**MARCH 1825**
**MARSEILLE, FRANCE**

*On February 2 we crossed* the equator for the fifth time. The winds veered from one rhumb-line to another, the sky was black and thunder rolled to our ears. Driven by a stiff breeze and navigating by sea swells, we crossed the Tropic of Cancer without delay and were abeam the Cape Verde Islands when our sails became coated with red dust, the wind now blowing from the land and bringing with it fine particles of the volcanic slag on these islands.

On March 22, we recognized Cape Sicié and the coast of Toulon. But in vain did we tack all that day, the wind being directly against us as if to provoke our rightful impatience and keep us pinned to this part of the much longed-for French coast. For two long days the entrance to the great harbor at Toulon seemed set on resisting us, giving us a foretaste of all the mishaps that were presently to dispel what few illusions we had remaining. It was then that Monsieur Duperrey determined on drifting ashore. So it was that we shortly entered the port of Marseille, whence we made sail a few days later for Toulon and the decommissioning of the corvette *Coquille* in the very same arsenal that had witnessed her departure on August 11, 1822.

*A cradle stood in the place of Napoleon's bed and his bed chamber was become a stable. Gone were the tapestries, gone was the wood paneling.*

Longwood House, final residence of Napoleon Bonaparte, Island of St Helena.

Facing-page: Detail from *Vue de l'île Oualan prise du havre de la Coquille* (a view of the Island of Oualan as seen from the anchorage of the *Coquille*).

IT ISN'T EASY MAKING A NAME FOR YOURSELF WHEN YOUR SURNAME IS BOUGAINVILLE AND YOU DREAM OF BUILDING A REPUTATION AS A NAVIGATOR LIKE YOUR FATHER BEFORE YOU. BUT HYACINTHE DE BOUGAINVILLE PULLED IT OFF PERFECTLY, BECOMING KNOWN FOR AN EXPEDITION THAT WAS RICH IN INFORMATION ALBEIT DISAPPOINTING AT TIMES. HIS ACHIEVEMENTS WERE THAT MUCH MORE COMMENDABLE CONSIDERING THE POLITICAL CONTEXT OF THE TIME. FOR KING LOUIS XVIII, SCIENTIFIC DISCOVERY CAME SECOND TO DIPLOMACY AND ESPECIALLY ESPIONAGE, THE AIM BEING TO HOIST THE TRICOLOR FLAG IN SEAS WHERE FRENCH TRADE WAS OBSTRUCTED BY ENGLISH DOMINANCE.

# Hyacinthe de Bougainville

**1781–1846**

Plate 49, George Wolfgang Knorr, 1766.

# THE AROUND-THE-WORLD VOYAGE OF THE *THÉTIS* AND THE *ESPÉRANCE*

**1824–1826**

**To fulfill his goals of reconnecting with China and Vietnam, forty-three-year-old Hyacinthe de Bougainville launched his expedition in the opposite direction to his father's, embarking on a complex diplomatic voyage in Asia at a time when the British Crown called the shots in the region and France had already missed her chance.**

Hyacinthe de Bougainville was born in Brest on December 26, 1781, the eldest of four brothers. His father being often absent, he was raised by his mother Marie-Joséphine; but he shared his father's love of mathematics, and in 1798 entered the École Polytechnique, which had been founded a few years earlier. But the call of the sea was stronger. Hungry for adventure, Hyacinthe wanted to see the world, put his learning into practice. His chance came with the Baudin expedition, which Hyacinth joined together with Louis de Freycinet. So it was that on August 16, 1800 (Thermidor 28, year VIII) he departed Le Havre aboard the *Géographe* on a mission to explore the southern lands. A forty-two-month expedition conducted in difficult conditions that led to many losses, but during which our budding navigator, then aged just nineteen, proved himself and learned hydrography in the process.

Promoted to ensign in 1803, he took part in all of Napoleon's maritime campaigns, whether on the banks of Newfoundland with the *Cybèle*, in the Indian Ocean or in the China Sea. In 1812, his frigate, the *Cérès*, was attacked by two English ships off Rio de Janeiro. He was arrested, taken to London and imprisoned (following a fine family tradition ...) Two years later he was released when peace was restored, tried before a council of war and acquitted. When he returned to France, the fall of the Empire had made way for the Bourbon Restoration, whose King he would serve as faithfully as he had served Napoleon. He was after all a sailor, and as a sailor he fought for France. So much so that King Louis XVIII entrusted him with the command of a new expedition ...

In March 1824, the frigates *Thétis* and *Espérance* left the port of Brest. Unlike his father, Hyacinthe headed east. But having obtained no passports from any foreign power, he knew he might be in for a fight—a daunting prospect made worst by having been forced by the King and the Minister of the Navy to take on so many passengers that his warship now resembled a liner. In April, he rounded the Cape of Good Hope and by May he was in La Réunion, an island Hyacinthe knew well and where his crew took a short break. From there, the frigates headed north to India. Pondicherry was his first diplomatic visit, a sobering experience for Hyacinthe who could not help but notice the growing influence of the English. The expedition then headed west towards the Strait of Malacca, and Penang, a small island off the coast of Malaysia where Hyacinthe thrilled to everything he saw. Charmed by the cosmopolitan nature of the island—its great mix of people and hundreds of ships come to trade spices, oils, fabrics and porcelain—he stayed for twelve days. In September 1824, the *Thétis* arrived in Singapore harbor, another place bustling with activity; Bougainville was in awe at how much Sir Thomas Raffles had achieved in just five years, turning Singapore into one of Asia's most beautiful cities, a free port and the base for British supremacy in this part of the world. Then it was off to China. After a stopover in Manila, a hurricane dismasted the *Esperance* and Hyacinthe was forced to set sail alone for Macao aboard the *Thétis*, leaving his second ship behind for repairs. His stay in China however proved short-lived, the Celestial Empire having lost none of its inward-looking attitude. By order of the King, the expedition then made a stopover in Vietnam, arriving in Tourane [Da Nang] in 1825, where Hyacinthe was to deliver a letter from King Louis XVIII to Emperor Minh Mang. It was a real lesson in diplomacy: the Emperor had him welcomed with great pomp and ceremony, but refused to grant him an audience or accept his gifts. Though his courtiers assured Bougainville of their Emperor's great love for France, our Frenchman strongly suspected that pleasing the English took precedence over everything else.

Leaving Tourane, the *Thétis* sailed south and explored new territories. Hyacinthe discovered and mapped the Anambas Islands, and met their inhabitants whose seafaring skills came as something of a surprise. Continuing south, our travelers skirted the coasts of Borneo, notorious for their head-hunters, and called in at Surabaya on the island of Java, where they had great difficulty obtaining supplies. Next stop was Sydney but only at the end of a long and particularly arduous crossing of the Indian Ocean that cost the lives of seven sailors, six who died from dysentery and one who fell overboard, the ship being meanwhile overrun by cockroaches. Having finally anchored in Sydney harbor, the *Thétis* remained there for three months, partly for refitting and partly to give the crew a well-earned rest. Hyacinthe was very well received by the Governor, Sir Thomas Brisbane, and marveled at the beauty of the city and its surroundings. As ever, he admired the British administrative system but feeling that more could be done to honor Lapérouse, he had a monument erected in his memory on a concession donated by Sir Thomas. Finally, in September 1825, Hyacinthe left Australia for South America, calling at Valparaiso and Rio de Janeiro, before heading for France.

On arrival in Brest on June 24, 1826, Hyacinthe was sad, disillusioned and already nostalgic. He knew that France had arrived too late in the Pacific and that the English would remain in place for generations. Australia would never be French. His expedition yielded a wealth of information that France would never take the trouble to use to her advantage. It was a poor reward for Hyacinth, materially and morally.

In 1830, the July Revolution seemed to bring his career to a halt. With little motivation, and under some pressure from his superiors, he set about writing an account of his voyage, which was published in 1837. By that time ten years had passed, and other navigators had since circumnavigated the globe. He received a few more honors, was appointed rear-admiral, took charge of the navy in Algiers and immersed himself in the technological revolutions that might improve the French fleet. He died a bachelor in Paris, at 3 Rue de la Cerisaie, on October 10, 1846.

**South American parrot brought back to Paris by Hyacinthe de Bougainville.**

Bessa del. Coutant sc.

LE CALLOCÉPHALE AUSTRAL, Mâle.

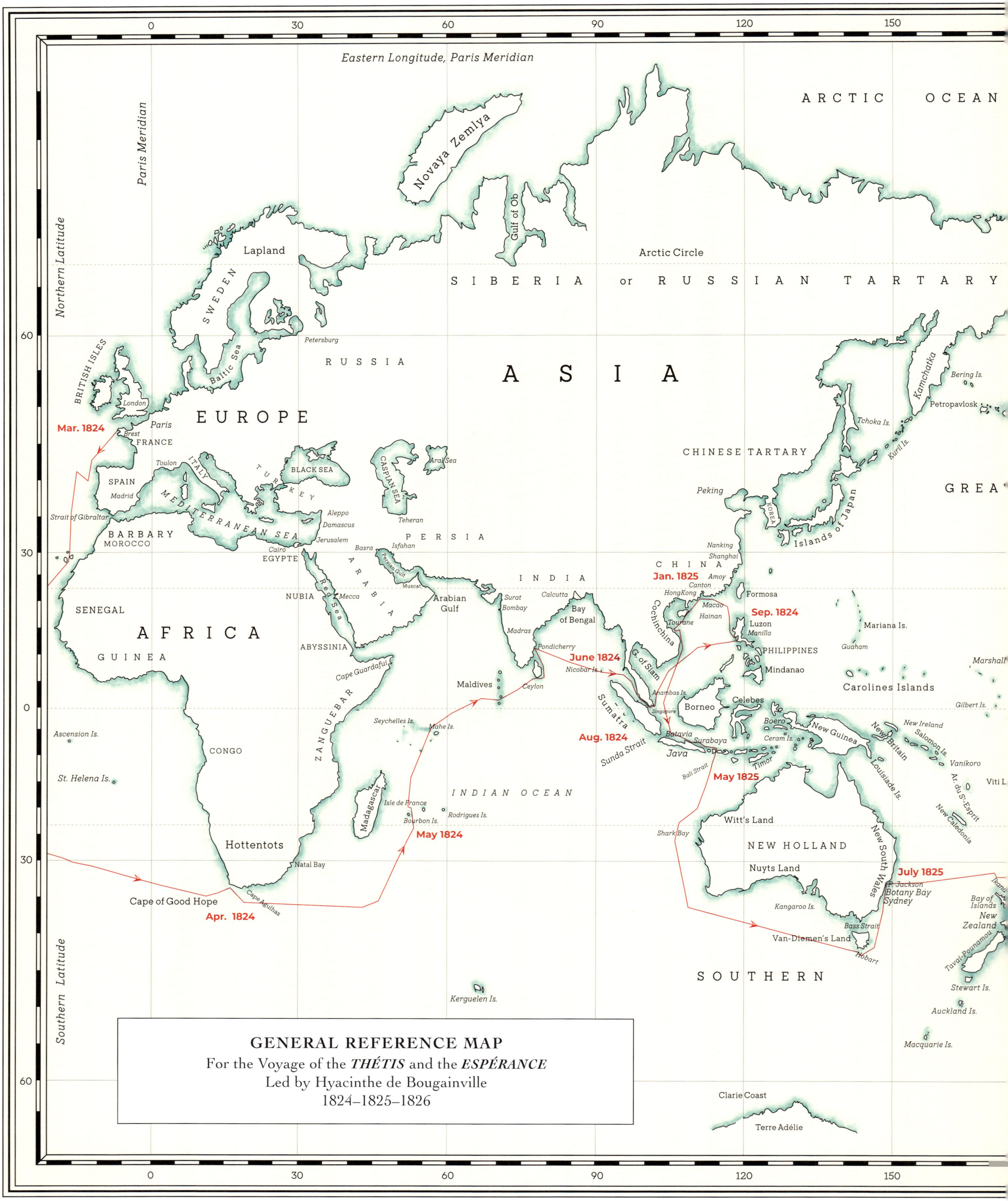


**GENERAL REFERENCE MAP**
For the Voyage of the ***THÉTIS*** and the ***ESPÉRANCE***
Led by Hyacinthe de Bougainville
1824–1825–1826

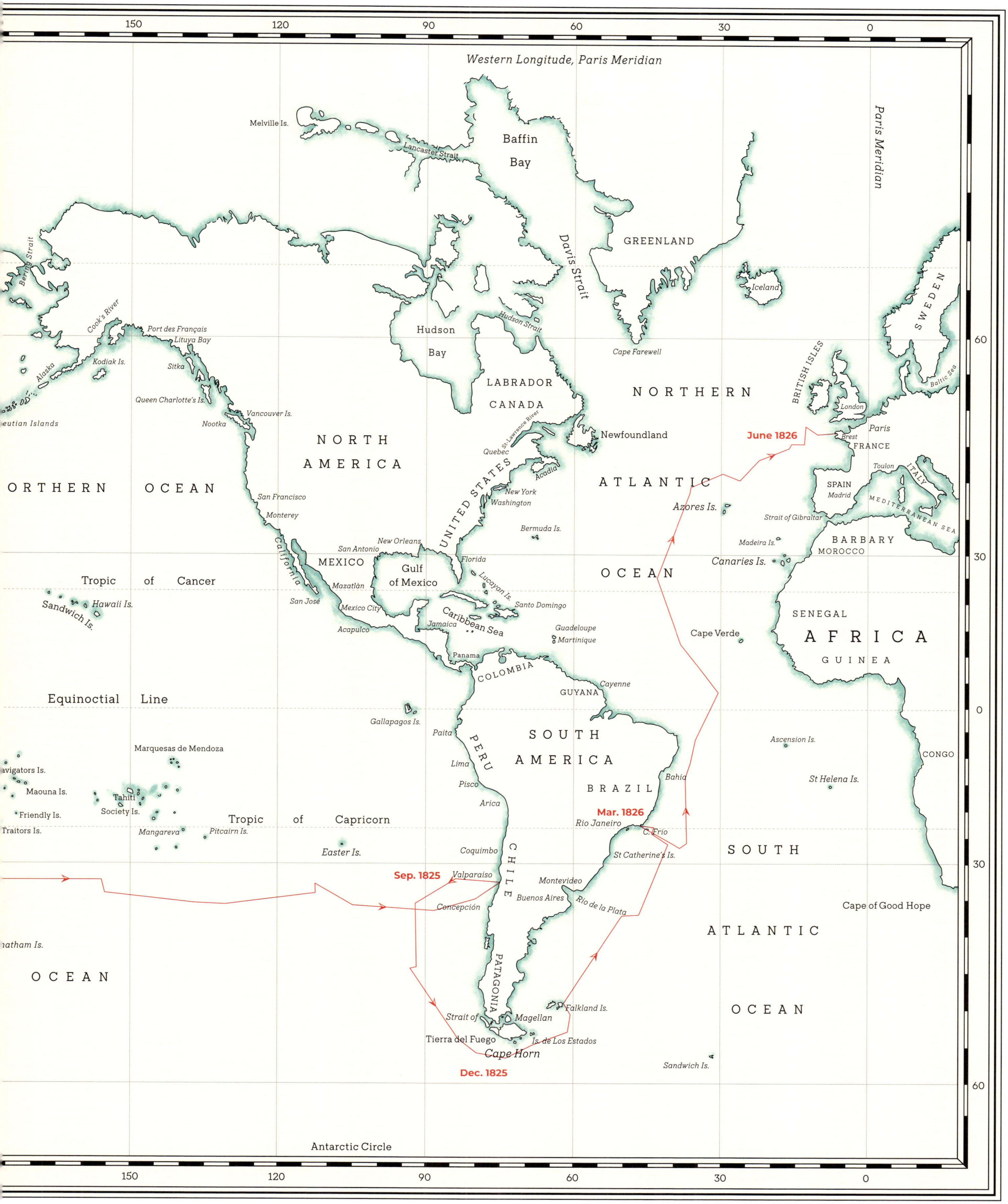
Western Longitude, Paris Meridian
Paris Meridian
150
120
90
60
30
0
Melville Is.
Lancaster Strait
Baffin
Bay
GREENLAND
Davis Strait
Iceland
Bering Strait
Cook's River
Port des Français
Lituya Bay
Hudson
Bay
Hudson Strait
Cape Farewell
SWEDEN
Alaska
Kodiak Is.
Sitka
LABRADOR
CANADA
NORTHERN
BRITISH ISLES
Baltic Sea
Queen Charlotte's Is.
Vancouver Is.
London
eutian Islands
Nootka
St-Lawrence River
Newfoundland
June 1826
Paris
Brest
FRANCE
NORTH
AMERICA
Quebec
ORTHERN
OCEAN
UNITED STATES
Acadia
ATLANTIC
Toulon
ITALY
SPAIN
Madrid
San Francisco
New York
Washington
Azores Is.
MEDITERRANEAN SEA
Monterey
Strait of Gibraltar
Bermuda Is.
California
San Antonio
New Orleans
Madeira Is.
BARBARY
MOROCCO
MEXICO
Florida
Canaries Is.
Tropic of Cancer
Gulf
of Mexico
OCEAN
Mazatlàn
Lucayan Is.
Sandwich Is.
Hawaii Is.
San José
Santo Domingo
SENEGAL
Mexico City
Caribbean Sea
Jamaica
Acapulco
Guadeloupe
Martinique
Cape Verde
AFRICA
GUINEA
Panama
COLOMBIA
Cayenne
GUYANA
Equinoctial Line
Gallapagos Is.
Paita
SOUTH
Ascension Is.
PERU
Marquesas de Mendoza
Lima
AMERICA
CONGO
avigators Is.
Bahia
St Helena Is.
Pisco
BRAZIL
Maouna Is.
Tahiti
Society Is.
Arica
Friendly Is.
Tropic of Capricorn
Mar. 1826
Rio Janeiro
Traitors Is.
Mangareva
Pitcairn Is.
C. Frio
Easter Is.
Coquimbo
SOUTH
CHILE
St Catherine's Is.
Sep. 1825
Valparaiso
Montevideo
Buenos Aires
Rio de la Plata
Cape of Good Hope
Concepción
ATLANTIC
hatham Is.
OCEAN
PATAGONIA
Falkland Is.
OCEAN
Strait of Magellan
Tierra del Fuego
Is. de Los Estados
Cape Horn
Dec. 1825
Sandwich Is.
Antarctic Circle
60
30
0
30
60

To sail around the globe was until a very few years ago a most hazardous enterprise. Indeed, less than half a century has elapsed since the time when an expedition of this nature sufficed to establish a certain reputation for the man at the helm.
The public interest and solicitude accompanied him across those so imperfectly known seas, a journey made more enchanting still by a prolonged absence and impossibility of communicating his news that cast a veil of mystery over his destiny.
Circumnavigation then was in its heyday and the dangers and privations suffered by the navigator were repaid a hundredfold when, laden with precious discoveries, he returned to salute the shores of his homeland.
Those hitherto much-dreaded seas are today an established theater of commerce between two worlds; of that profusion of islands and skerries scattered over the surface of the Great Ocean, there is nary one that has not been visited and reconnoitered by the ships of the Atlantic.
Various circumstances have deferred the publication of this journal and the manuscript in its entirety is now long abandoned. Revisited here is only that part concerned with the nautical aspect of my journey, the only part that may yet prove useful.
The major part of what I had begun to write about the lands that we visited is now redundant, having lost the novelty that was its chief merit.
It would be unseemly to present such information in the wake of those more interesting accounts rendered by officers who have more lately visited the same countries.
Thus is this publication but largely a logbook; and since it is principally written for seafarers, I have not shrunk from including details that albeit apparently trifling may at times prove useful to whomsoever may frequent those climes that we ourselves traversed.
Based on the foregoing, it is plain that this recital of the voyage of the *Thétis* and the *Espérance* will have little if any appeal to worldly folk, but I felt it my duty to warn them to forestall such reproaches as they might have should they ever have a mind to open this volume.

HYACINTHE YVES PHILIPPE POTENTIEN,
BARON DE BOUGAINVILLE

**MARCH 2, 1824**
**BREST, FRANCE**

*In the event, the mission* of the *Thétis* and the *Espérance* was not to carve out new paths or explore uncharted coasts but simply to protect the commerce of France, to fly the French colors, as they used to say, and verify the geographical positions of the principal reference points of those seas of which the English, drawing advantage from the proximity of their colonies, have so far advanced the exploration.

We departed Brest harbor on March 2, 1824, the *Thétis* armed with forty-four canon and crewed by three hundred seamen. We also carried a prodigious number of passengers for despite my efforts and the minister's goodwill toward me, our frigate was not only a transport ship but also had to do duty as a passenger vessel.

**JUNE 1824**
**PONDICHERRY, INDIA, FRENCH TERRITORY**

*Toward midday we sighted Pondicherry,* which a wooded mound of reddish earth would suffice to make known from afar were it not for its white stucco clad houses that catch the eye first. Two hours later, the division dropped anchor in the harbor, twenty days after leaving Bourbon and having noticed on its passage but one single ship off the coast of Ceylon.

The appearance of the place before which we were moored could not but give me pause for thought. For if Coromandel is not so devoid of attractions as to be incapable of rejoicing the eye of the wave-weary navigator, it falls far short, in my opinion, of the expectations it arouses in most Europeans. There is no Frenchman, even less a French seafarer, who on seeing these deserted beaches and this sea bereft of vessels and movement would not bitterly recall what Pondicherry once was and what it could have been—its glory and its downfall. But what use is it to recall a time we will never see again when the fate of Pondicherry is now sealed beyond redemption? The colossal power of the English is too firmly cemented in India and its jealousy of rival colonies too well known for there to be any hope of this region's ever enjoying more prosperous times. Afflicted by the past and wanting consolation for the future, we must consequently resign ourselves to speaking only of the present.

**JULY 1824**

**PONDICHERRY, INDIA, FRENCH TERRITORY**

*The Indian population* of the coast of Coromandel is divided into two classes: the Right Hand caste and the Left Hand caste. The origins of this division date back to a Nabob against whose authority the people rebelled, those remaining faithful to the prince bearing the Right Hand distinction and the others the Left Hand. These two great tribes account in near equal measure for the entire population and are constantly quarreling over the ranks and privileges originally granted to the prince's supporters. The latter meanwhile retain those positions in the gift of the Government while the others are relegated to the roles of shopkeepers and tradesmen.

The Right Hand and Left Hand are subdivided into eighteen castes, or trades, riddled with pretensions and prejudices that several centuries of association with Europeans has done nothing to diminish. Hence the rivalry and contempt felt by these men, a mutual disdain that would be the source of bloody battles were it not for the Indians' horror of bloodshed and their recoiling from violence in all its manifestations. Mild manners and dissent explain a political phenomenon whereby fifty million men remain under the yoke of twenty-five thousand foreigners.

*Arbre pétrifié de l'ancienne forêt de Trivicaré – Pondhichery* (petrified tree from the ancient forest of Trivicare, Pondicherry).

*French ships that travel the world in pursuit of trade carry a cargo of junk that gives a false and most damning impression of our manufacturers.*

**SEPTEMBER 1824**
SINGAPORE

*Singapore so captured the attention* of Sir Thomas Stamfort Raffles that he is said to have thought of making the island the seat of all English possessions on the Malay peninsula, with himself as governor general under the direct command of the ministry and the admiralty. From the military and commercial points of view alike, Sir Thomas judged Singapore to be admirably well situated, lying as it does at the midpoint between India's principal markets and the fertile islands of the Malay Archipelago; Singapore is located on the route of the Chinese trading vessels whose passageway it could open or close at it pleased, and adjoining the kingdoms of Siam and Cambodia.

It is impossible to know whether or no Sir Thomas did have such ambitious plans in mind. What is certain, however, is that the brilliant results achieved in the first years of the island's colonization more than testify to the excellence of his judgement. From the outset, he declared the port of Singapore to be "a free port, the trade thereof being open to ships and vessels of every nation, equally and alike to all." In the midst of negotiations, he put his time to good use attracting merchants, vessels, emigrants—whomever and whatsoever might give his yet nascent establishment the appearance of a colony. So successful were his efforts in this regard that the English Government ultimately found itself obliged to inform the Amsterdam cabinet that the expenditures being so considerable and the reclamations so tardy, it was no longer possible for the East India Company to divest itself of this possession without substantial compensation being paid, a sacrifice that the Dutch were not disposed to make. The island consequently remained in the hands of the English.

*"How good it is to preach the gospel to the natives, for, should we only teach them as much Christianity ...*

**OCTOBER 1824**
**MANILLA, PHILIPPINES**

*I stayed in Manilla but long enough* to acquit myself of some obligatory visits, of which I shall only mention the first, paid to Captain Don José Médinilla y Pinéda, Governor of the Mariana Islands at the time of Monsieur de Freycinet's voyaging. No Spaniard in the Philippines had a greater right than him to the attentions and gratitude of the French, and there is not an officer nor a seamen of the *Uranie* who has not tender recollections of the care and succor that he lavished upon them when the corvette dropped anchor in Guam, wanting of everything and with forty men laid low and fighting for their lives. Throughout their three months' stay, Monsieur de Médinilla gave them his own house, had delivered a profusion of refreshments of every kind and would not rest but that he had tended to their every need with a delicacy equaled only by its consistency. But so simple and natural did this noble conduct appear to him that he seemed really quite surprised by the warmth we evinced in our reminding him of it.

*... as is required to walk fully clothed, it would be of great benefit to British manufacturers."*

**DECEMBER 1824**
**MACAO, CHINA, PORTUGUESE TERRITORY**

*For the well-nigh six hundred years* that China has been known to Europe, we have dealt with her so curiously and with such constancy that it would seem whatsoever might arise from this country must necessarily be of interest to us. China has been presented to us from every point of view, whether its government, laws, usages, ceremonies, inland navigation and commerce, fruits of the soil and industry, masterpieces of art, inventions or discoveries. All of these things have been amply described by voyagers, embassy workers and especially missionaries, the latter being best placed to form a sound judgement of those men and matters that they are at leisure to study. We know our Chinese by heart. We know about their pride and ceremonious politeness, their cowardice and duplicity and their address and patience no matter how testing the transaction and business at hand. We know about their talent for imitation in the mechanical arts, their sobriety and their resignation, their gravity of demeanor and their moral depravity. We know about the concupiscence of the Mandarins, the roguery of Chinese salesmen and the impertinence of all, lesser or greater, toward foreigners and European foreigners above all!

**DECEMBER 1824**
**MACAO, CHINA, PORTUGUESE TERRITORY**

*Among the dominant traits* of the Chinese character, if there were one that would seem calculated to shock, it is the sovereign contempt that they affect for Europeans, of which they never let slip an opportunity to give us proof. In our eagerness to reach out to the Chinese from afar, persevering in our efforts to strengthen the ties between us despite the impertinent reception we receive, we have no doubt contributed to their disdain by appearing unable to do without our Chinese friends. Whatever the destiny reserved for this immobile empire (and in the interests of China, long may it remain so) its contempt for Europeans, albeit inherent in its people, is no less intolerable and one cannot conceive of any nation willingly exposing itself to treatment of this kind. Such however is the history of the colony in Macao, whose benefits to Portuguese trade are now non-existent and whose humiliation at the hands of the insolent Chinese is a source of undying shame to the House of Braganza.

Top: *Les idoles chinoises de la grande pagode à Macao* (the Chinese gods of the great pagoda at Macao)

Facing-page: *Costumes et cases d'indiens Tagals, près Manille* (Tagalog house and Tagalogs in traditional dress, near Manila).

**JANUARY 1825**
**MACAO, CHINA, PORTUGUESE TERRITORY**

*Here as elsewhere however,* commercial success essentially turns on the appointment of a consul, in this case in Canton, and the care taken in his selection. The position of consul in China requires a person of great prudence and tact capable of holding his own amidst foreign business agents with a long experience of dealing with the whims and trickery of the local population. The consul must be an educated man of flawless character and with impeccable credentials. He must inspire confidence by abstaining from any form of trading on his own account, this last point being essential no matter how difficult to achieve. Hence the importance of his being allocated a salary that allows him to serve as a worthy ambassador and bear such costs as may be required of any consul. Our consul to Holland receives twelve thousand piasters, his vice-consul half that amount. Proceeding on this basis would seem expedient as there can plainly be no question of rivaling the opulent East India Company, whose highest-paid workers can afford to put on a great show of luxury however trifling the occasion, the while setting aside a goodly sum for themselves.

**JANUARY 1825**
**TOURANE [DA NANG], COCHINCHINA [VIETNAM]**

*All of these people climbed aboard* upon receiving permission, and Captain Longueville escorted the Mandarins to the poop deck where Monsieur du Camper and I were waiting. I bade the first of these Mandarins to sit beside me on the sofa. He was a man of some fifty years, of pleasing countenance and wondrous solemnity, said by the interpreter to be the chief of justice. Pali-Tokinguang, who accompanied him, was shown to an armchair facing the armchair of the Captain of the *Espérance*, and a dozen valets bearing the personal effects of their Cochinchinese lords took their places at our feet. These preliminaries completed, the door was closed and the conference commenced, our having paid the greatest attention to every little detail, however trivial, that was likely to hinder our dealings with these men who strut more than they walk and whose comportment is ceremonious in the extreme. The description given here is necessarily informed by the preoccupation I felt at the time. The sitting underway, more than a quarter of an hour elapsed without any person opening their mouths. I was reminded of the story of the Chinese envoy and the Governor of Java, who having vied for supremacy in gravity and prudent diplomacy, remained in each other's presence for twenty-four hours without saying a word before departing in like manner.

► *Réception faite au commandant Bougainville par les mandarins de l'empereur à Tourane* (welcome ceremony for Commander Bougainville staged by the Mandarins of the Emperor of Tourane).

▸ *Roussette à tête cendrée* plate (grey-headed flying fox plate).

▸ Facing-page: Route of the *Thétis*, and discovery of the Anambas Islands, Indonesia.

*Upon their return our pinnaces were enchanted by the discovery they had made of a port, to which I gave the name of Clermont-Tonnerre.*

**MARCH 1825**
ANAMBAS ISLANDS, DUTCH EAST INDIES, [INDONESIA]

*The lands that lay before us* and whose northern border we had skirted for quite some distance that very morning, in no way resembled the Anambas Islands indicated on the English map of China, published in 1823. And for good reason since according to Horsburgh, these islands have never been explored and rarely if ever visited. The obscurity in which they languish to this day had led me to believe them of little importance, there being small chance that lands as well situated as these would have been left uncharted had they provided navigators with essential provisions. Imagine my delight therefore, when a host of islands and islets came into view as we approached, making up quite a large archipelago that likely offered excellent moorings at the time of the winter and summer monsoons.

Off to the southwest were two round islands fringed with white sand that we named Les Frères.

Still sailing upwind, we perceived a small, isolated island that appeared to stand in the center of a circular basin; this island we named for the *Thétis*.

**MARCH 1825**
BORNEO, DUTCH EAST INDIES, [INDONESIA]

*These natives, who they call the Dayaks,* are as brave as they are industrious but their natural bloodthirstiness and atrocious customs have led to their being dubbed headhunters. The torch of Hymen like the funerary torch is only ever lit for them when the head of an enemy, a stranger to the tribe or a miserable slave, has been severed by the warrior, the High Priest or the assassin. We should no doubt rejoice that such monsters are now confined to skulking in the mountains, having fled into the dense forests with only one another to tear apart. But alas! the people who have routed them from the coast are not much better. The natural treachery and rapacity of the Malays, all of whom engage in piracy, render them an enemy at least as daunting for the Europeans who ply their trade along these same coasts.

The Dayaks are not the only natives in this archipelago that daily offend against Nature in its most noble and sacred form. The Bataks, on the island adjacent to Sumatra, and probably many others whose ways are unknown to us, rival the Dayaks in ferocity and likewise feed on human flesh. The Bataks go even farther in

Gravé par Ambroise Tardieu, d'après les dessins de M. Bigot de la Touane, Lieutenant de Vaisseau.

PRIX. Un Franc.

Pl. 48

indulging their horrible taste for such feasting, having included in their tribal code a law that condemns certain criminals to be eaten alive piece by piece! But let us avert our eyes from such disgusting practices and thank Providence for our having been born in the Old World, in this Europe of ours that all said and done remains the best place in the world, whatever certain melancholy and troubled souls may say, who would no doubt grow in wisdom and amenability were they ever to visit the land of the Dayaks and the Bataks.

### APRIL 1825
### JAVA SEA, DUTCH EAST INDIES, [INDONESIA]

*Borneo, by virtue of its size* and central position, is second only to Java as a candidate to become Holland's most prized overseas possession. In Sulawesi, the principal Dutch establishment is Makassar, which commands four trading posts and is protected by Fort Rotterdam, whose high walls are studded with artillery. It nonetheless surrendered without a fight when invaded by an English division in 1810.

Ambon is the seat of the Government of the Moluccas and the center for all the clove islands under its jurisdiction. Every year in the month of January, a ship arrives from Java to take on a cargo of some fifty tons of cloves. In 1810 the island was captured by fewer than five hundred men from an English division of just two frigates and a corvette.

Ternate is the second most important Dutch settlement in the Moluccas and is protected by a fortified stronghold called Fort Oranje, manned in 1824 by garrison infantry. Seram and Buro are two Ambon dependencies. Seram has never submitted to the yoke of the Dutch and once traded in spices on behalf of the English with Ambon, which in exchange received merchandise from India worth a fifth of the price of the exported cloves.

Banda Neira is the capital of the ten islands in the Banda Sea, for which they are named, and the center for all the nutmeg harvested there. In January the same ship that calls in at Ambon also calls in here to export some five hundred thousand pounds of nutmeg and one hundred thousand pounds of mace (the nutmeg pericarp). Banda Neira boasts a large harbor, protected by two forts, Fort Belgica and Fort Naussau, that hinder the entrance of ships. The channel leading to the harbor does not exceed a mile in breadth and is of unfathomable depth.

► Plate showing the different vessels encountered along the way by the *Thétis* and the *Espérance*.

**GUIDALO**, bateau de passage de Cavite à Manille.

Grande barque de pêche de la Baie de Manille.

Bateau pilote chinois de Macao.

Bateaux de pêche chinois dans les parages des Iles Ladrones.

CHAMPANGS de l'Empereur de la Chine en rade de Macao.

de la Baie de Tourane.

Bateau pilote hollandais de la Rade de Sourabaya.

Pirogue des naturels des iles Anambas.

une JONG destinée aux voyages du Japon.

Petit navire de l'ile Célèbes.

◄ Pages 218-219:
*Le Cañada, promenade publique de Santiago, Chili* (the Cañada esplanade in Santiago, Chile).

▼ Bottom:
*Volcan du Broumo, dans les montagnes de Mallang, Java archipel de la Sonde* (Bromo volcano in the Malang mountains, Java, Sunda Islands).

► Facing-page:
*Têtes et groupe de sauvages du Camden-shire, Nouvelle Galles méridionales* (busts and group of aboriginals in Camden, New South Wales).

**MARCH-APRIL 1825**
SOURABAYA [SURABAYA] DUTCH EAST INDIES [INDONESIA]

*For the foreigner arriving in Surabaya* for the first time, a visit to the Sultan of Madura, in his Bakalam residence, is as obligatory as visiting the Palaces of Versailles and the Trianon for the provincial newly arrived in the French capital. Four-horse carriages were waiting for us, together with a breakfast served in a vast warehouse designed as a first stopping place for visitors. Tea, biscuits and jams did not delay us long but the paths were so detestable and the horses so bad that we all but became mired in mud twenty times. Our saving were the men stationed at intervals along our route who gave our wheels a push, then escorted us to His Highness (for such is the title given him by the Dutch) who received us at the entrance of the main block where he resides, in the fifth courtyard of the palace.

The first moments were devoted to etiquette. The Sultan having bade us sit beside him, his relatives, who were seated on their heels on the floor, received permission to approach and be introduced. And so they came, prostrating themselves before the Sultan and bringing both hands to their foreheads such that they almost crawled to take their places below some officers. The conversation was in English and the toasts came thick and fast, the prince drinking to our health from some bottled tea that he served himself in place of Madeira. As the head of religion in his states, he adheres rigidly to the letter of the Koran, never drinks wine and spends a large part of his time at the mosque. But he is no less a good host and his conversation is none the worse for the austerity that may be expected from such a strict way of life. The truth is that his time is not entirely spent in prayer, and the scenes to which we were privy gave an altogether different idea of his morals.

*Such was the ignorance that persisted until lately of the interior of the Dutch possessions in the East Indies.*

**JULY 1825**
**NEW SOUTH WALES, NEW HOLLAND, [AUSTRALIA]**

*When speaking earlier of Singapore,* I remarked that the ideas we form about distant countries spread today at astonishing speed. There can be no more striking example of this than New South Wales. It was but yesterday that Botany Bay was a place of exile and punishment, exclusively populated by undesirable citizens banished from Great Britain. Today it enjoys top ranking among the most important of her colonies, already aspiring to self-government and perhaps even quietly hoping to break free of the mother country.

How has this yet recent creation acquired such impetus, almost unbeknown to Europe so to speak? And how has it suddenly become an object of general curiosity, the theme of so great a number of publications that Australia, to call her by the name recently imposed on New Holland, is presently as well-known to many people in France as the majority of our own provinces?

*This genius belongs to Great Britain, to the country with an understanding of colonization that far surpasses that of any other nation.*

**MARCH 1826**
**RIO DE JANEIRO, BRAZIL**

*The Emperor was away* on a tour when we arrived and his return was the occasion for all manner of feasts and receptions that animated the otherwise dull existence of this city, the saddest and gloomiest city in the world for foreigners. The environs are all the same blessed by Nature to a very great degree, and the immense harbor is the very picture of activity, bustling with the innumerable vessels of the trans-Atlantic trading nations that converge on this port. Ships coming and going, craft of every type passing each other—an almighty racket made yet more deafening by the gun salutes received and returned by the forts and warships, or celebrating the anniversary or feast day of some such saint. There is also the incessant exchange of civilities between the officers of foreign navies as they pay each other visits, and between the diplomatic envoys of these great powers to the court of Rio.

Yet it is not even thirty years since the capital of Brazil, far from being one of the world's richest repositories as it is now, was but the miserable vassal of an inconsequential European power. And so it would remain today but for its enviable position on the route of ships passing by way of the three oceans, and for the terror struck into the hearts of the sovereign and his court by the conqueror that put them to flight. At the beginning of this century, the coast of Brazil was forbidden to foreigners and as my father's account makes clear, it was a chilly welcome that awaited those navigators forced by circumstances to drop anchor in Rio de Janeiro.

**JUNE 1826**
**BREST, FRANCE**

*The corvette, whose crew* until now had enjoyed perfect health (a miracle considering the quality of the accommodation), was suddenly hit by an epidemic that laid low some twenty men. The cause was initially attributed to the defective coating of certain cauldrons, which I made all necessary arrangements to repair, and to half of our remaining poultry and refreshments. The malady nonetheless continued to play havoc with the intestines, such that we were condemned to a period of quarantine that sorely tested what little remained of our patience after such a prodigiously long crossing. So it was that we were assailed by one vexation after another, though the windless conditions did eventually cease in favor of winds that bore us all the way from the Azores to Brest. On the twenty-third day at six o'clock, harbor pilot Nicolas Malgorn came aboard, the very same as had escorted me into Brest in 1823 and with whom I was most satisfied on both occasions.

On the morrow, a decision of the health authorities saw us exiled to the vicinity of Tréberon: the *Thétis* to undergo forty days of quarantine under observation (despite its three hundred men being as well as the day they had departed), and the *Espérance* to remain in isolation until further notice. Our freedom having been restored at last, we unfurled our sails one final time and at the moment of regaining the port, following twenty-eight months of absence, difficulties, vexations and champing at the bit, everything was forgotten. Pretty soon, the voyage and the voyagers were forgotten too.

# Auguste Bernard Duhaut-Cilly

**1790–1849**

DID EUROPEANS TRAVEL TO SATISFY NEEDS
THAT THEY CREATED FOR THEMSELVES?
THIS WAS A QUESTION THAT OBSESSED THE
SWASHBUCKLING AUGUSTE BERNARD DUHAUT-CILLY,
A SAILOR, PHILOSOPHER AND SOCIOLOGIST
ALL ROLLED INTO ONE.
CHARGED WITH FINDING NEW OUTLETS
FOR FRENCH TRADE, HE EMBARKED ON A COMMERCIAL
MISSION ABOARD THE *HÉROS* WITH A MYSTERIOUS
ADVENTURER AT HIS SIDE ...
ALTHOUGH TRAVELING IN THE REIGN OF KING CHARLES X,
HE HIMSELF WAS A CHILD OF THE REVOLUTION;
AND AS HE SAILED FROM PORT TO PORT MEETING
DIFFERENT PEOPLE, HE NEVER CEASED
TO QUESTION THE REPUBLICAN MOTTO:
*LIBERTÉ, ÉGALITÉ, FRATERNITÉ.*

Plate 5, George Wolfgang Knorr, 1766.

# JOURNEY TO CALIFORNIA AND HAWAII ON THE *HÉROS*

**1826–1829**

**He had been a disciple of the privateer Robert Surcouf since the age of seventeen, and always a great admirer of the navy. At the age of thirty-six, Duhaut-Cilly then sailed to Hawaii along the west coasts of the USA and Mexico, trading his cargo of goods in a series of improbable encounters ...**

Born in Saint-Malo on March 26, 1790, Duhaut-Cilly earned his spurs in the Indian Ocean and joined the navy in 1810. That same year, in August, he distinguished himself at the Battle of Grand Port off Mauritius, one of Napoleon's great maritime victories against the English (it features on the Arc de Triomphe in Paris.) Three years later he faced the English again off the coast of Guinea, this time under the command of his brother Malo, sailing aboard the *Artheruse*, which gave chase to the English frigate *HMS Amelia*. A fierce battle ensued, with many deaths on the English side but no clear victory for the French. Throughout his years as a sailor, Duhaut-Cilly remained unswervingly loyal to France and her navy, and was filled with admiration for her officers. But he also felt an immense respect for their English counterparts and was even slightly jealous of their prowess at sea.

In 1815, with peace restored, Duhaut-Cilly joined the merchant navy. It was a carefully considered decision, as evidenced by the reflections that illustrate the account of his voyage to the French colonies, Brazil, Havana and Buenos Aires. Around 1824-1825, he crossed paths with a number of bankers, among them Jacques Lafitte who was struck by Duhaut-Cilly's military and commercial experience. It was at this time that a certain Jean-Baptiste Rives landed in London in the company of the then rulers of the Sandwich Islands [Hawaii], King Kamehameha II and his wife. Rives had met French explorer Freycinet in Kealakekua and acted as his interpreter but without making much of an impression—rather the reverse. With the sudden death of the monarchs, "M. R.", as Duhaut-Cilly called him, tried to convince the English to mount an expedition to take control of the islands. They refused, as did the French when asked, the government having already paid dearly for financing those many dubious adventurers who returned from the islands at that time with their heads full of ideas of conquest. However, arguing that he was in possession of an "exclusive trade treaty", Rives managed to convince Lafitte and his colleagues to mount an expedition, which they entrusted to Duhaut-Cilly. Rives was to sail with him, together with the remains of King Kamehameha and his Queen that he was to return to Honolulu, the while assessing the situation there and trading along the American coast. Duhaut-Cilly accepted his mission, but remained on his guard...

The *Héros* left Le Havre in April 1826, her holds full of merchandise. After a stopover in the Canary Islands and Brazil, she rounded Cape Horn in very difficult conditions, pausing in Valparaiso in August to make necessary repairs before heading up the American west coast. Duhaut-Cilly coasted from Lima to San José del Cabo, doing good business selling his cargo along the way, and then further north to San Francisco for eighteen months. He then headed back down to Mexico as far as Mazatlán, then north again to Monterey. There, Rives deserted (Duhaut-Cilly had been right to be wary) and would die in poverty in Mexico City a few years later. Turning south again, the *Héros* headed for Lima, before finally returning to San Francisco. At each stopover our navigator took the opportunity to meet the locals, study their customs and talk to the residents, prominent families and missionaries. His account of those experiences, though certainly moving, pulls no punches, often amounting to a furious excoriation of the missionaries for mistreating and enslaving the Indians and failing to honor their commitments.

In September 1828, the *Héros* entered Honolulu harbor, where Duhaut-Cilly learned, not surprisingly, that Rives had totally failed to influence the authorities since the disappearance of King Kamehameha. He also notes, much to his dismay, just how strong are the Americans in Honolulu, thanks partly to the power of the Methodist missionaries but also to America's virtual monopoly of the sandalwood trade (vital to her dealings with the Chinese). Duhaut-Cilly berated what he saw as a reckless deforestation for the sake of trade that would no doubt have dire consequences for the environment. Continuing his world tour, he arrived in China in December 1828 where, like so many of his predecessors, all he understood was that everything remained to be understood. The *Héros* then sailed on to Java, Mauritius and finally the island of St. Helena, where Duhaut-Cilly paid his respects at the tomb of Napoleon Bonaparte (who died in 1821, and whose ashes were not returned to France until 1840).

By the time the *Héros* finally arrived in Le Havre on July 10, 1829, Duhaut-Cilly was exhausted. In 1834, he published the first volume of his *Voyage Autour du Monde*, each chapter of which begins with a statement of his profound philosophical convictions. In particular, he sincerely believed that "pure happiness" was only achievable by respecting what he called the "natural order" and that to flout that law was to court "disaster and misery." After a few years' rest, he takes command of the corvette *Ariane* in the Pacific.

In 1837, he crosses paths with Abel du Petit-Thouars, and in 1838 his ship anchors in Valparaiso harbor alongside the *Astrolabe* and Dumont d'Urville's *Zélée*, which had just returned from a grueling mission to Antarctica. Surprised by the slander circulating in Paris about Dumont d'Urville's expedition, Duhaut-Cilly pored over the *Astrolabe's* maps and other discoveries and was amazed by the results that the crew had already achieved. So much so indeed, that he readily provided the six seasoned sailors required by Dumont to continue his extraordinary circumnavigation and return to France.

Duhaut-Cilly died of cholera near his home town on October 26, 1839, aged 60.

► *Vue de la côte du Brésil, vis à vis de l'île de Sainte Catherine* (view of the coast of Brazil from Santa Catarina Island).

GENERAL REFERENCE MAP
For the Voyage of the *HÉROS*
Led by Auguste Bernard Duhaut-Cilly
1826–1827–1828–1829

150
120
90
60
30
0
Western Longitude, Paris Meridian
Paris Meridian
Melville Is.
Lancaster Strait
Baffin
Bay
Davis Strait
GREENLAND
Iceland
Bering Strait
Cook's River
Port des Français
Lituya Bay
Kodiak Is.
Sitka
Alaska
Queen Charlotte's Is.
eutian Islands
Vancouver Is.
Nootka
Hudson
Bay
Hudson Strait
Cape Farewell
LABRADOR
CANADA
BRITISH ISLES
SWEDEN
Baltic Sea
London
N O R T H E R N
St-Lawrence River
Quebec
Newfoundland
Apr. 1826
Brest
Paris
FRANCE
N O R T H
A M E R I C A
UNITED STATES
Acadia
A T L A N T I C
Toulon
ITALY
SPAIN
Madrid
MEDITERRANEAN SEA
O R T H E R N
O C E A N
Jan. 1827
San Francisco
July 1827
Monterey
New York
Washington
Azores Is.
Strait of Gibraltar
Apr. 1827
Bermuda Is.
BARBARY
MOROCCO
Madeira Is.
New Orleans
San Antonio
MEXICO
Gulf
of Mexico
Florida
Canaries Is.
O C E A N
Sep. 1828
Tropic
of
Cancer
California
Oct. 1826
Mazatlàn
Lucayan Is.
Sandwich Is.
Hawaii Is.
San José
Santo Domingo
Mexico City
SENEGAL
May 1827
Caribbean Sea
Jamaica
Guadeloupe
Martinique
Cape Verde
A F R I C A
Acapulco
GUINEA
Panama
COLOMBIA
Cayenne
GUYANA
Equinoctial
Line
0
Gallapagos Is.
Paita
PERU
S O U T H
Sep. 1826
A M E R I C A
Ascension Is.
CONGO
Marquesas de Mendoza
Lima
avigators Is.
Oct. 1827
Pisco
Bahia
BRAZIL
St Helena Is.
Maouna Is.
Tahiti
Society Is.
Arica
Friendly Is.
Tropic
of
Capricorn
Rio Janeiro
Traitors Is.
Mangareva
Pitcairn Is.
C. Frio
June 1826
Easter Is.
Coquimbo
CHILE
St Catherine's Is.
S O U T H
30
Valparaiso
Montevideo
Aug. 1826
Buenos Aires
Rio de la Plata
Concepción
Cape of Good Hope
A T L A N T I C
hatham Is.
O C E A N
PATAGONIA
Falkland Is.
Strait of
Magellan
O C E A N
Tierra del Fuego
Is. de Los Estados
Cape Horn
July 1826
Sandwich Is.
60
Antarctic Circle
150
120
90
60
30
0

In 1824, King Rhio-Rhio of the Sandwich Islands [Hawaï]
was driven by curiosity to visit England aboard a whaler
that had dropped anchor off one of his islands.
He was accompanied by his wife, his minister and a Frenchman
named R., who served as his interpreter and secretary.
On arrival in London, he was an object of curiosity
and ridicule, shown off like an exotic animal
by the nobility who invited him to visit and derided by journalists
ever on the look-out for a good story.
Capitalists vested hopes in his ignorance that were never
fulfilled, and the Government covered his every expense
without turning a hair, seizing the opportunity to curry favor
with a foreign potentate who might be induced
to favor English trade. But before ever any negotiations
were got underway, Rhio-Rhio and his wife fell victim
to smallpox and succumbed to death.
Monsieur R., having remained in London, sought to profit
from his position by spreading the rumor that
he had been empowered by the dying Rhio-Rhio to enter
into an exclusive trade agreement in the name of the latter.
The expedition that I am about to relate
was undertaken by Messrs Javal and Laffitte,
bankers by trade, whose love for the common welfare
more than a desire to augment their immense fortune
persuaded them to furnish such capital as might
be required to fulfill their hopes of opening up new
opportunities for French commerce.
An agreement was duly drawn up whereby Monsieur R.,
acting for the Sandwich Islands, granted significant advantages
to the interested party. It was signed in Paris in late 1825,
whereupon I was invited to take charge of this expedition
together with command of the ship destined thereto.
I betook myself to Bordeaux where I acquired a charming
vessel of 370 tons that was christened the *Héros,*
which I fetched to Le Havre to complete its outfitting and load
its cargo. The Naval Minister, who had appeared
to take a certain interest in the expedition,
had promised to supply me with such instruments as might
further its progress. In the event, my entreaties to make good
his promise fell on deaf ears and I departed
with no more than a special passport and a collection
of nautical maps that I handed back on my return.

AUGUSTE BERNARD DUHAUT-CILLY

**APRIL 1826**
LE HAVRE, FRANCE

*On the morning of April 10, 1826,* we prepared to make sail. The wind having first come from the east abruptly veered west but remaining moderate withal and bringing not a single cloud. Every thing was in readiness to welcome aboard Monsieur Martin Laffitte and his family and friends, for whom a breakfast had been laid out. Among the guests was the celebrated François Joseph Talma, who had eagerly accepted the invitation to enjoy a meal at sea and whose presence in my opinion was the main attraction for the curious onlookers who thronged the piers. Many of them were still enthusing about his gifted performance the previous evening in the title role as the magnificent Sylla; the ship's elegant lines, its gun batteries and even its destination paled into insignificance alongside Europe's most celebrated tragedian and it was to him that all eyes were drawn. The crowd seemed far more intent on distinguishing him among the company assembled on the bridge than on observing the ship's maneuvering.

**SEPTEMBER 1826**
PAITA, PERU

*After many months at sea,* the sailor who first contemplates the land that he has been skirting for some time is assailed by the strangest ideas. Distance does indeed lend enchantment to the view, creating an overall picture of velvety smoothness that changes from moment to moment as the vessel approaches, such that novelty is enhanced by variety.

It is then that the inexperienced young man may fall prey to the illusions conjured by this moving and too often misleading tableau. In place of impenetrable thickets, he sees slopes carpeted in lush grass; in place of woodlands strewn with decaying tree trunks and bedeviled by thorny lianas he sees charming groves skillfully laid out to include shady nooks and leafy retreats. He dreams of pleasing dwellings and bucolic wildernesses swathed in greenery, with not a town or a ship in sight. Giddy with delight, he promises himself to fulfill his dreams, reach out to those kindred spirits just waiting to break bread with him and include him in these illusory pleasures.

Imagine his disenchantment when all he hears is talk of politics, trade and seafaring! And should he persist in wanting to see for himself and venture in search of this garden of Eden, gone are the charming landscapes, enchanting woodlands and verdant plains. Now he sees only thickets and marshes infested with reptiles and insects, dried-up riverbeds choked with sand and pebbles, and coarse wild grass growing between the stones and the crevasses beneath his uncertain feet.

*This ungrateful soil is watered by the sweat of those unhappy wretches who toil for the enrichment of their oppressors.*

What he actually discovers are sugarcane, maize, cotton and coffee plantations, worked by wretches to the whistle of the cane wielded by the overseer. He recoils in horror at the sight of their misery and returns to town having seen too much, plagued by fatigue, defeated by despair and forever cured of his folly.

**Top:**
Benedictine monastery in the Bay of Rio de Janeiro, 1830.

**Pages 232-233:**
Scene from everyday life, Sandwich Islands (Hawaii).

**OCTOBER 1826**
**SAN JOSÉ DEL CABO** [LOS CABOS],
**BAJA CALIFORNIA**

*It was at daybreak* on the twenty-seventh of the month that we first caught sight of the land of California. It might be worth knowing whether overseas travel comes naturally to man or whether he has been driven to it out of necessity by civilization. That there was certainly a time when God inspired man with an irresistible longing to spread his race far and wide is plainly evident from the peopling of the Great Ocean. For it would be absurd to posit a separate Act of Creation for each of those islands.

The most enlightened voyagers have everywhere been struck by the resemblance between the different dialects spoken by these islanders. We know, for instance, that the inhabitants of the Sandwich Islands and the Marquesas understand each other perfectly, and that the latter likewise understand their counterparts in the Society Islands. Manners, idiom, physique: there is every indication that these populations are the result of successive emigrations by the same race.

However, these emigrants having once settled in these different lands, there is no record of their having invaded other places or lived otherwise than in peace, no doubt because the purpose of Providence had been accomplished.

The same goal does not drive we Europeans to venture to the ends of the Earth. No, our designs lie in persecution, curiosity, lust for glory, ennui and most especially the needs we have created for ourselves for all manner of different objects, gold and silver above all. Such are the motives that induce us to abandon our homeland and leave our loved ones, forswearing the comforts of the family home and the conjugal roof.

## NOVEMBER 1826
### SAN JOSÉ DEL CABO [LOS CABOS], BAJA CALIFORNIA

*There are no longer* any free indigenous people: they are all taken into the service of missions and private households. Despite the changes occasioned by revolution and the establishing of independence, the natives are no less considered and treated as slaves by the neophytes who manage the missions, and by the servants who manage the private households. As new missions were established, the missionaries would fill it with as many Indians as could be accommodated. No steps were taken to further their emancipation.

Having no personal interest in work, they became lazy; and virtue itself being powerless to guide them in the pursuit of grateful activity, they succumbed to wickedness and the diseases that killed off the greater part of their population, the birth rate being much diminished by servitude and in no way compensating for the death rate.

So it was that for more than a century the politics of the missionaries perpetuated a system tantamount to slavery, depriving these poor wretches of all the blessings of civilization.

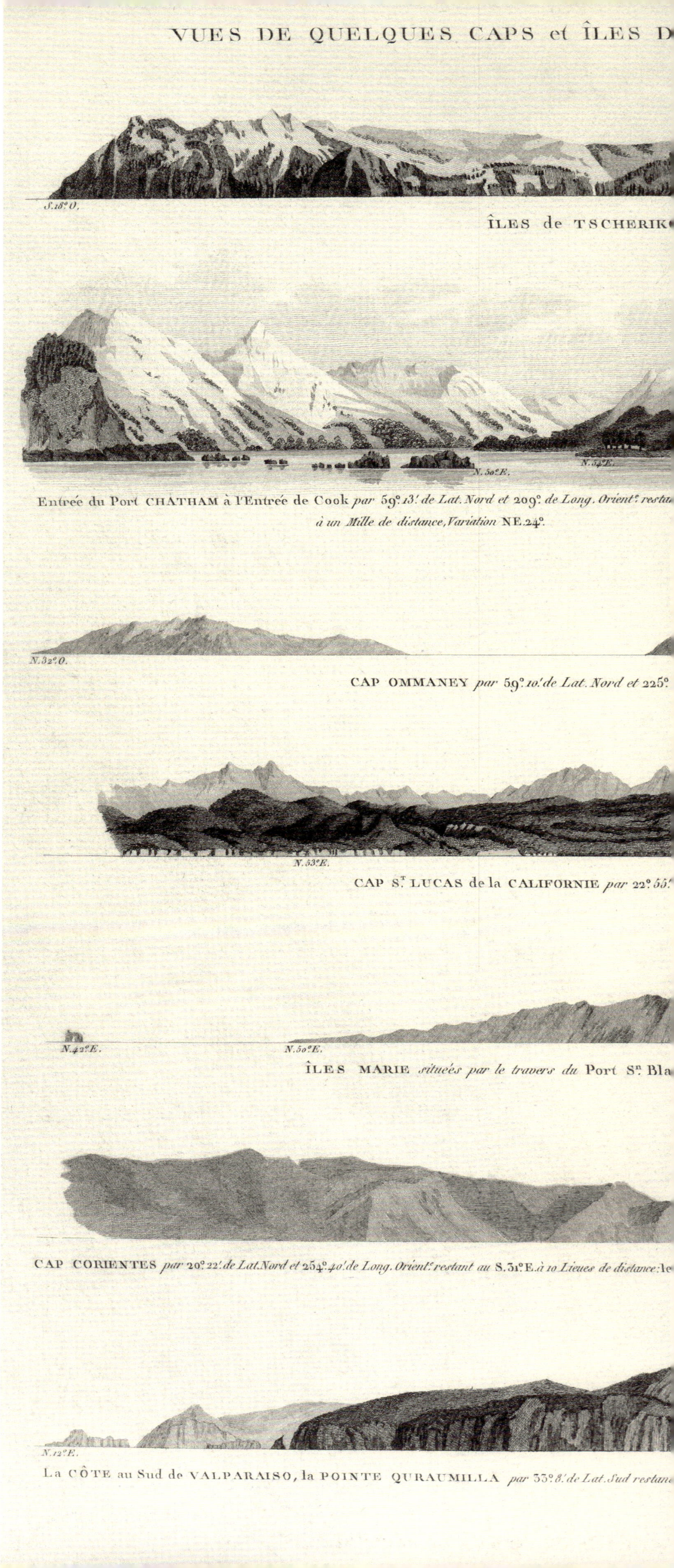


► Views of the American coastline as seen from the *Héros*.

A CÔTE DU NORD OUEST et DES CÔTES DU SUD-OUEST et DU SUD DE L'AMÉRIQUE.

55°. 49'. de Lat. Nord et 205°. 4'. de Long. Orient.e restant du S. 18° O. au S. 80° O. du Compas à la dist.e de 4 Milles. Var.on NE. 23°. 30'.

ÎLE de WINGHAM restant au N. 12° E. la Pointe Ouest de L'ÎLE de KAYE ou N. 23° E. et le CAP HAMOND par 59°. 48'. de Lat Nord 215°. 49' de Long. Orient.e restant au N. 48° E à 4 Lieues de dist. V.on NE. 29°. 20

E. au N. 54° E

g. Orient.e restant au N. 32° O. la Pointe Sud de l'ÎLE CORONATION au N. 11° O. à la distance de 5 Lieues; le MONT CALDER au N. 13° E. V.on NE. 25° 30'.

rd et 250°. 16'. de Long. Orient.e restant au S. 67° E. à la distance de 4 Lieues, la Terre la plus proche au N. 53° E. à 4 Milles de distance. Variation NE 7°.

XIQUE. la Pointe Ouest de l'Île la plus Nord par 21°. 43'. de Lat. Nord et 253°. 29'. de Long. Orient.e restant au N. 50° E à 3 Lieues de dist. Variation NE. 7°. 30'.

oches au S. 33° E. à 4 Milles de dist. V.on NE. 6°.

L'Îlot des DIEGO RAMIREZ au Sud du Cap de Horn par 56°. 30'. de Lat. Sud et 68°. 58'. de Long. Occid.e restant à l'O. à 5 Milles de dist.e Var.on NE. 22°

E. la plus proche Terre à l'ENE à 3 ou 4 Milles de dist.e la Pointe escarpée au Sud de la Montagne de la Selle au S. 62° E. Var.on NE. 14°. 45'.

*Suite de la Vue ci-dessus.*

**JANUARY 1827**
**SAN FRANCISCO, ALTA CALIFORNIA**

*We distinguished the entrance* to San Francisco from a distance of some three leagues. The view was exactly as described by George Vancouver. Having rounded the rocky spit of land that juts into the entrance, we came upon a second, higher, point upon which stands an ancient Spanish fort and there before us was a group of houses that we took for a farm. On closer inspection and having consulted the accounts of navigators Vancouver and Roquefeuil, I recognized it as the *presidio*. We dropped anchor some two hundred yards from the beach.

I found the San Francisco mission to be much changed since the time of Vancouver's visit in 1794. In those days, it consisted of a chapel and a house forming the two sides of a square. In recent years, it had grown to become one of the most substantial missions in California, as much for the wealth of its products as for the number of its Indians. By the time of our arrival in 1827, all that remained of its former splendor were the many buildings made necessary for its operation, most of them already in ruins. Of the seven hundred Indians still present when Roquefeuil visited the mission in 1816, there remained just 260. Having too few hands to maintain its standards, the mission had rejoined the ranks of the most indigent establishments on the entire coast. It had taken but two consecutive, talentless and feckless missionaries to set this once flourishing concern on the road to ruin.

**APRIL 1827**
**MONTEREY, CALIFORNIA, ALTA CALIFORNIA**

*Freedom! Freedom!* Over the past half-century, this oft-repeated word has been uttered by mouths that appear to belong in heads that ignore its significance, or rather reduce it to no significance at all. For every person who cries "we are free!", there are ten others who cry "we are oppressed!". Where only yesterday there was thought to be too much freedom, that much more again is required today. The fact is that freedom means different things to different people and it is impossible to create a freedom that suits one and all. The freedom to pocket public money with impunity is not, after all, every body's idea of freedom. No more than is the freedom to secure preferments, accumulate sinecures or enjoy fat payments for non-existent services.

*Navigators! Fellow risk-takers! Let us beware fortune, that capricious and flighty goddess ...*

It has been shown that no one agrees as to the meaning of political freedom and nor is it political freedom that I wish to discuss here. No, what I have in mind is the God-given freedom of Nature: the freedom that is understood by all living creatures if not by all men. It is this freedom that society is forced to take away from the convicted criminal; but it is also this freedom that is wrested from the miserable slave at the mercy of injustice and brutality, the freedom forever lost to the unfortunate Indians that I was to escort to San Diego at the behest of Don Ignacio Martinez.

### MAY 1827
### MAZATLÁN, MEXICO

*It came as no great surprise* to discover that every inch of this vast republic was in the throes of a great turmoil. The entire country was divided into two parties, *Los Escoseses* and *Los Yorkinos*. The latter formed a terrorist faction that was suspected of fomenting plots to take extreme and bloody measures against the Spaniards and even against all Europeans. Unfortunately, their views and some of their principles had initially been shared by certain members of the government in the mistaken belief that *Los Yorkinos* were no more than over-zealous patriots. The veil concealing their hideous designs was however soon lifted, and the government felt confident that it had stopped the storm in its tracks. It was not hard to see that in kicking all the rich Spaniards out of Mexico, or better still cutting off their heads, there was no other goal than to seize their fortune. Is there ever any other reason for mass proscription?

People take fright very easily amid the heady excitement of new found freedom. They fly into cruel rages on the least little suspicion, and it is easy to convince them of the perfidy of those we would ruin. In Guadalajara, the cathedral had been looted of its treasure, and a monk named Arena had just been executed for having allegedly dabbled in a conspiracy against the Republic. Discontent, fear and exasperation were the orders of the day. The government mistrusted its functionaries and most especially its bookkeepers. The corruption I witnessed in Mexico pretty much beggars belief. Only those born of revolution as we are, know that there is but small chance of any barrage, no matter how strong, prevailing against such raging torrents.

◄ Facing-page: :
*Vue de Monterey dans la haute Californie prise de la rade* (view from the harbor of Monterey, alta California).

▼ Bottom:
*Vue de l'établissement russe de la Bodega, à la côte de la Nouvelle Albion en 1828* (view of the Russian establishment of Bodega, on the coast of New Albion in 1828).

*... who will weary of favoring us long before we weary of following her.*

*The freedom to slander, vilify that which is most deserving of respect, is this a taste of freedom? It is an abuse and desecration of its sanctity.*

Top:
*Vue de la mission de San-Luis-Rey en Californie* (view of the mission of San Luis Rey, California).

Facing-page:
Chinese traditional dress in 1904.

**AUGUST 1827**
**CALIFORNIA**

*There is a total neglect of agriculture* in California and it is certainly the case that a country so devoid of laws is not propitious to the successful pursuit of farming. Only a landowner can wish to improve the land he works, yet there is not a single person here who has a legal title to his land. Only the missionaries and the government enjoy, and can transfer, the right of possession. I spoke earlier of those big farms or *ranchos* occupied by private citizens; I mentioned the gardens in the *pueblos* of San José and Los Angeles. The occupants here can be forced out at any moment, having built even their houses on land that is not legally theirs. Every concession hitherto made to the Californians can be revoked. More than half a century of occupation by the same family does not as yet confer any legal entitlement. We have seen inheritance disputes over this kind of property that ended in its being provisionally granted to a party with no right thereto, having first been seized in the name of the government and granted to said party to the mutual dissatisfaction of all the plaintiffs.

These are therefore no more than fiefdoms that can be withdrawn on a whim or at the lord's pleasure. Never has the government or the missionaries either donated or sold the least little piece of land to the people who lived there.

**OCTOBER 1827**
**CALIFORNIA**

*The same system that oversaw* the establishment of the missions in Baja California by the Dominicans was applied by their Franciscan brethren in Alta California. And neither one nor the other honored their side of the bargain, which is to say that at the end of the ten-year period they did not feel it appropriate to distribute the lands to the neophytes by restoring them to freedom. Several of the missions picked a few of the most intelligent Indians, who were established with their wives and children in lands of more than sufficient size for each family. They were furnished with enough livestock, plowing implements and food to see them through their first year. In short, their every need was supplied and fortune was theirs for the taking.

But far from bettering themselves, they succumbed to indolence and vice. Their gardens became choked with grass and their houses fell into ruin, such that within a few years they were obliged to return to the mission lest they starve to death. There, they could at least be sure of wanting for nothing, forced to work certainly but fed and given a roof over their heads. What are we to make of all this? Simply that these people are not born to be farmers. Theirs is the existence of the native, precarious no doubt but true to their nature. Forced into farming, they pine and die.

**OCTOBER 1827**
**CALIFORNIA**

*Slavery, for all that it may masquerade* as humanity dedicated to the betterment of mankind, is still slavery. As such, it is a state incompatible with the intellectual spirit that animates our being. The will of a people to be free can be quashed but never extinguished. The chain that prevents a man from falling into the abyss is no less of a chain in his eyes, no less of a shackle that he would break even at the risk of perishing.

What do the *padres* ask of the Indians of Alta California? A little work in exchange for plentiful food, warm clothing and the blessings of civilization. But even such tangible advantages as these are not enough to suppress an instinct for freedom that cries out to the Indians to forswear the comfort of monotony in favor of the impoverished, uncertain existence of the wilderness dweller.

**DECEMBER 1828**
**CANTON [GUANGZHOU], CHINA**

*We reached Whampoa* [Huangpu] where we dropped anchor in six fathoms of water, surrounded by twenty-five to thirty ships of various nationalities but mainly English and American. What can I say about Canton that has not been said one thousand times before? The Chinese nation is essentially immobile and hostile to progress of any kind. Or at least, for centuries now it has been government policy in this vast country to put every conceivable obstacle in the way of innovation. Were the very first voyagers to return to Canton today they would find it pretty much unaltered since they were last here three centuries ago. The manners and usages of the people you see today are the same as they were then and the only changes to the town would appear to be the European-style houses built here and there by the foreigners who now reside along the river in settlements known as Factories.

**JANUARY 1829**
**CANTON [GUANGZHOU], CHINA**

*The objects, industries and usages* found on all sides of China are entirely comparable to those found in Europe but practically all of these objects, industries and usages are the opposite of ours. Firstly, while we play battledore with the hand, they play with the foot. Their streets like ours are hung with a multitude of signs, but whereas ours are horizontal theirs are vertical. We sell dead fish (in France at least), they sell live fish. The roofs of our houses are convex or domed, theirs turn up to form gutters. Our boats are tapered at the bow to reduce wind and air resistance, theirs finish above the waterline in a tympan or vertical plane that

*To visit Canton is not to visit China. A foreigner who spends two months in this town lives with his fellow foreigners.*

Top:
Mandarins and soldiers, China, early nineteenth century.

Facing-page:
An audience with King Tammeamea [Kamehameha] in Ovayhi, Sandwich Islands.

is perpendicular to the keel. We write horizontally from left to right, they write vertically from right to left. We count by numbers, they count with beads. For us the Earth is round, for your low-born Chinese it is square with China accounting for half of that square. The Chinese have achieved practically the same degree of civilization and industry by virtue of processes that almost always seem exactly opposite.

**MARCH 1829**
**CHINA SEA**

*As we approached the Sunda Strait* we encountered a contrary wind but the current was so strong and so favorable that our zigzagging route was as good as any straight course we might have followed sailing into the same wind but with no currents. Only a zigzagging route could have given us the opportunity to observe the coastlines and islands of the strait from so many different aspects. The nature of the land had changed. Gone were those low-lying, wooded islands whence we came. These lands were of nobler, more severe countenance. Once past the bare, rocky acclivity of the Pointe Saint-Nicolas, the shore that drew closer with each tack was fringed with a pretty beach of bluish sand that likely consisted more of fine particles of lava and basalt rather than those particles that form our yellow European sands. Next came a wide, straight ribbon of land studded with coconut palms, their fronded crowns swaying gently in the breeze and so uniform in height that they could be seen shimmering from the tops of our masts, as green as the grain fields of Beauce in the month of June. Rising up beyond this sea of leafiness were hills of every shape, robed in forests or carpeted with yellow grass dotted with clumps of trees. So rugged was this terrain and so varied Nature's production, that the parks and gardens of Great Britain's wealthiest aristocrats appeared absurdly small by comparison. This was a parkland to make the English gardening enthusiast despair.

**MAY 1829**
**LE HAVRE, FRANCE**

*Desirous to take on fresh water,* we passed by way of the Island of St Helena where such provision is easily found. Thus did we drop anchor there for two hours before preparing to make sail anew. Another French ship arrived at the same time as us, whose passengers requested permission to spend a day visiting Napoleon's tomb. This being now a place of pilgrimage, there is practically no captain today on the homeward journey from the Indies who would refuse such a request from his passengers. So it was that we were invited to join the latter as they set off on horseback; but having been absent from France now for more than three years, every hour was precious and we declined despite the very real interest aroused by the project. But oh! what inexpressible sadness was ours at the sight of this rock, this infamous island that will forever stand as a monument to misfortune, shameful injustice and petty revenge.

Toward the Azores, a few days of becalmed conditions provided an opportunity to paint and refurbish our ship, such that on arrival in Le Havre on July 19, 1829 it was as fresh and clean as on our day of departure. This much to the astonishment of a good many people who imagined that a ship just returned from a tour of the globe must surely look no better than a hulk.

# Cyrille Pierre Théodore Laplace

1793–1875

INTELLIGENCE HAS ALWAYS BEEN THE MUSCLE OF WAR;
AND IN THE MIND OF KING CHARLES X OF FRANCE
THERE WAS NO BETTER MAN TO PROVIDE THAT INTELLIGENCE
THAN THE ZEALOUS AND RESOURCEFUL
CYRILLE PIERRE THÉODORE LAPLACE.
AS SUCH, HE WAS THE KING'S FIRST CHOICE
TO CONTINUE THE DIPLOMATIC MISSION ENTRUSTED
TO HYACINTHE DE BOUGAINVILLE.
A SPY PAR EXCELLENCE, LAPLACE LOOKED,
ASKED QUESTIONS AND TOOK NOTES WHEREVER HE WENT.
BUT LIFE AS A SAILOR IS MERCILESS
AND LAPLACE'S OWN SHIP, THE *FAVORITE*,
ONLY SURVIVED THE JOURNEY THANKS TO THE KINDNESS
OF LOCAL POPULATIONS. ILLNESS AND DAMAGE
MIGHT OTHERWISE HAVE BEEN HER UNDOING.
AS IT TURNED OUT, LAPLACE HAPPILY LIVED
TO LEAD ANOTHER AROUND-THE-WORLD VOYAGE
A FEW YEARS LATER...

Plate 55, George Wolfgang Knorr, 1766.

# VOYAGE AROUND THE WORLD, VIA INDIA AND CHINA

**1829–1832**

**For this first expedition, thirty-seven-year-old Théodore Laplace was charged with protecting French interests in the Pacific, most especially India, China, Vietnam and the Philippines. It was a voyage of discovery for this keen observer, in the course of which he faithfully reported to the King of France all that the English and Dutch were doing there.**

It is anyone's guess why Anne Eulalie de Ballon was at sea aboard the *Rebecca* on November 7, 1793; but it was on this day that she gave birth to a son, Cyrille Pierre Théodore Laplace. After spending his childhood in Saint-Domingue, Laplace joined the Imperial Navy as a *mousse* (ship's boy) in 1809, at the age of 16, just as Napoleon was stepping up shipbuilding to resume hostilities against the English. Laplace fought in one battle after another, demonstrating a courage in action, seamanship skills and a talent for leadership that saw him promoted from ensign in 1812 to lieutenant in 1819. His flawless maritime career continued under King Charles X (crowned in 1824) with the awarding of the Cross of the Royal and Military Order of Saint Louis in 1825, and promotion to corvette captain in 1828. Freycinet, Duperrey and Dumont d'Urville had already published their adventures or had just returned from their circumnavigation of Oceania. But King Charles X required another expedition for the sake of intelligence gathering. With France gradually regaining influence at sea and French merchants sailing the oceans, the government needed to know what the English were doing, to protect French citizens and to raise the Tricolor around the world. Laplace was the ideal candidate to lead such a mission and complete the work left unfinished by Hyacinthe de Bougainville ...

The *Favorite* departed Toulon on December 30, 1829, headed for the Cape of Good Hope and Laplace's first port of call, Gorée in Senegal, a place notorious for its slave trade. She then dropped anchor in the Port-Louis roadstead, Mauritius at the time being jointly owned by the English and the French, and the scene of constant internecine conflict that a furious Laplace took as a sign of undue British influence. He feared that the history of Quebec would be repeated and that the island would fall into the hands of these strutting English who were already behaving as masters of the seas. Throughout his expeditions and life as a sailor, one of Laplace's great challenges was to ensure that France had a strong and powerful navy.

The corvette then headed north towards La Réunion, followed by the Seychelles and Mahé in particular, where the crew encountered a "friendship and benevolence" that added to the charms of the stopover. Sailing northeast, the *Favorite* headed for India, where Laplace was commissioned to examine what had become of the French trading posts. In Pondicherry, he spent ten days filling his notebooks with information on commercial activities, taxes, duties and influential families. He pursued this work in Madras, where he mainly acquired anchors and chains to replace those lost at sea. But on the night of June 29, 1830, in complete darkness, the *Favorite* ran aground on the mud. It took Laplace eight days and the help of several Indian boats to refloat her. Equipped with a brand-new rudder and yet another new anchor, the *Favorite* crossed the Bay of Bengal and reached the Strait of Malacca.

Laplace spent just one week in Singapore, recording his observations of the city and its administration, trying to understand how the English had made their mark so quickly. In the process, he correctly analyzes the trade mechanisms put in place by Sir Thomas Raffles that led to the fall of Batavia's [Jakarta's] and Malacca's economic dominance and the concomitant rise of Singapore to become one of the richest places on Earth. By September 4 the *Favorite* was in Manilla where for Laplace, fever, cholera and hurricanes were as nothing compared to the disgraceful behavior of the monks.

The *Favorite* anchored in Macao on November 21, 1830. Here again, Laplace observed and took notes on local customs and the governance of the Mandarins, before heading for Canton. He found China fascinating but sadly misunderstood by Westerners, so he made a detailed description of the prevailing customs in his logbook. Before his eyes in Canton was "an immense people, who had long before us reached the highest point of industry and commerce, and yet who have nothing in common with Europeans but the hemisphere they inhabit." He was shocked that Europeans could be so "glaringly oblivious to the manners and customs of the people among whom they spend their lives." And he was convinced that the reason the Chinese were so distrustful of Europeans was precisely because the latter had never made the slightest effort to understand them, even after spending years on their territory. For Laplace, there could be no question of behaving in China as explorers and missionaries had behaved in the islands of Oceania. On the contrary, what was needed here was understanding and respect.

After twenty-one days in the Celestial Empire, it was time to return south. The stopover at Tourane [Da Nang], in Cochinchina [Vietnam], was a diplomatic flop, the Emperor going all out to frustrate negotiations, egged on by the local authorities. Laplace was convinced that the authorities were too afraid of antagonizing the already too-powerful English to accede to the demands of French diplomats. He could also see that those in the Emperor's entourage so resented British dominance that they had come to hate all foreigners, with the exception of the Chinese, thus laying the foundations for future alliances between Vietnam and China. But his notebooks also contained a great many positive observations that would prove invaluable.

All in all, it was with no regrets that Laplace departed Tourane and headed for Australia, first making a stopover in Java. On arrival in Dutch-controlled Surabaya, he was surprised to learn that during the July Revolution (the *Trois Glorieuses*) the monarchy had once again been overthrown in France. King Charles X had capitulated and fled Paris for England. With European alliances now in turmoil, the *Favorite* found herself in a port that no longer regarded France as a friend but as an enemy.

In July 1831, in the middle of the Indian Ocean, cholera, dysentery and endless storms took a severe toll on the crew. More than seventy men fell ill, reducing the rest to a very low ebb. The *Favorite* found refuge in Hobart, Tasmania, where the sick were cared for in a hospital that was put at their disposal, then the vessel continued on to Australia. Sydney was a fast-developing city at the time, arousing admiration in Laplace for the English economic model, but also doubt as to France's future in this part of the world ... In his opinion, France simply didn't understand the process of expatriation, leaving the field too wide open for the English who were experts at this game. His crew was meanwhile delighted by the splendor of Sydney, its enchanting appearance and the hospitality of the Australians.

The frigate then headed east again, crossing the Pacific and stocking up on supplies in Valparaiso. This was the age of whaling, an industry dominated by the Americans and the English. Once again, Laplace takes a lucid look at the conditions that had led to Anglo-Saxon success and French failure. He contrasts the great freedom of trade enjoyed by the English with state-supervised French industry, whose subsidies were misappropriated for personal gain by unscrupulous shipowners. The Anglo-Saxon system of government, which had confidence in and encouraged its major merchants, was in sharp contrast to the French system based on overarching regulatory oversight that was all too often counter-productive.

From Chile, the *Favorite* rounded Cape Horn and returned to Toulon on April 21, 1832, after thirty months at sea. The time had come for Laplace to write up his journey.

*To be continued*

► *Marché de Pondichéry* (Pondicherry market).

PRISON GENERALE

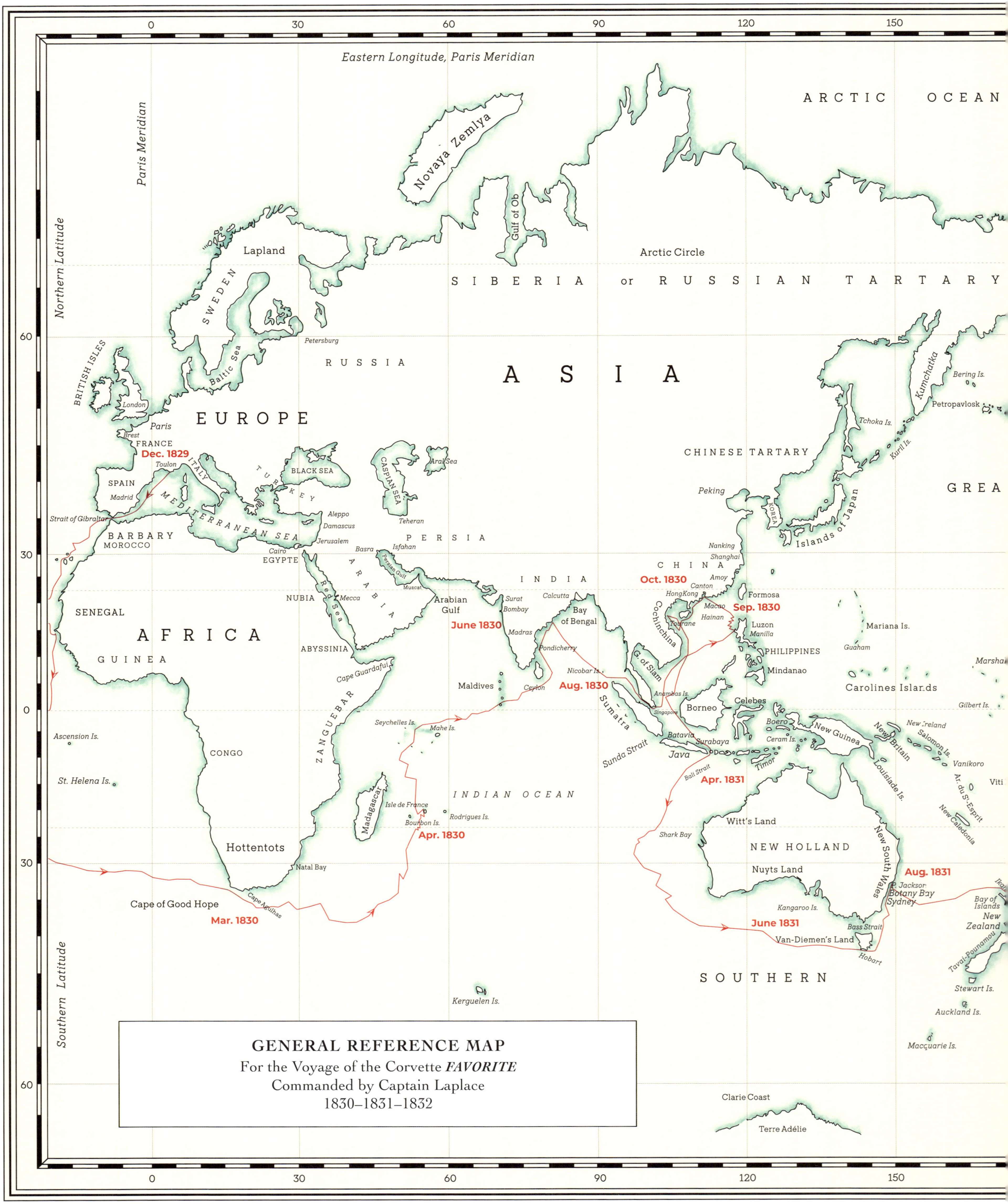


# GENERAL REFERENCE MAP

For the Voyage of the Corvette ***FAVORITE***
Commanded by Captain Laplace
1830–1831–1832

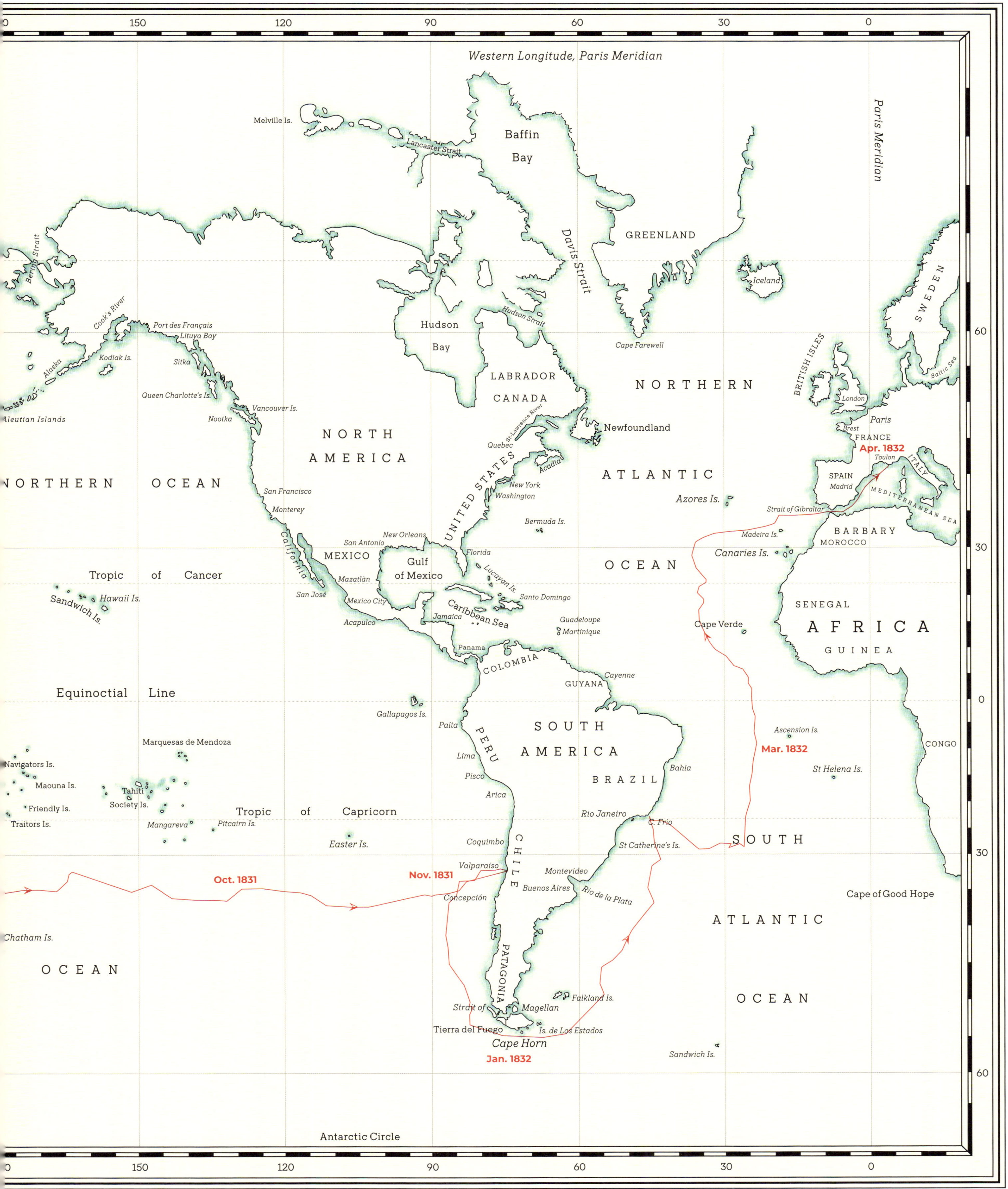

Western Longitude, Paris Meridian
Paris Meridian
150
120
90
60
30
0
Melville Is.
Lancaster Strait
Baffin
Bay
Davis Strait
GREENLAND
Iceland
Cape Farewell
Hudson Strait
Hudson
Bay
LABRADOR
CANADA
Bering Strait
Cook's River
Port des Français
Lituya Bay
Kodiak Is.
Alaska
Sitka
Queen Charlotte's Is.
Vancouver Is.
Aleutian Islands
Nootka
NORTH
AMERICA
St-Lawrence River
Quebec
Newfoundland
Acadia
UNITED STATES
New York
Washington
NORTHERN
ATLANTIC
OCEAN
NORTHERN
OCEAN
San Francisco
Monterey
California
San Antonio
New Orleans
MEXICO
Gulf
of Mexico
Florida
Bermuda Is.
Azores Is.
Mazatlàn
San José
Lucayan Is.
Santo Domingo
Mexico City
Caribbean Sea
Jamaica
Acapulco
Guadeloupe
Martinique
Panama
COLOMBIA
GUYANA
Cayenne
Tropic of Cancer
Hawaii Is.
Sandwich Is.
Equinoctial Line
Gallapagos Is.
Paita
PERU
Lima
Pisco
Arica
SOUTH
AMERICA
BRAZIL
Bahia
Rio Janeiro
C. Frio
St Catherine's Is.
Marquesas de Mendoza
Navigators Is.
Maouna Is.
Tahiti
Society Is.
Friendly Is.
Traitors Is.
Mangareva
Pitcairn Is.
Tropic of Capricorn
Easter Is.
Coquimbo
CHILE
Valparaiso
Montevideo
Buenos Aires
Rio de la Plata
Concepción
Oct. 1831
Nov. 1831
Chatham Is.
OCEAN
PATAGONIA
Falkland Is.
Strait of Magellan
Tierra del Fuego
Is. de Los Estados
Cape Horn
Jan. 1832
Sandwich Is.
SOUTH
ATLANTIC
OCEAN
Cape of Good Hope
Ascension Is.
Mar. 1832
St Helena Is.
CONGO
Cape Verde
SENEGAL
AFRICA
GUINEA
Canaries Is.
Madeira Is.
MOROCCO
BARBARY
Strait of Gibraltar
MEDITERRANEAN SEA
SPAIN
Madrid
FRANCE
Brest
Paris
Apr. 1832
Toulon
ITALY
BRITISH ISLES
London
SWEDEN
Baltic Sea
Antarctic Circle

This account was written and published by order of the Government and is the work of a naval officer who went to sea as a boy and has remained there for close to all his life. As such, he was somewhat of a stranger to the task expected of him but no less determined to speak the language of truth and apprize his readers of the present condition of the many distant countries visited by the *Favorite*. I hope to have achieved both these designs thereby atoning for such stylistic errors as may be discovered. It certainly is not my intention here to contest the views of others, even less impose any opinion of my own. My goal, I repeat, is solely to inform or at the very least arouse interest by speaking the truth; whether or not I have succeeded is for others to judge. Those persons good enough to read this account shall find it devoid of any nautical details or scientific observations likely to interfere with subjects of greater interest to civilian readers and accordingly obscure those subjects. My wish is to erect a monument to the memory of the works of my companions. This task may well have deserved a more practiced hand than mine; but I felt there was no person better able than their captain to do justice to the devotion of the officers and seamen of the *Favorite* throughout this long and difficult campaign. None but their captain might better bear witness to their selfless performance in the service of their country, or speak about them with more affection to their fellow countrymen.

CYRILLE LAPLACE, TOULON, 1833

**DECEMBER 1829**
**TOULON, FRANCE**

*Having received my final instructions* from the minister, the *Favorite* took sail on December 30, 1829, at two o'clock in the afternoon, driven by a light northwest breeze in uncertain weather and furnished with everything that might ensure the success of the mission that it was to undertake. The snow-covered land and sun hidden by clouds exuded a melancholy akin to our own sentiments in bidding adieu to our beautiful homeland. None of us would have wished to change his destiny nor forego the fine future that beckoned; but many of us knew by experience that we were perhaps leaving our families and friends for all time.

Thus was it with a heavy heart that we watched the coastline of France slowly receding into the distance until suddenly it sank below the horizon and finally disappeared from sight for a long time to come.

**APRIL 1830**
**ILE BOURBON** [LA RÉUNION]

*Far be it from me* to assert that France should forswear forming colonies or building settlements in distant countries. On the contrary I believe that her growing population and inability to provide for this new generation will soon leave her with no other option but to encourage emigration. How else is she to meet the clamor for change and an easy existence than by following the example of England, the country that having overtaken her European neighbors in civilization and commercial influence was first brought to resort to colonization—an expensive measure certainly, but essential to avoid great misfortunes.

Looking back at the past, we see Canada wrested from France despite a dogged defense due to our sadly neglected navy being no longer a match for the enemy. In the last of the Napoleonic wars, and for the same reason, we lost all our colonies one after another; and the same fate awaits our settlements old and new for as long as their wellbeing is divorced from that of our navy. Until such time as the French navy receive the promotion befitting such a great and noble institution, it will remain the object of new recriminations every year, and the interests of maritime trade and the protection essential for its thriving shall ever bow to the exigencies of the moment.

Only when this is no longer the case shall our belle France be ready to follow England's example and establish colonies that she can protect: colonies that will provide her ever-growing population with new homelands, her manufactories with new outlets for their products and her naval forces with waters under French jurisdiction.

*England will profit from her colonies just as she has always done, reaping the fruits of her rival's labours and rendering all her sacrifices in vain.*

**JUNE 1830**
**PONDICHERRY, INDIA, FRENCH TERRITORY**

*Pondicherry, though poor* and sparsely populated, remains a place of merrymaking and pleasure. Madras, with its palaces, treasures and population, is a place of sadness and ceremony. What Frenchwoman would willingly pay the price of such opulence, exchanging the pleasures of European society for isolation and especially the frigid rules of etiquette to which the English lady in India appears to devote her life? A drive to the sea in her dazzling carriage brings to a close a day spent almost entirely indoors. Exhausted by the heat and enforced idleness, she hopes that the cooler evening air will revive her energies; but nay, she returns more bored than fatigued to appear at the dinner-table as pale as she was in the morning, with her appetite equally jaded. The dessert course is her opportunity to escape and leave the men to get drunk while she withdraws to her apartments or takes up a now month-old invitation (etiquette oblige) to a soirée where the stiff and starchy demeanor of all those present will speak of a duty more than a distraction. From beginning to end, this will be a ball devoid of warmth, gaiety and spontaneity, an event attended with no pleasure whatsoever and no regrets when the time comes to leave. In vain did I study the physiognomies of these English ladies in search of a glimmer of joy; all that their pale, pinched faces evinced was ennui and disgust.

**AUGUST 1830**
**SINGAPORE**

*The cold afternoon wind* having lost much of its strength, we fought against the current as we navigated the Détroit du Gouverneur [Singapore Strait]—a passage made more difficult still by the current's changing direction twice in the year depending on the season. Rounding the rocks and reef bordering the island of Singapore was a struggle; but the *Favorite* eventually dropped anchor on August 17 at eleven o'clock at night, standing off the English colony of the same name.

Pages 250-251:
*Échouage de La Favorite* (shipwreck of the Favorite).

Top:
*Intérieur du port de Singapour* (the port of Singapore).

This town stands as a shining example of what the maritime trade of a great nation can achieve when it is encouraged and conducted by enlightened and appropriate institutions. The Dutch grew rich in Batavia by virtue of their monopoly over the local produce. Two centuries appeared to have bent the Malays to Holland's will, these people enduring endless vexations at the hands of their colonial masters, not the least of which was the imposition of duties as unjust as they were exorbitant. England heard the Malays' cries for help and immediately understood the advantage to be gained from going to their rescue.

Singapore is now the entrepot for Europe's formidable trade with this part of Asia, its fine, safe harbor bedecked with the flags of every trading power in the world.

**SEPTEMBER 1830**
**MANILLA, PHILIPPINES**

*This colony,* having languished in the darkness of ignorance and superstition for the past three hundred years, is now slowly awakening to the first glimmers of European civilization. The progress of the Enlightenment, however, must be introduced with caution to these almost wild people who are only kept in order by the righteous influence of the monks, and for whom the word "freedom" is synonymous with disorder and unrest. Ignorance and fanaticism are rife even among the upper echelons of this society, who though also deserving of pity constitute one further obstacle to be surmounted before ever a change of system might exert its happy impact, this mass of men being incapable of appreciating its advantages.

*Singapore became a flourishing city, a free port open to ships of every nation at no charge, and Batavia faded into insignificance.*

The monks of Luzon enjoy a power without limits, reaping the rewards of their many good works and dogged perseverance in the face of many dangers. The monks are the conquerors of these ferocious people but, in their courageous and admirable zeal to evangelize, far from showing Christian humility the brothers sought to unite spiritual influence with temporal power, such that the convents are now the real masters of the Philippines.

The monks are the enemies of innovation and use every means at their disposal to keep out foreigners, making them a target of hate for the natives, who only the monks can control.

**OCTOBER 1830**
**MACAO, CHINA, PORTUGUESE TERRITORY**

*The taxes here being very low,* the emoluments paid to the mandarins are relatively modest considering their obligations. So it is that, as if by tacit agreement between the Government and the depositories of its power, a right of peculation

*The ruin of our merchant navy has ultimately extinguished any french desire for expatriation.*

has been established. Every provincial governor, as a fully paid-up official of the Imperial court, is under the obligation to impose unlawful taxes on the people to fulfill onerous and ignoble engagements whose benefits are only secured at the cost of new sacrifices every year. The great Mandarins put pressure on their more lowly counterparts, who in turn put pressure on the population by the most iniquitous means. Together this multitude of petty officials comprise a kind of network that extends pretty much throughout the country and takes a heavy toll on the affluent classes.

Personal taxation however is low to non-existent, China's immense proletariat working as day laborers and being freer and happier here than anywhere else in the world. The Government handles them with great care, ensuring their subsistence and warding off anything likely to displease or stir up so great a mass of people; the fear inspired by the masses, indeed, is the only bulwark against that spirit of peculation that seems to animate the authorities here, regardless of rank.

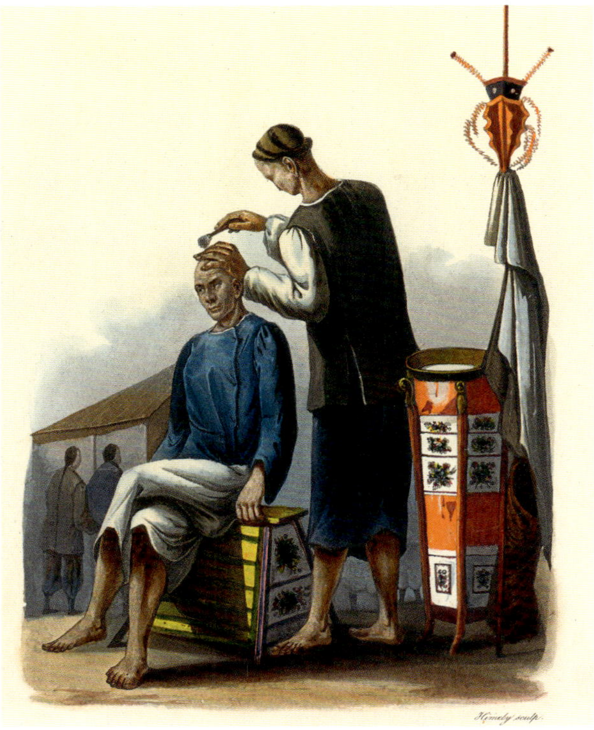

**Facing-page, top:**
Filipino couple in Manila, circa 1830.

**Facing-page, bottom:**
*Port de Singapour* (the port of Singapore).

**Top:**
*Église dans un faubourg sud de Manille* (church in the southern suburbs of Manila).

**Left:**
Chinese barber, Canton, China.

**Pages 256-257:**
Chinese village near Canton.

**NOVEMBER 1830**
**CANTON [GUANGZHOU], CHINA**

*The* Favorite *having dropped anchor* in Macau harbor, we made the necessary arrangements to provide the crew with fresh food and such distractions as might make them forget the fatigues of the crossing they had just endured. At that point I could turn my mind to Canton, an object of curiosity for me and the city that I was most earnestly entreated to visit by the French Consul to ensure the success of negotiations with the Chinese government.

Before 1826, no European and most especially no naval warfare officers (who were strictly forbidden from entering the country), could proceed to Canton without obtaining a costly permission from the Mandarins. Hotly debated ever since, this requirement has now been dropped on the grounds that it constituted an impediment to trade and was often obtained by fraudulent means. Now, charming schooners flying English or American flags transport passengers to Canton at great speed. It was aboard one of these delightful little vessels, chartered by two employees of the English company who impressed us with their warm welcome, that my two companions and I left Macao on the morning of November 29 and made toward our new destination.

**DECEMBER 1830**
**CANTON [GUANGZHOU], CHINA**

*The European who is prepared* to overlook the venality and rapacity of the Mandarins and observe only the manners and character of the Chinese people will see that their good points outweigh the bad. In truth, the Chinese have all the requisite qualities to become the greatest nation on Earth if only they were well-governed. They are patient, hard-working and intelligent with a gift for trade and an industrious disposition. They live off rice, fish, poultry and vegetables and keep their homes and their persons scrupulously clean. Peace and tranquility reign in the family home; wives remain faithful to their marriage vows; and the children have unbounded affection and respect for their parents. In addition, the poorest members of society are treated with humanity and kindness by their richer counterparts, rich and poor alike being united by bonds of generosity and gratitude.

The Chinese are accused of being cowardly, defiant and suspicious; they are reproached for having been as often enslaved as attacked by the Barbarians. But did these same Barbarians encounter greater resistance when they unseated the Later Roman Empire and subjected our ancestors, almost without

*What importance can a Chinese man of inferior rank attach to insults and blows ...*

*... when he sees the Empire's premier Mandarin being flogged by order of the Emperor himself?*

a fight, to the yoke of feudalism? In those distant times, industry and civilization always gave way to brutality. Time and circumstances, more than violence, have put freedom within our grasp. Not so in China, where the people remain subject to their victors: a government bereft of any principle of loyalty, whose agents live off plunder the while ruling these vast lands with an iron rod. What noble and elevated sentiments can be expected from men who have endured such tyranny for six hundred years?

**JANUARY 1831**
MACAO, CHINA, PORTUGUESE TERRITORY

—*Among Chinese women of a certain class,* usually quite pretty, well-made and neatly dressed women, some are of mixed Chinese and European parentage. Most however were sold as children by their parents to Portuguese settlers who have made of them Christians as miserable as they are themselves. The girls are almost invariably condemned to debauchery by the meanness or cupidity of their masters and live by doing business with foreigners. The more fortunate ones are bought by Chinese men whose concubines they become, in the hope that they will produce the much-desired male heir. Others still are shipped out to the provinces for the sordid pleasure of the local population. Those who remain in Macao lead a life of vice and enjoy a certain reputation as the fillers of the opium pipes smoked in debauched parties by men reclining on divans, slowly smoking themselves into dreamy stupefaction. Drawing up the smoke through the pipe's short stem, they inhale air rich with the fumes of the poppy seeds that have been rolled into a tiny pellet and inserted by means of a silver needle into the very narrow orifice at the tip of a conical red clay pipe. Holding the pellet over the flame of a spirit lamp causes the black sticky gum to catch fire, a single inhalation sufficing to consume its fumes. This operation is repeated until the smoker fall into a senseless delirium that must be very agreeable to judge by the passion it arouses in opium devotees. Over time, the effects of opium are equally deleterious to mind and body but are only felt through repeated practice, or so I am told by those many Europeans who say that their flirting with opium left them unmoved physically or mentally.

**Facing-page:**
Canton street in 1830.

**Top:**
*Procession chinoise à Macao* (Chinese procession in Macao).

**FEBRUARY 1831**
TOURANE [DA NANG], COCHINCHINA

*It was unfortunate that France* had overlooked certain advantages offered by the Philippines and one cannot help but regret that she abandoned Cochinchina at a time when the future looked promising; unlike now, when everything seems to conspire against us and shatter whatever hopes we retained.

The past few years have indeed seen a dramatic change in circumstances, our own merchants now being at a disadvantage compared to our tireless rivals who will stop at nothing to seize valuable assets that we have failed to put to good use. Since the foundation of Singapore—the port that like some great lighthouse casts its light on all the countries roundabout—the English have been aware of, and intent on exploiting, the commercial potential of Cochinchina. When they disembarked in Tourane, they were admitted to the capital where they received a magnificent welcome. Our own trade with the inhabitants meanwhile became subject to innumerable barriers and as many exactions. Eventually the court of Hué-Fou [Hué] opted to keep out Europeans altogether, treating all foreigners alike in its hatred for the English and anxious above all not to give the governor of Singapore further cause for complaint. And so things stand today, the court of Hué-Fou doing everything in its power to prevent Europeans setting foot in Cochinchina. Such was the sorry state of France's affairs when the *Favorite* appeared in the Bay of Tourane.

**JUNE 1831**
AT SEA, SOUTH OF NEW HOLLAND

*Dysentery was rife aboard ship,* affecting several of our officers including me, though I was not so indisposed as the worst of them and so much the better as we were about to embark on a long and difficult crossing of the stormy Southern Ocean in winter.

Such were the gloomy auspices when the *Favorite* cast off from the Sunda Strait on the morning of June 1. A contrary wind alternated with calms, so forcing us to zig-zag between the shores of Java that are sandy and deserted in these parts, and the dark and mountainous coastline of Baly [Bali].

On our left to the southeast stretched the immense expanse of New Holland, whose long coast bears the names of the navigators who contributed to the exploration of this fifth part of the world. These coasts are very irregularly shaped and terminate toward the south in the Bass Strait that separates them from Van Diemen's land [Tasmania], a region that was then only lately settled by Europeans. Thither did the *Favorite* make sail, heading for the English settlement of Hobart where we hoped to be provided with shelter for our ship and succor for our increasingly epidemic-stricken crew.

**JULY 1831**
HOBART, VAN DIEMEN'S LAND [TASMANIA], NEW HOLLAND

*Toward the end of the day* I sighted the first dwellings of the European settlers. Fairly sparse and uninteresting at first, they became far more numerous and much finer as we approached that arm of the sea called Derwent on which stands the English settlement. Those among our poor sick who had managed to drag themselves to the bridge gazed longingly at these houses, orchards and herds scattered along the coast. Thinking to see their homeland again, hope sprang anew in their hearts. Who knows how many of them would have died had we arrived but a single day later!

The pilot came alongside us in the place where the Bay of Storm converges with the canal and is joined by the river Derwent. He dispelled all our uncertainties and conducted our corvette before nightfall to the mooring in Hobart, where our safe arrival was greeted by a crowd of townspeople aboard pleasure boats bedecked with flags and banners in three colors. It was a heartwarming reception that spoke of allies and friends. Indeed the very next morning, all our sick were transported to the colony hospital where they were treated with all the tender care befitting the great nation on whose hospitality we were imposing.

Facing-page: *Grotte et pagode de Tourane* (grotto and pagoda in Tourane).

Bottom: Woolloomooloo in Port Jackson, Australia.

Pages 262-263: Kororareka beach, New Zealand.

**AUGUST 1831**
**SIDNEY [SYDNEY], NEW HOLLAND**

*Among their many charms,* young Australian women exhibit a gay disposition and taste for pleasure that make them very like Frenchwomen. Added to this, however, is that strength of will and wifely devotion so often displayed by Englishwomen who abandon their parents and a life of ease to follow their husbands to the Indies, to China and even into the forests of the Antarctica. How many times have I not been astonished by the apparent indifference with which the ladies of Sidney or Hobart have told me of their forthcoming departure for Europe, from which they expected to return having been absent for less than a year! This although the journey would take them right around the world, passing the Horn and the Cape of Good Hope with apparently none of that dread felt by the seamen themselves. The mere idea of such a crossing would be enough to scare a Frenchwoman to death.

Not so her English counterpart, who embarks without batting an eye, takes possession of her cabin, remains entirely oblivious to everything happening on board, and looks after her children and herself but certainly not her fellow passengers. I have often seen even the longest crossing undertaken in this manner by very young English ladies who by birth and fortune plainly belonged to the cream of British society.

**OCTOBER 1831**
**IN THE PACIFIC OCEAN**

*Having had our fill of scenes of horror,* it was not the cannibals of New Caledonia nor the Viti Levu (Fiji) and Salomon Islands that excited our curiosity; rather it was our desire to visit the charming archipelago of the Friendly Islands [Tonga] where the navigators of the past century had received such an hospitable welcome. We knew, of course, that we would no longer find there those good-looking Tahitian natives whose combination of loose morals and childlike innocence elicited such glowing descriptions from the straitlaced Cook and our witty Bougainville. In vain does today's traveler look for those pretty huts shaded from the sun by groves of fruit trees, or those lovingly tended fields separated by fragile fences. Gone are those men full of confidence and affection. Gone are those women full of grace and beauty whose tasteful apparel was as neat as their demeanor.

All you see today are near-deserted lands, stripped of trees and overgrown, populated solely by wretches: thieves, drunks, letches, impure women infected with venereal diseases through contact with European seamen, and English and American missionaries who live in holy leisure off the backs of these unfortunates. Such is the picture presented by most of the Polynesian islands. The Sandwich islands are now pretty much a British colony and the Friendly Islands, likewise overrun by Anglican ministers, are well on their way to becoming an annex of New South Wales.

▲ The *Favorite* arrives in Valparaiso.

**NOVEMBER 1831**
**VALPARAISO, CHILE**

*By the thirteenth day of the month* our eyes were scouring the horizon in search of the snowy peaks of the Andes that can be seen from a distance of forty leagues when sailing in clear weather. But so thick was the mist shrouding the mountains that at eleven o'clock, within but six leagues of land, our lookouts had yet to sight anything.

Our journey of exploration ended here, in this land so well-known in the Old World for its trade and revolutions. It is difficult to describe our pleasure at seeing warships and merchant ships flying our national colors. The sort of moral isolation that had been ours since our departure from Bourbon was finally at an end. Our motherland was still very far off, but we could speak about her with fellow citizens and friends; and oh! what music to the ears of the seafarer is his home language! How harmonious and expressive are the words of his fellow native speakers and how interesting are the accounts of what has been happening in France in his absence!

**APRIL 1832**
**TOULON, FRANCE**

*With mounting excitement* we approached the entrance to Toulon only to encounter a last-minute problem at the end of our journey. To our great dismay, the breeze dropped in the afternoon and by the evening we were becalmed beneath Cap Sicié. Imagine our relief when the wind picked up again at night, such that at nine o'clock we were able to regain the same mooring whence the *Favorite* had weighed anchor on December 30, 1829.

On the morrow the decks of the corvette were practically deserted, officers and men being now released from duty and busy regaling their families and friends with the story of our adventurous expedition. By the time the *Favorite* had been decommissioned, most of them had been paid for their good services by the Government and I could leave them to their new-found freedom and repair to Paris to render account of my mission.

There ended the task of the marine officer and here endeth that of the historian. I have fulfilled one and the other with the same zeal and devotion. During our twenty-eight months of navigation, I did everything in my power to protect our merchants and uphold the honor of France in all parts of the world. Throughout that time, I meanwhile applied myself with unstinting perseverance to the penning of my observations, a task as alien to me as it was arduous, and undertaken for one purpose only: to open the eyes of my fellow citizens to the decline of France's maritime trade and demonstrate the necessity of promoting and expanding our navy.

◄ Banyuwangi, Java, Indonesia.

# Auguste Nicolas Vaillant

**1793–1858**

UNDER THE JULY MONARCHY AND THE GOVERNMENT OF KING LOUIS-PHILIPPE I, THE FRENCH NAVY WAS INCREASINGLY CALLED UPON TO EXPAND FRANCE'S DIPLOMATIC PRESENCE AND PROTECT HER TRADE, A ROLE IN WHICH SKILLED SAILOR AUGUSTE NICOLAS VAILLANT EXCELLED. IT WAS A STRANGE WORLD TOUR FOR THIS FUTURE MINISTER OF THE NAVY—A MAN WHO WOULD PONDER INDEPENDENCE AND THE EMANCIPATION OF SLAVES THE WHILE DISEMBARKING FRENCH DIPLOMATS AT HIS VARIOUS PORTS OF CALL AND FILLING HIS HOLDS WITH ANIMALS OF ALL KINDS.

Plate 43, George Wolfgang Knorr, 1766.

# VOYAGE OF THE *BONITE*

**1836–1837**

**Manila, Canton, Macao, the Sandwich Islands ...
As a worthy representative of the government, the forty-two-year-old August Nicolas Vaillant sailed his ship the *Bonite* to faraway shores where France's Tricolor had been absent for too long.**

Vaillant was born in Paris on July 2, 1793, just a few months before Laplace. He too entered the navy as a novice, but in Brest in 1808, and was assigned to a succession of gunboats, then sailed with convoys along the coasts of Brittany and Normandy. In the Napoleonic Wars, he fought on the coasts of Flanders and Holland, before being promoted to midshipman first class in 1813 and heading for the West Indies. He then embarked on the *Hermione* and later sailed the English Channel aboard the *Durance*. Briefly dismissed at the time of the Restoration, Vaillant was reinstated in 1817 and promoted to ensign in 1818, in which capacity he was sent to French Guiana to explore the possibility of establishing a colony there. He took charge of Cayenne and made a name for himself by drawing up a very precise map of the Maroni valley. He embarked on a series of battles and expeditions, defending French interests on the oceans and repelling pirates—in short, Vaillant was every inch a sailor. In 1827, after the battle of Navarin in the Peloponnese, he became chief of staff to Admiral Henri de Rigny on the *Conquérant*, and having been appointed frigate captain in 1831 he continued to sail the high seas.

In July 1833, Rigny became Minister of the Navy and took Vaillant on as his aide-de-camp. Then came a return to Paris where King Louis-Philippe, who titled himself King of the French instead of King of France, was turning his attention to regions of the world where the foundations of a new empire might be laid ... It was in this spirit and to protect trade, particularly whaling, that the *Bonite* expedition was set up and entrusted to Vaillant. Preparations only lasted a few months, which is not very long for a circumnavigation, and Vaillant would have occasion to complain about his crew's lack of experience, even teaching some of them to read and write during the Atlantic crossing. The Académie des Sciences invited itself on board, in the persons of three scientists— a hydrographer, a naturalist and a botanist— who would gradually transform the *Bonite* into a menagerie over the course of her many stopovers and wanderings.

The corvette set sail from Toulon on February 6, 1836, carrying sixteen passengers, nine consuls and seven servants—quite a lot for a crew made up of novices but Vaillant personally saw to it that everyone was kept busy. So busy indeed, that what with climbing to the top of the mast and back down again, sea training and learning to fire a cannon, by the time the *Bonite* reached Rio de Janeiro, the crew was fully trained and ready to round Cape Horn.

On her arrival in Valparaiso in June 1836, the new consul disembarked, the scientists began collecting data and the diplomatic mission could begin. In the bay, however, were four French ships in need of Vaillant's help, among them a whaler called the *Geneviève* whose crew had deserted. No more sailors, no more whales. The *Bonite* continued on to Peru, where Vaillant met Morand, commander of the Peruvian troops. Despite the warm welcome, he was dismayed by the state of the country, its poverty and its inability to cope with the departure of the Spaniards, who had left nothing but ashes in their wake.

In October 1836, Vaillant left the American coast for the Sandwich Islands [Hawaii], where religious tensions remained rife. He met the king, displayed his cannon and by dint of smiles and threats made himself clear to his majesty. But the American presence was already well established, and Vaillant suspected that Hawaii's future was already mapped out. The *Bonite* then headed west, entering the port of Manila to land a new consul, before continuing on to Macao and Canton.

In the China Sea, scientists caught hundreds of fish still unknown in Paris. In December 1836, when Vaillant explored the city of Canton, nothing had changed: the Middle Kingdom was still closed to foreigners, Canton was off-limits to foreign wives, and Chinese politics remained the same. Following his mission order to visit Cochinchina [Vietnam], Vaillant then anchored the *Bonite* at Tourane [Da Nang]. But the new Emperor Minh Mênh was overtly hostile, hating all Westerners and the French among them, courtesy of England's now ruling the waves and making themselves the masters of India.

In February 1837, in the Straits of Malacca, the *Bonite* called at Singapore, whose destiny Vaillant had already guessed, like other great navigators before him. Heading north across the Bay of Bengal to Calcutta, he arrived to undertake an altogether different kind of mission.

Convinced that certain species of silkworm could be acclimatized in France, specialists had asked him to find cocoons and bring them back to France intact. It was a mission that spoke to Vaillant, and he carried it out with great success. With famine raging in Pondicherry, he also took on board a large number of sacks of rice but his progress along the Indian coast was slow due to the worst hurricane of his whole career.

The *Bonite* rounded the Cape of Good Hope in August 1837, sailed up to St. Helena Island and arrived in Brest on November 6, 1837. The expedition had been a success, with no loss of life and a great deal of information gathered, particularly in the form of live animals that would enrich the cause of science: twenty large mammals, nine hundred reptiles and amphibians, six hundred birds, four hundred fish and crates of shellfish.

Promoted to Captain, Vaillant pursued his career as a sailor in Mexico (where he took part in the battle of San Juan de Ulúa in 1838), Uruguay (where he was present in 1845 during the blockade of Río de la Plata and the occupation of Montevideo) and the West Indies. His successes earned him a rear-admiral's commission. Back on land, he became maritime prefect of Rochefort and a member of the Admiralty Council, then briefly Minister of the Navy in 1851. With his health deteriorating, he was denied permission to take part in the Crimean War and left active service with the rank of vice-admiral. He died in Paris on November 1, 1858.

► Pagoda in Tourane, Cochinchina.

**GENERAL REFERENCE MAP**
For the Voyage of the Corvette ***BONITE***
Led by Captain Auguste Nicolas Vaillant
1836–1837

150
120
90
60
30
0
Western Longitude, Paris Meridian
Paris Meridian
Melville Is.
Lancaster Strait
Baffin
Bay
Davis Strait
GREENLAND
Iceland
SWEDEN
Bering Strait
Cook's River
Port des Français
Lituya Bay
Kodiak Is.
Sitka
Alaska
Queen Charlotte's Is.
Vancouver Is.
Nootka
Hudson Bay
Hudson Strait
Cape Farewell
LABRADOR
CANADA
NORTHERN
BRITISH ISLES
London
Baltic Sea
Paris
Brest
FRANCE
Feb. 1836
Toulon
ITALY
St Lawrence River
Newfoundland
Quebec
Acadia
NORTH AMERICA
ORTHERN OCEAN
eutian Islands
UNITED STATES
New York
Washington
ATLANTIC
SPAIN
Madrid
MEDITERRANEAN SEA
San Francisco
Monterey
Azores Is.
Strait of Gibraltar
Bermuda Is.
BARBARY
MOROCCO
Madeira Is.
California
San Antonio
New Orleans
MEXICO
Gulf of Mexico
Florida
Canaries Is.
OCEAN
Tropic of Cancer
Mazatlán
Lucayan Is.
Sandwich Is.
Hawaii Is.
San José
Mexico City
Santo Domingo
SENEGAL
Caribbean Sea
Jamaica
Guadeloupe
Martinique
Cape Verde
AFRICA
Oct. 1836
Acapulco
GUINEA
Panama
COLOMBIA
Cayenne
GUYANA
Equinoctial Line
Gallapagos Is.
0
Paita
SOUTH AMERICA
PERU
Ascension Is.
July 1836
Lima
CONGO
Marquesas de Mendoza
avigators Is.
Pisco
BRAZIL
Bahia
St Helena Is.
Maouna Is.
Tahiti
Society Is.
Arica
Friendly Is.
Tropic of Capricorn
Rio Janeiro
Traitors Is.
Mangareva
Pitcairn Is.
C. Frio
June 1836
Mar. 1836
CHILE
Coquimbo
Easter Is.
St Catherine's Is.
SOUTH
Valparaiso
Montevideo
Buenos Aires
Rio de la Plata
Concepción
Cape of Good Hope
Aug. 1836
ATLANTIC
atham Is.
OCEAN
PATAGONIA
Falkland Is.
Strait of Magellan
Tierra del Fuego
Is. de Los Estados
Cape Horn
Sandwich Is.
May 1836
60
30
0
30
60
Antarctic Circle
150
120
90
60
30
0

V VIAU

Letter from the Minister of the Navy to Monsieur Vaillant,
in command of the corvette *Bonite*.
Reference: the object of your mission
Paris, December 28, 1835

Sir, the cargo sloop *Bonite (corvette de charge)* is hereby entrusted to your command for the purposes of conveying to Chile, Peru and Manilla various consular agents lately appointed by His Majesty the King.
As an adjunct to this primary goal, it is in the interests of French trade that you fly the national flag in the Sandwich Islands, on the coasts of China and Cochinchina and in several trading posts of India.
Over the course of this long voyage you will visit a great number of places in which our national flag appears only at intervals. You will fully appreciate, Sir, the advantages of inquiring into the situation of French trade on these distant shores, and of garnering all such intelligence as may apprise our merchants of the potential benefits and risks of expeditions to such destinations.
The Sandwich Islands, Manilla, Canton and Macao are of particular relevance in terms of this information-gathering, partly to supplement research by your predecessors but also to identify such changes as may have occurred since then and provide an accurate picture of the present state of these countries in both political and commercial terms.
You shall apply yourself, Sir, to deepening our existing ties of friendship with the regions you visit and to forging new ones. The French flag must ever be seen as a friendly flag and France herself as a strong and powerful nation but no less committed to welcoming overtures of friendship.
To facilitate the accomplishment of your mission, you will be furnished with an assortment of objects suitable as gifts for the authorities of the countries you visit, and as rewards for such services as may be rendered to you by the people in those places where you drop anchor.
Such expenses as you may incur in mooring places where we have established agents shall be settled by means of bills of exchange drawn on the royal treasury.
However, to provide for the ship and its crew in places where this expedient is not available, you have my authority to withdraw a sum of between one thousand two hundred piasters and one thousand five hundred piasters.
I well know, Sir, what we may expect of you in this regard. Your previous services are the measure of those that I am confident you will provide on this fine mission that is entrusted to you, and I have no doubt that by your efforts and your dedication this voyage shall prove as useful in all respects as it brings credit upon yourself.

VICTOR GUY BARON DUPERRÉ, ADMIRAL AND PEER OF FRANCE
AND MINISTER OF THE NAVY AND THE COLONIES
PARIS, DECEMBER 28, 1835

**FEBRUARY 1836**
**TOULON, FRANCE**

*On February 6 at half past eight* in the morning the corvette cast off under a press of sail and a cloudless sky. There was a fine day in prospect, the wind initially light and blowing from the east but soon freshening.

Even the most seasoned sailor cannot help but be moved when like the captain of the *Bonite* he says to himself: "This land that I am perhaps leaving never to see again is the land of my loved ones: parents and devoted friends who grieve to see me go and wish me a safe return. Would that their wishes may be answered! But alas! Who can know what dangers and mishaps lie ahead on this long journey on which I am embarked? Nay, this fine corvette that is mine to command across the perilous sea is not destined to perish! It shall fulfill its mission and bear me home to our belle France, filled with happiness at returning to you. Adieu then and au revoir all of you who love me; keep me in your hearts as I will keep you in mine. May happiness and peace be yours while I am away on duty, striving to earn new rights to your affection and your esteem."

**MARCH 1836**
**RIO DE JANEIRO, BRAZIL**

*For the Frenchman who has never left* Europe, there is something repellent about the sight of these poor wretches, reduced to the level of brutes far more by slavery than the native life from which they are now excluded. One cannot help but feel a profound sadness when one reflects that though Nature might not have endowed these people with the superior intelligence owed to civilization, they are for all of that no less human beings.

God forbid that I should join in the philanthropic rantings of those zealous but imprudent souls who seek the sudden emancipation of all slaves in our colonies. We should think very carefully before embarking on such a difficult reform, for in hastening to liberate the vassal we may succeed only in bringing down his master to the ruin of them both.

*In Brazil these men bought at market like pack animals receive the same treatment as the horse or mule sold alongside them.*

It is true to say indeed that the lot of the native slave is not always an unhappy one. I have seen some arrive in France only to reject the freedom being forced upon them by law, complaining about what they call their "abandonment" by the masters they have come to love. It is as if in leaving them to their own devices, they are being forcibly excluded from the family of which they form part.

However, while the natives in certain colonies and especially certain colonial households may forget or even cherish their enslavement, such does not appear to be the case in Brazil. Slaves here do exactly as they are told, whether as porters, craftsmen, boatmen or any other job reckoned to amortize their purchase price. They also bear children, in accordance with Nature's bidding and much to the satisfaction of their masters for whom these children born into slavery represent a further source of wealth.

**JULY 1836**

LIMA, PERU

— *Who would reproach the United-States* for having shaken off the yoke of English oppression? Those who first incite a whole country to expose herself to the hazards of revolution shoulder great responsability. Peru, before independence, was the jewel in the Spanish crown and the richest of all Spain's New World possessions. Lima, self-proclaimed "city of kings", was conspicuous on the American coastline for a splendor unrivalled by any of her European counterparts. All that has changed today. Only time will tell whether this yet nascent civilization will ever enjoy that wise governance and rule of law essential to her becoming a great and powerful empire.

In modern times, the colonies established overseas by European powers remained the offshoots of a common trunk, from which they drew life, strength and sustenance but on which they depended and to which they owed everything.

◄ **Pages 276-277:**
The port of Toulon in 1835.

▲ **Top:**
*Eglise de la Gloria à Rio-Janeiro, Brésil* (church of La Gloria, Brazil).

Colonies founded on that principle are but the extensions of their mother country. They can increase her power and wealth but the day she neglects them, the day they realize they can survive without her and have a lot more to gain alone rather than remaining in her orbit, that is the day that they will break free. Sooner or later, any overseas European possession worth its salt is bound to reject the mother country. Not that this will mark the beginning of a growing prosperity for our newly independent nation. Nay, she has a long way to go before then. The path to future greatness is strewn with trials and tribulations, always proceeding by trial and error in pursuit of that peaceful development that will bear fruit, though certainly not for the instigators of independence.

Spain's former colonies know this by bitter experience and for Peru that experience is bitterest of all.

**AUGUST 1836**
**LIMA, PERU**

*French trade has often been accused* of being the architect of its own downfall, of failing since the beginning to prepare properly for its expeditions or win the confidence and sympathy of the foreigners with whom she would be treating. I dare say that these accusations are not entirely undeserved. We have seen too many of our ships depart loaded almost exclusively with exports that our own warehouses have rejected. It is as if foreign trade were but a means of disposing of products so badly made as to be unfit for local consumption.

Proceeding in this manner may suit the books of manufacturers and warehousers with sub-standard products on their hands and, in the first instance at least, may prove extremely profitable for those who have undertaken expeditions on this basis. What it most certainly has not done is contribute to a favorable idea of our products abroad.

**Top:**
*Une rue de Payta, Pérou*
(street in Paita, Peru).

**Facing-page, top:**
*Une rue de Valladolid à Lima (*Pérou),
(street in Valladolid, Peru).

**Facing-page, bottom:**
*Vue de Callao, Pérou*
*w*(view of Callao, Peru).

*Speculation on the basis of the real or presumed ignorance of the other party is always misguided.*

*Chinese policy with respect to foreigners is to deter any interaction likely ...*

**OCTOBER 1836**
SANDWICH ISLANDS [HAWAII]

*In a country that hardly ever sees* a French warship, a visit by the *Bonite* was unlikely to pass unnoticed. Our brief was to make France in all its power and glory known to a semi-civilized king and his rather kinglier counsellors, good will and even justice being often a measure of a nation's ability to command respect. It was hoped to bring the king and his court to look favorably upon those French merchant ships and French travelers who happened to alight upon these shores, where, it must be said, practically no one had ever heard of France.

No-one here suspected that the Americans and the English might not be the only great nations. While the former were hailed as masters wherever they went and the sovereignty of the English was recognized by the King of the Sandwich Islands himself, France could expect a lukewarm reception at best in Honolulu, an island ever at the mercy of its capricious chiefs and the vested interests of their rivals.

Monsieur Vaillant suggested giving the government of the Sandwich Islands a clearer idea of the grandeur of our country, taking advantage of his future relationship with Taméaméa III [Kamehameha] to inspire the prince with a healthy respect for the power of France, the while treating the latter in such a way as to inspire a certain sympathy for the French.

**DECEMBER 1836**
MACAO, CHINA, PORTUGUESE TERRITORY

*Chinese politics* have always striven to uphold those feelings of pride that make Chinese people see themselves as superior to every other nation on Earth and maintain an aversion toward foreigners that borders on contempt.

The fact is that the Chinese government has good reason for such a policy. Its prime motive appears to be fear of the Europeans, whose successive conquests in India and the great Asian archipelago leave China in no doubt as to the conquering and controlling genius of the

*... to nurture closeness between themselves and the subjects of the Celestial Empire.*

Occidentals. Hence her careful avoidance of anything likely to provoke a collision. This explains why European warships are explicitly forbidden from entering any imperial port and why those few Europeans admitted with great difficulty onto Chinese soil are confined to one corner of a Canton suburb.

All that these measures go to show is that the Chinese Government has no intention of falling prey to the same fate as those Indian princes who having allowed a few simple merchants to establish themselves in their lands, finished subservient to them at best and at worst totally dispossessed by them of all their states.

As a further deterrent to prolonging their stay in China, Europeans are forbidden from receiving their womenfolk in Canton. Foreigners wishing to reside there must therefore agree in advance to foreswear the charms of female company and resign themselves to living like monks. Married men are obliged to live apart from their families, who can reside in Macao at best. These are hardly circumstances conducive to planting roots in China and all those who come here in the hope of making their fortune cannot wait to leave.

**JANUARY 1837**

**MACAO, CHINA, PORTUGUESE TERRITORY**

*And yet, we French are relatively* well-thought-of in China compared to other European nations, and more likely than the English to be treated fairly by the authorities— or so say the merchants who reside there.

There are two reasons for this. On the one hand, China has no cause to believe that we will take advantage of her concessions to seek a foothold in her empire. On the other hand, she has no cause to doubt the might of France, having heard talk of our wars and our conquests and learned by example what happens to those who insult our flag. The Chinese authorities also know that in the event of such insults occurring, they would have nothing to mitigate France's revenge, unlike the English who have traditionally had a vested interest in remaining in China.

Facing-page:
Port of Honolulu, Sandwich Islands, Hawaii

Top:
*Scène de danse, aux îles Sandwich* (dancing scene, Sandwich Islands).

Pages 284-285:
*Place des factoreries européennes à Canton, Chine* (area of Canton, China, reserved for the European "factories", meaning accommodation and trading posts for foreigners).

**FEBRUARY 1837**
**TOURANE, COCHINCHINA** [DA NANG, VIETNAM]

*The captain had nothing to hide:* "He was conducting a tour of the world by order of his most Illustrious and Mighty Majesty the King of France, for the purposes of protecting French people wherever they might be, ensuring respect for their persons and their goods, demanding satisfaction for such wrongs as they might complain of and exacting a blazing revenge on whomsoever has insulted or molested them. His most August Majesty was particularly desirous that the expedition should pay a visit to Cochinchina, the French being the sincerest and the oldest friends of the Cochinchinese and the expedition officers being charged with conducting certain works of science here as in those other points of the globe that the *Bonite* had already passed through."

The Mandarins listened attentively to the foregoing and took issue with one point only: the tri-color flag that they saw flying above their heads. "If you are French, why are you not flying the French flag?" they asked, meanwhile showing us a large placard displaying a white flag dotted with many fleur-de-lis. Monsieur Vaillant replied that the tricolor flag was that of Louis Philippe 1, the present King of the great French nation, as so admirably represented in the full-length portrait that adorned the great gallery of the corvette and before which the Mandarins now bowed in respect. The fact that visitors to the *Bonite* were always much impressed by this portrait led Monsieur Vaillant to conclude that it would be useful for all ships on a similar mission to display a richly framed portrait of the French sovereign in ceremonial robes amid the trappings of power. Nothing, indeed, could speak with more eloquence to the minds of barely civilized people.

**MARCH 1837**
**SINGAPORE**

*Despite the ancient rivalry* that has always existed between France and England, at the individual level French and English alike set aside national interests and observe a punctilious civility that is the measure of their mutual respect and sympathy. Nowhere is this more evident than in the communications that take place between the officers of these two nations in countries remote from Europe. As the representatives of the world's two most advanced cultures, they seem to form a single noble family amid that great disparity of families from less privileged backgrounds in terms of power and especially education. I have lived in a French colony. The arrival of an English warship

► Beach scene, Tourane, Cochinchina.

or a few isolated officers from that nation was always an occasion for celebration and everyone was at great pains to make them feel welcome, not only local officials but also society in general. It is the same for French officers who arrive in an English colony. Considering the circumstances, there is never any talk of politics and people confine themselves to polite intercourse, as befits well-behaved persons bent on honoring one another as friends whatever their opinions may be.

**MAY 1837**
**CALCUTTA, BRITISH INDIA**

*I would only be repeating myself* in speaking of the interesting work that the expedition naturalists and observers were to execute in Calcutta and did indeed execute with great success. But I must mention one mission that was particularly recommended to Monsieur Vaillant. Its specific purpose was to collect various species of the mulberry trees that grow in Bengal, together with the eggs of all the varieties of silkworm that feed off their leaves. Several species of silkworms and mulberry trees have long been known to acclimatize well to conditions in France and her colonies alike.

At first glance, employing a navy warship for this purpose hardly appears necessary when merchant ships come and go between Europe and Calcutta all the time. On closer inspection however, things are not as simple as they seem, the trees in particular requiring storage conditions that are almost impossible to provide aboard a merchant ship. The eggs too must be very carefully protected to prevent their spoiling or hatching prematurely under the varying effects of long sea travel, not least the sudden transitions from one climate to another.

**NOVEMBER 1837**
**BREST, FRANCE**

*Heavy seas,* pouring rain and ever stronger wind gusts: weather conditions until October 21 conspired to plague our progress, taking a heavy toll on the health of the crew. Scurvy, already detected by our surgeon on his weekly visits to the crew, now struck the men down in their droves. The sick bay was soon overflowing and the corvette gun deck looked like a hospital.

The wind eventually shifted in our favor, accelerating the ship's progress but further fatiguing her passengers. The bad weather meanwhile continued, as feared by the captain, and that for several difficult days. On November 5 the storm did at last cease, the weather brightened up and on the very next day we sighted the Ouessant lighthouse.

A cry of joy resounded from one end of the *Bonite* to the other, all our woes suddenly forgotten! On the night of November 6 the *Bonite* dropped anchor in Brest harbor to a chorus of seamen's voices chanting: "We shall go around the world, the *Bonite* shall not perish." Words endlessly repeated throughout the journey, repeated now for the very last time.

**Top:**
*Mosquée aux environs de Calcutta, Inde* (mosque on the outskirts of Calcutta, India).

**Facing-page, top:**
*Pagode de Dina Marlinga sur les bords de l'Hougly près de Chandernagor, Inde* (Pagoda of Dina Marlinga, on the banks of the Hooghly River, near Chandernagore, India).

**Facing-page, bottom:**
*Une mosquée aux environs de Pondichéry, Inde* (mosque on the outskirts of Pondicherry, India).

**Pages 290-291:**
*Pagode aux environs de Canton, Chine* (pagoda on the outskirts of Canton, China).

*France asks nothing of China and has nothing to lose from an eventual war, unlike England that depends on China for its supplies of tea.*

A FEW MONTHS AFTER VAILLANT,
ABEL AUBERT DU PETIT-THOUARS WAS ALREADY
EMBARKING ON A NEW CIRCUMNAVIGATION.
POLITICS AND DIPLOMACY WERE THE FIRST ORDERS
OF THE DAY FOR THIS GENIAL WARRIOR;
SHOWING THE FRENCH FLAG IN KAMCHATKA
AND PROTECTING WHALERS CAME SECOND.
A STRATEGIC MISSION CALLED FOR A SUBSTANTIAL SHIP
AND THE *VÉNUS* DID NOT DISAPPOINT.
WITH ACCOMMODATION FOR MORE THAN FOUR HUNDRED
MEN AND ARMED WITH FIFTY-TWO CANNON,
THE FRIGATE WAS A SIZABLE VESSEL
BY ANYONE'S STANDARDS. LÉON DE LAPÉROUSE,
NEPHEW OF THE FAMOUS EXPLORER, WAS AMONG
PETIT-THOUARS' HANDPICKED CREW.

# Abel Aubert du Petit-Thouars

**1793–1864**

Plate 50, George Wolfgang Knorr, 1766.

# VOYAGE OF THE *VÉNUS*

**1836–1839**

**Du Petit-Thouars took his mission to heart. To obey the King's orders to protect French interests in Oceania, our forty-three-year-old explorer was ready to declare war on a queen and aim his cannon at a king. Thanks to his expedition, Polynesia would remain forever French.**

Du Petit-Thouars was born into a family of famous sailors on August 3, 1793, in Maine-et-Loire. His uncle Aristide, most notably, fought for the Americans in the War of Independence; he was one of the first to go in search of Lapérouse, and died a hero at the Battle of Aboukir. Quite an act to follow. Du Petit-Thouars himself joined the navy as a *mousse* (ship's boy) at age eleven, enlisted at age fifteen and became an ensign at age twenty-one. At age twenty-six he made lieutenant (navy) before being appointed Head of Mission in Haiti then serving on the Brazilian naval station. His first command was a dispatch boat during the Spanish campaign.

Promoted to frigate captain in 1827, du Petit-Thouars played a decisive role in preparations for the Algiers expedition, proving himself such a consummate strategist that his battle plan was readily adopted by King Charles X. And with that, du Petit-Thouars went into battle to put his theory into practice. King Louis-Philippe's accession to the throne did not affect his career; rather the reverse. It was under the July monarchy that he was sent to the Pacific naval station, then to Peru where he played the role of diplomat to perfection. Returning to France at the end of 1835, he wrote a note on the importance of a military presence in the region to defend French interests. No sooner had he read it, than the Minister of the Navy, Claude du Campe de Rosamel, approved the funding for a circumnavigation of the globe to be led by du Petit-Thouars, with the blessing of His gracious Majesty. This would be the second such expedition of 1836, Vaillant having just made a somewhat hurried departure from Toulon in February. No such haste for our own, rather more relaxed navigator, who took all the time he needed and obtained altogether different results.

The *Vénus* left Brest on December 29 and, after weathering a mighty storm, she entered the bay of Rio de Janeiro in February 1837, where the crew rested for three weeks in preparation for rounding Cape Horn. Du Petit-Thouars was struck by the number of begging slaves, "bought at the market like beasts of burden". In April 1837, he arrived in Valparaiso, a port bustling with ships of all stripes— fifteen Chilean ships preparing to wage war on Peru, at least six French ships (including the corvette *Ariane* under the command of Duhaut-Cilly), American whalers, English traders and more. After putting in a brief appearance, the Vénus headed north again.

After a stopover in Peru and before crossing the Pacific, du Petit-Thouars anchored in Hawaii in July. On arrival, he learned of the unjust fate of two missionaries, one English, one French, imprisoned by order of King Kamehameha III. Together with Edward Belcher, an English captain anchored in the bay at the time, he demanded a treaty from the king guaranteeing the protection of their fellow citizens. With two corvettes in his harbor and one hundred cannon aimed at his city, Kamehameha was hardly in a position to refuse.

On August 25, 1837, Russia was in sight—a part of the world unvisited by a French ship since Lapérouse and duly explored and mapped by the *Vénus*. In Petropavlovsk-Kamchatski, du Petit-Thouars was received with all due honors by the Russians and took part in festivities to celebrate the emperor's birthday, punctuated by numerous toasts and cannon salutes. Being introduced by the Governor to a Russian sailor who had known Lapérouse was a particularly emotional moment for du Petit-Thouars.

He then headed due east to arrive in California, where he deplored the chaotic state of affairs created by the decline of the monasteries. Next stop was Acapulco, followed by a detour to Easter Island for hydrographic surveys, then stopovers in Valparaiso and Peru to stock up with provisions for the Pacific crossing.

In August 1838, at anchor in the Marquesas, du Petit-Thouars noted that the inhabitants "have lost their savage ways and with them, their originality." After a short rest, the *Vénus* continued on to Tahiti where, as in Hawaii, the English missionaries had too much influence, and French people were being mistreated. Du Petit-Thouars was having none of it and promptly dispatched what amounted to a declaration of war to Queen Pōmare IV. Also as in Hawaii, a treaty was signed to guarantee the protection of the French. The *Vénus* then set sail for New Zealand and Australia where our Frenchman, like his predecessors, was both admiring and jealous of England's achievements, France's own efforts at colonization paling somewhat by comparison. Then it was back to France.

The *Vénus* entered Brest harbor on June 24, 1839, bringing back, in addition to the two treaties, chests filled with thousands of multicolored shells. Still hungry for adventure, du Petit-Thouars suggested to Louis-Philippe that France strengthen her presence in the Pacific by taking possession of the Marquesas and Society Islands. His proposal was accepted, and our man was promoted to rear-admiral, taking command of the naval division in Oceania and crisscrossing the Pacific islands aboard the frigate *Reine Blanche*. On April 28, 1842, he had the various Marquesas chiefs sign an act of protectorate, distributed uniforms and installed a garrison: the tricolor flag now flew over the Marquesas. The *Reine Blanche* then made sail for Tahiti, showed her teeth to the English pastors and signed a new treaty of protection with Queen Pōmare in September 1842.

Du Petit-Thouars then continued his missions in the Pacific and on the American coast. But he was in for an unpleasant surprise when he returned to Tahiti a year later, the Union Jack now flying in place of the Tricolore. Du Petit-Thouars asked three corvettes to join him and had the longboats armed. Early in the morning, the cannon thundered and the English missionaries kept their heads down. Du Petit-Thouars and his men took possession of the town and the royal house and annexed Tahiti, without a drop of blood being shed. By November 1843, Tahiti had returned to France's fold; but du Petit-Thouars was recalled to France where King Louis-Philippe, under pressure from the English, disavowed him. The "Tahiti Affair", as it become known, inflamed public anger but though the French population came to du Petit-Thouars' defense, the King did not ratify the treaty and even went so far as to compensate the English missionaries. Pure treachery!

Du Petit-Thouars was nevertheless promoted to vice-admiral and continued to serve France until his death in Paris on March 16, 1864. As for Queen Pōmare, she voluntarily abdicated in favor of France in 1880, and Tahiti and her islands became a French possession.

Chilean horseman in 1835.

Costume Chilien.

## GENERAL REFERENCE MAP
For the Voyage of the *VÉNUS*
Led by Captain Abel Aubert Dupetit-Thouars
1836–1837–1838–1839

150
120
90
60
30
0
Western Longitude, Paris Meridian
Paris Meridian
Melville Is.
Lancaster Strait
Baffin
Bay
Davis Strait
GREENLAND
Iceland
SWEDEN
Bering Strait
Cook's River
Port des Français
Lituya Bay
Kodiak Is.
Sitka
Alaska
Hudson
Bay
Hudson Strait
Cape Farewell
LABRADOR
CANADA
NORTHERN
BRITISH ISLES
London
Baltic Sea
Queen Charlotte's Is.
Vancouver Is.
Nootka
utian Islands
St-Lawrence River
Quebec
Newfoundland
Dec. 1836
Paris
Brest
FRANCE
NORTH
AMERICA
UNITED STATES
Acadia
Toulon
ITALY
ORTHERN
OCEAN
San Francisco
Monterey
New York
Washington
ATLANTIC
SPAIN
Madrid
MEDITERRANEAN SEA
Azores Is.
Strait of Gibraltar
Oct. 1837
Bermuda Is.
California
New Orleans
San Antonio
Madeira Is.
BARBARY
MOROCCO
MEXICO
Gulf
of Mexico
Florida
OCEAN
Canaries Is.
Tropic of Cancer
Mazatlán
Lucayan Is.
San José
Mexico City
Santo Domingo
Sandwich Is.
Hawaii Is.
July 1837
Caribbean Sea
Jamaica
Guadeloupe
Martinique
SENEGAL
Cape Verde
AFRICA
Acapulco
GUINEA
Panama
COLOMBIA
Cayenne
GUYANA
June 1838
Equinoctial Line
Gallapagos Is.
Paita
SOUTH
AMERICA
PERU
May 1837
Ascension Is.
CONGO
Marquesas de Mendoza
Lima
Aug. 1838
vigators Is.
Maouna Is.
Tahiti
Society Is.
Pisco
BRAZIL
Bahia
St Helena Is.
Arica
Friendly Is.
Tropic of Capricorn
Rio Janeiro
C. Frio
raitors Is.
Mangareva
Pitcairn Is.
Coquimbo
Easter Is.
CHILE
St Catherine's Is.
SOUTH
Feb. 1838
Valparaiso
Montevideo
ct. 1838
Buenos Aires
Rio de la Plata
Concepción
Cape of Good Hope
ATLANTIC
atham Is.
OCEAN
PATAGONIA
Falkland Is.
OCEAN
Strait of Magellan
Tierra del Fuego
Is. de Los Estados
Cape Horn
Sandwich Is.
Apr. 1837
Antarctic Circle

I have tried to write an account of the long journey
that commenced on December 31, 1836 and ended
on June 24, 1839, sailing there and back aboard the frigate *Vénus*.
In pursuit of this task, I resolved to include several new
observations that I believed might prove useful
to navigation and the sciences.
I also wished to include a few words of advice on the best places
to go whale-hunting, the potential for expanding this most
interesting branch of maritime industry and the trade relations
that might be established with those parts of the world
visited by us but otherwise heretofore largely unfrequented.
Since the voyage of the celebrated Lapérouse in 1786,
not a single state vessel nor a single French merchant ship
had visited Kamtschatka [Kamchatka].
Our national flag had therefore never flown in sight of those
frozen lands and distant shores where it was known
only by reputation. To be the first to display its full glory
was for the seamen of the *Vénus* among the major
attractions of this campaign.
In addition to providing incentives for French trade
there was the yet more important imperative represented
by whale hunting: that adventurous industry protected
and favored by the government by means of special regulations,
bonuses and every other means at its disposal.
Despite all the sacrifices, however, whale hunting remained
a "static" industry so to speak for the simple reason
that whaling crews did not feel duty bound and often mutinied,
wrecking the expedition if not sinking the ship herself.
It was therefore decided to put an end to these disturbances
by placing whalers under the supervision of state vessels
charged with enforcing the law in all four corners of the world.
The *Vénus*, as a worthy representative of France,
was to fly the national flag in all such parts of the world
as might rarely have seen it before.
In short, nothing was overlooked in the attempt
to make her tour as beneficial as possible.

ABEL AUBERT DU PETIT-THOUARS
PARIS, 1840.

**DECEMBER 1836**
**BREST, FRANCE**

*With the crew of the frigate* *Vénus* now complete thanks to the good offices of Vice Admiral and Minister of Marine Affaires Jean-Baptiste Grivel, we took advantage of an east-northeast wind to cast off from Brest harbor, anxious to avoid such delays as might be occasioned in this season by the return of on-shore winds.

At the moment of departure, the harbor was almost deserted and the weather was dark and cold. The coasts were cloaked in snow that continued to fall heavily, but the wind was so light that our progress was slow to begin with; as the breeze picked up, so too did our speed.

By midday, the frigate was across the fairways and in the Iroise Sea. I dismissed the pilot, who departed with our letters to families and loved ones—our last marks of affection for those who we would not be seeing for a very long time to come. The wind was changeable at first but eventually freshened, speeding us on our way.

**MARCH 1837**
**VALPARAISO, CHILE**

*Valparaiso is home* to trading companies of different nationalities, among them French, English, American and German. Each of them is managed by a sufficient number of foreigners to function independently of one another, in accordance with the customs and tastes of its particular nationality. This faithfulness to patterns of behavior rooted in family and education is especially conspicuous in the English, who cling to their traditions to the point of obstinacy. Hence those influences that over time have proven as favorable to the politics as to the trade of the mother country.

From the moment the port of Valparaiso opened to foreign trade, it attracted merchant ships of all nationalities, all of them equally keen to profit from the end of Spanish dealings with her former colony. Thus almost to the exclusion of all other countries have the North Americans been able to supply Chile with most of the furniture now in use in this country.

The English, by dint of a growing abundance of products at cheaper prices, have meanwhile succeeded in furnishing Chile with the greater part of her porcelain, durable wares and other everyday items and basic household goods.

British industry has maintained its leadership and so come to monopolize world trade and create a need for goods made in the English style.

*A Frenchman who marries in Chile becomes Chilean. An Englishman who marries in Chile changes not—it is his wife who changes, to become English.*

**APRIL 1837**
CHILE

*Chilean society is most advanced* and agreeable, having made such rapid progress since independence that there is now little to distinguish a salon in Santiago from a salon in London or Paris. Education is widespread and plainly regarded as very important to judge from the care taken in its expansion. Santiago boasts a tolerably good school, and many young Chileans are sent away every year to study in Europe.

Female education enjoys even greater attention. In the period 1818-1820 the capital welcomed a boarding school for young persons, who received as good an education here as any that might be provided in Europe. Its director was a Frenchwoman and former pupil of Madame de Campan, who saw to it that her young charges were schooled in good manners. Most of them left her care proficient in music, able to speak or understand French and skilled in every kind of needlework befitting the fairer sex. There are few houses in Chile, whether in Santiago or Valparaiso, that is not graced by a piano. Chilean ladies follow the French fashions with even more enthusiasm than their counterparts in many parts of France.

**JULY 1837**
SANDWICH ISLANDS [HAWAII]

*Note to the King of the Sandwich Islands*

*I have the honor to inform the King that being sent by my Government to visit several parts of the world, I am charged with overseeing the safety of my nation's commerce and that of all French people now established in foreign countries.*

*It is my duty to receive their requests and complaints and seek justice on their behalf for such demands as may be founded on human rights and fairness, said rights and said fairness guaranteeing the freedom and property of all foreigners for so long as they comply with the laws and regulations of the land wherein they reside.*

*On arriving in the Sandwich Islands, I expected to have only thanks to offer to the King for the gracious promises he had been good enough to make to Captain Vaillant, captain of the corvette* Bonite, *to take under his immediate protection all such Frenchmen as might alight in the islands under his dominion.*

*Imagine my unpleasant surprise, therefore, to learn of the violent and inhumane treatment only lately meted out to a Frenchman who had been driven by necessity to pass by way of Honolulu, seeking an opportunity to return to his homeland!*

▲ *Marchande de fruits à Lima, Pérou* (lady fruit seller, Lima, Peru).

Bottom:
*Vue d'une rue d'Honoloulou, capitale des îles Sandwich - La Reine Kinau revient du temple des étrangers accompagnée de ses dames d'honneur* (street scene in Honolulu, capital of the Sandwich Islands: Queen Kina'u returns from the foreigners' temple escorted by her ladies-in-waiting).

Facing-page:
*Assemblée des chefs des îles Sandwich en conférence avec le commandant de la Vénus* (the chiefs of the Sandwich Islands in conference with the Captain of the Vénus).

Pages 304-305:
*Vue de Pétropawlowski (Kamtschatka), Russie* (view of Petropavlovsk, in Kamchatka, Russia).

*I fear, however, that the King may have been misled in this matter by certain persons who have abused their influence.*

*Hence my request that, in order to forestall any such disputes as might arise in the future and throw our harmonious relations into disarray, the King may be pleased to join me in signing a convention similar to that lately concluded, on behalf of Great Britain, by Lord Edward Russell, captain of the corvette* Acteon. *I claim no special privileges or indulgences for my compatriots, merely that they be granted the same treatment as the most favored foreigners, a treatment which is their right and cannot be refused by the King without reneging on the promises he himself hath made.*

*A. du Petit-Thouars*

*Aboard the Vénus, Honolulu, July 20, 1837*

**JULY 24 1834**
**HONOLULU, SANDWICH ISLANDS [HAWAII]**

*Agreement* between ship-of-the-line-captain Petit-Thouars, commander of the frigate *Vénus*, and the King of the Sandwich Islands, Taméhaméha [Kamehameha] III

*Article 1 (sole article)*

*That there shall be perpetual peace and friendship between the French and the inhabitants of the Sandwich Islands.*

*That the French shall be free to come and go as they please in all the states that make up the governance of the Sandwich islands; that they shall be received nd protected therein and enjoy the advantages due to the subjects of a most-favored nation.*

*That the subjects of the King of the Sandwich Islands shall likewise be free to visit France, and shall be received and protected there as most-favored foreigners.*

*Signed, Taméhaméha III*

*A. du Petit-Thouars*

**SEPTEMBER 1837**
**PÉTROPAWLOWSKI [PETROPAVLOVSK-KAMCHATSKY], RUSSIA**

*His excellency the governor* proposed a toast to the health of the King of France; I in turn proposed a toast to His Imperial Majesty the Emperor of all Russia. These were Champagne toasts, raised to the accompaniment of a twenty-one-gun

*A promise made by a King has always been as good as a treaty and I hereby demand that the King fulfill his promise.*

salute, with the frigate *Vénus* contributing to the solemnity of the occasion. No sooner was the meal finished than his Excellency the Governor came to talk to me, amid other conversations taking place in the salon. Music played as we whiled away the hours until ten o'clock when we retired to rejoin the frigate. The entire day had been filled with an uninterrupted succession of new sights that interested us greatly. After our visit to the Sandwich Islands, we counted ourselves fortunate to encounter a society so far from France whose polite manners and discerning affability recalled our absent friends, and whose warm welcome made us all but forget that we were strangers to each other.

### OCTOBER 1837
### MONTEREY, ALTA CALIFORNIA

*Under the Spanish dominion,* the two Californias were ever neglected. The government of mainland Spain seemed to consider this fine possession of secondary importance, occupying it only to arouse jealousy and to prevent its being possessed by any other nation. The government thus appeared to underestimate the immense value of California in terms of agriculture, and to be oblivious to the advantages to be gained from a fertile land blessed by an idyllic climate. It contented itself with appointing governors but otherwise sent no aid or succor of any kind.

Until 1823 then, Alta California was very far from being well administered but even that seems preferable to the present deplorable state of affairs. Every day since independence has seen the mission populations, like their income, decrease at an alarming rate. The country finds itself in that difficult period of transition that follows a change of sovereignty, its entire social edifice having collapsed with the destruction of its religious foundations. The former monastic order has crumbled and instead we find nothing but anarchy. The powers that be are toothless and indecision reigns supreme. A new political order will no doubt emerge from this chaos; but in the meantime, this largely ungoverned country is home to a disconnected society of insecure, unhappy people.

### NOVEMBER 1837
### MONTEREY, ALTA CALIFORNIA

*The Russians, confined* in their agricultural settlement of La Bodega, appear to have their sights set on grabbing the port of San Francisco—or so it would seem from the words that inadvertently escaped the lips of a few high-ranking officers. No-one can really say today which nation will eventually gain possession of this fine port and two fertile basins. What seems certain, however, is that in the present political state of Europe and the New World, any power bold enough to seize this plum by means of a *de facto* annexation is unlikely to face much resistance.

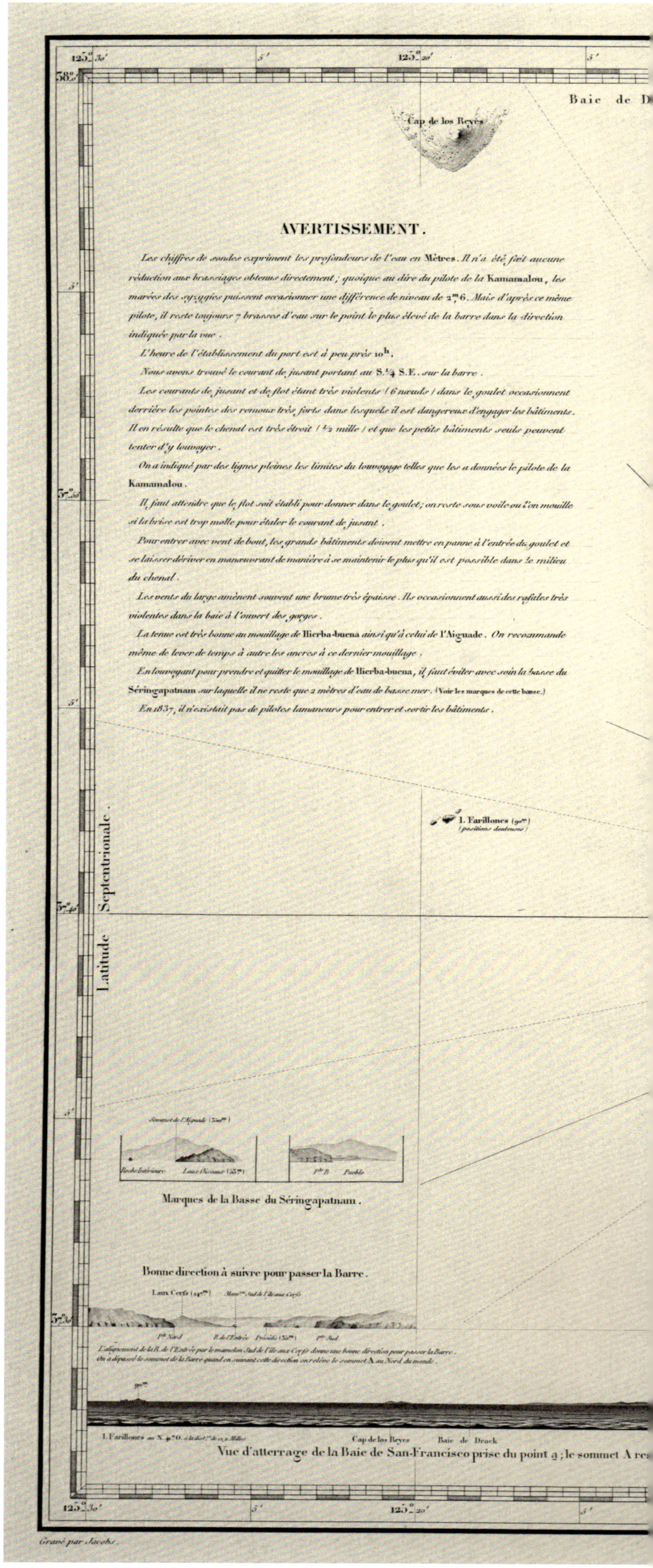


Route of the *Vénus* through the Bay of San Francisco.

Table de réduction des Mètres en Brasses et en Pieds.

| Mètres | Brasses | Pieds | Mètres | Brasses | Pieds | Mètres | Brasses | Pieds | Mètres | Brasses | Pieds | Mètres | Brasses | Pieds | Mètres | Brasses | Pieds |
|---|---|---|---|---|---|---|---|---|---|---|---|---|---|---|---|---|---|
| 1 | 0,6 | 3 | 6 | 3,7 | 18 | 11 | 6,7 | 34 | 16 | 9,8 | 49 | 30 | 18,5 | 92 | 80 | 49,2 | 246 |
| 2 | 1,2 | 6 | 7 | 4,3 | 21 | 12 | 7,4 | 37 | 17 | 10,4 | 52 | 40 | 24,6 | 123 | 90 | 55,4 | 277 |
| 3 | 1,8 | 9 | 8 | 4,9 | 25 | 13 | 8,0 | 40 | 18 | 11,0 | 55 | 50 | 30,8 | 154 | 100 | 61,6 | 308 |
| 4 | 2,4 | 12 | 9 | 5,5 | 28 | 14 | 8,6 | 43 | 19 | 11,7 | 58 | 60 | 36,9 | 185 | 200 | 123,1 | 615 |
| 5 | 3,1 | 15 | 10 | 6,1 | 31 | 15 | 9,2 | 46 | 20 | 12,3 | 61 | 70 | 43,1 | 215 | 300 | 184,7 | 923 |

Pl. 6. N° 1002.

PRIX. Un Franc.

According to the statutes of the California missions, Indians who had completed a ten-year stint with the priests were entitled to receive a plot of land sufficient to their needs and those of their families, together with the seeds and plowing implements required to farm those lands. In practice, however, this was seldom implemented and almost always with disappointing results. This may have been because the Franciscan priests who managed the missions in Alta California were at odds with Government policy; or because they felt it ran counter to the interests of the community; or because, as they claimed, the Indians were incapable of appreciating the benefits of such civilization. Those Indians who were emancipated fell into idleness and drunken depravity; and having squandered everything they possessed in gambling, which they love to distraction, returned to the missions for want of that subsistence that they were incapable of providing for themselves through hard work and good husbandry.

### FEBRUARY 1838
EASTER ISLAND, CHILE

*We sighted Easter Island* at midday on February 25, 1838. Seen from our direction of approach the island appeared to be divided into two. The extremities being at a much higher elevation than the center, it was only as we drew closer that the middle of the island became visible and corrected our false impression. At seven o'clock in the evening, we were within nine or ten miles of the eastern tip and could perfectly distinguish that whitish aridity that first strikes the eye, followed by the rugged formations of the island. We tacked all night with mainsail reduced to remain upwind of the island, and at daybreak we realized that the currents had not materially altered our position.

On the morning of the twenty-fifth we sailed within two miles of the eastern tip where we made a first stop to gauge the shape of this land. We then continued northward along the coast, always remaining within one or two miles of land to reconnoiter that part of the shore left unexplored by Lapérouse.

### JULY 1838
O-TAÏTI [TAHITI]

*France had long been* justified in complaining of the conduct of the government of Tahiti, under the rule of Queen Pomare [Pōmare IV], toward all those French people who had passed through or stayed in the islands, whether seafarers, merchants, industrialists or other persons. Only lately two French missionaries and a carpenter-mechanic whose business had brought them to Tahiti had been the victims of unprecedented outrages that constituted a violation of human rights.

*For these peoples the mighty warships they behold are the sole measure of the relative power of distant nations.*

So ill-disposed, indeed, was the government of these islands toward we French that not long ago a French pearl diver in Valparaiso had been refused passage to Pomotou [Tuamotu] on a ship flying the Tahitian flag, the captain arguing that should he return to Tahiti with a Catholic on board he might himself be barred from landing, or ordered to return whence he came by the head missionary Monsieur Pritchard, and at the very least be ordered to pay a fine. The situation was intolerable, amounting to nothing less than the outlawing of the French, whose ships henceforth would think twice before dropping anchor in the bays of Taïti even when they had no other choice.

The French Government, having received timely warning of this flagrant flouting of the law, had sent dispatches to Valparaiso ordering the *Vénus* to call in at Tahiti on her return from Kamchatka and put things in order. In particular we were to insist on Queen Pômare's making appropriate reparations and paying damages for the mistreatment suffered by Messrs Laval and Carret in being forced to take ship and return whence they had come.

**AUGUST 1838**
O-TAÏTI [TAHITI]

*Communication from the Captain* of the *Vénus* to the Queen of Tahiti

Aboard the French frigate *Vénus*, in Papeïti [Papeete] bay, August 30, 1838, at ten o'clock in the morning

To the Queen of Tahiti

*Madame,*

*The King of the French, being rightly angered by the outrage caused to his country by the wicked and shameful treatments inflicted on several of his subjects who betook themselves to Tahiti, most notably the apostolic missionaries Messrs Laval and Carnet in 1836, has sent me to request and if necessary insist upon, prompt reparations to a great and powerful nation that has been so grievously and gratuitously insulted.*

*The King and His Government demand:*

*1° That the Queen of Tahiti write to the King of the French apologizing for the violence and other affronts inflicted on His subjects, whose honorable conduct had in no way deserved the treatment meted out to them.*

Facing-page: *Chasse de l'ours sur les bords de l'Avatsha, Kamtschatka* (bear hunt on the banks of the Avacha River, Kamchatka).

Top: *Danse à cloche-pied des indigènes de l'île de Pâques à bord de La Vénus* (Easter islanders performing a "hop dance" aboard the *Vénus*).

Top:
*Village de la baie de la Madre de Dios, Îles Marquises* (village in the Bay of Madre de Dios, Marquesas Islands).

Facing-page:
*Vue de la baie de Huaheiné, Îles de la Société* (view of the Bay of Huahine, Society Islands).

Pages 310-311:
The *Vénus*, Papeete harbor.

*The Queen's letter shall be written in Polynesian and in French, and both documents shall be signed by Her. Said letter shall be officially conveyed to the Captain of the* Vénus *within twenty-four hours of receipt of the present communication.*

*2° That the sum of two thousand Spanish pieces of eight shall be deposited, within twenty-four hours of the present communication, in the coffers of the frigate* Vénus, *in compensation to Mssrs Laval and Carret for the damage occasioned to them by their mistreatment.*

*3° That the French flag shall be flown on September 1 at midday on the island of Motu-Uta and honored with a twenty-one gun salute from the queen's fort.*

*I hereby declare to Her Majesty that in the event of failure to comply with the conditions herein within the time allowed, I shall regrettably be obliged to declare war and commence hostilities against all the States under Her dominion, said hostilities being continued by every warship passing by way of these islands thereafter, until such time as France receive satisfactory reparations.*

*I remain, with the most profound respect,*
*Your Majesty's most humble servant,*
*A. du Petit-Thouars, Captain of the frigate* Vénus

Letter from Her Majesty Pōmare, Queen of Tahiti, to His Majesty the King of the French
Tahiti, August 31 1838

*To the King,*

*May peace be with you; here is what I would make known to Your Majesty: I acted wrongly in opposing the residency of two French citizens. May His Majesty be not too angry for what I have done and may peace be restored. I am but the sovereign of a small and insignificant country. May knowledge, glory and power be forever with Your Majesty; may your anger cease and may you forgive the error that I have committed.*

*May peace be with Your Majesty.*
*Signed: Pōmare*
*To the King of the French*

Agreement between His Majesty Louis-Philippe I, King of France, represented by ship captain Abel du Petit-Thouars, Officer of the Legion of Honor and captain of the frigate *Vénus*,
And Her Majesty Pōmare, Queen of Tahiti

*French people, of whatever profession, shall be free to come and go as they please, establish businesses and conduct trade in all the islands that make up the governance of Tahiti; they shall be received and protected there as most-favored foreigners.*

*The subjects of the Queen of Tahiti shall likewise be free to visit France; they shall be received and protected there as most-favored foreigners.*

*Written and agreed at the palace of the Queen of Tahiti, in Papeete, on September 4, 1838*

*Signed: Pōmare, Queen of Tahiti,*

*A. du Petit-Thouars, Ship Captain*

**OCTOBER 1838**

**KORORAREKA [RUSSELL], NEW ZEALAND**

*For the New Zealanders,* as for pretty much all barely civilized peoples, clemency counts not as a virtue but rather as an effect of weakness or powerlessness. No-one here would consider clemency as a sign of humanity, even less feel any respect for a nation that allowed itself gravely to be insulted without seeking retribution for the offense inflicted. So true is this that one is tempted to conclude that had we severely chastised a few of the tribes in the Bay of Islands, where the crime was committed, they would have held us in high esteem instead of mistrusting us. But then, our treatment at the hands of such foreigners would not be half so prejudicial were we to make ourselves better known in these remote parts by flying our flag here more often. Our absence provides a breeding ground for every version of France imaginable, every one of them given credence since there is no one to say otherwise. The throwing of mud at France extends to all aspects no matter how petty or puerile the cause may appear to be. It has even been maintained that France has no warships in the vessel or frigate class— an opinion widely held, to judge from the number of toy boats sporting the names of French warships. It is true that said vessels have rarely been sent to Western Oceania but even we were not expecting the astonishment evinced at the appearance of the *Vénus* in those places where she showed herself.

*In Tahiti and in the Sandwich Islands, the Queen is but a figurehead—a passive instrument in the hands of the Methodist missionaries.*

**DECEMBER 1838**
**SYDNEY, NEW HOLLAND, AUSTRALIA**

*On seeing this brand-new town,* it is incredible to think that barely fifty years ago there was not a single native hut on this spot. Today, everything about this flourishing city arouses admiration, from the elegant faubourg of Woolloomooloo to its public institutions, and the parade of sumptuous stores that line its broad avenues, constantly bustling with passers-by and vehicles crossing back and forth. Fifty years ago, no-one had heard of the River Parramatta except for a few brutish natives who sailed their shapeless canoes on its waters. Today their crude handiwork has made way for passing ships of gross tonnage and steamboats sailing up and down river by day and by night. From these erstwhile deserted waters now rises a forest of masts that merge with the still virgin forests along the riverbanks, adding to the already considerable charms of this picturesque tableau. All honor therefore to the Government behind this magnificent creation; or to be more precise, to the trading genius that has spawned such wonders.

Dame de Monterey.

Costume de la Haute Californie.

**JANUARY 1839**
**SYDNEY, NEW HOLLAND, AUSTRALIA**

*It is worth mentioning here* the most unusual meeting that took place between Captain Arthur Phillip, first governor of New South Wales, and Captain Lapérouse. The occasion was the moment that Captain Phillip and his fleet were leaving Botany Bay headed for Port Jackson, and Lapérouse's division was making toward the bay in search of a mooring and rest and recuperation. Captain Phillip was initially on his guard, France and England being then at war, but he recovered his composure on recognizing the scientific expedition of the *Boussole* and *Astrolabe*, led by the celebrated French captain. Putting culture and peace before the exigencies of war, Phillip immediately offered his services to Lapérouse, whereupon their respective officers shook hands, repeatedly enjoyed visits aboard their respective vessels and exchanged innumerable courtesies. We know that this was the last ever meeting between Lapérouse's division and other Europeans, the expedition having gone missing thereafter. Forty years later it seems that Captain Peter Dillon may have found traces of the lost Lapérouse expedition on the Malicolo islands [Vanikoro], where he was sent for this purpose aboard the *Research,* a ship outfitted by the East India Company.

**JUNE 1839**
**BREST, FRANCE**

*At daybreak on June 23,* at the entrance to the Gulf of Gascony we sighted several ships crossing our path, among them a brigantine flying French colors that we realized in passing by her was none other than the *Henriette*. The captain gave us all the latest French political news and the upshot of our quarrel with Mexico. We thereupon changed our route and made toward Brest, where we arrived a day later, at six o'clock on the night of June 24, 1839. So ended a campaign of almost thirty months, during which we had had the good fortune not to suffer a single fatal accident; we returned with our crew intact save for six men lost to illness or desertion.

**Top, left and right:**
Traditional dress in 1838.

**Facing-page, top:**
*Village de Kororaréka, Nouvelle Zélande* (Village of Kororareka, New Zealand).

**Facing-page, bottom:**
*Cases de Paha Village de Kororaréka, Nouvelle Zélande* (native huts in the *paha* [village] of Kororareka, New Zealand).

**Pages 316-317:**
The *Vénus* in Sydney harbor, with view of Fort Macquarie, New Holland, 1838.

*For these unenlightened peoples the mighty warships they behold are the sole measure of the relative power of distant nations.*

## Cyrille Pierre Théodore
# Laplace

**1793–1875**

AGENT LAPLACE HAD A NEW MISSION!
KING LOUIS-PHILIPPE, TAKING ADVANTAGE OF THE YEARS OF PEACE, WISHED TO EXPAND FRANCE'S PRESENCE OVERSEAS. IN HIS SIGHTS WERE HIS LONG-STANDING ADVERSARIES, THE ENGLISH, WHO INCREASINGLY RESENTED SEEING THE FRENCH FLAG FLYING IN OCEANIA. WOULD THEY INVADE CHINA AS THEY HAD INVADED INDIA? IT WAS LAPLACE'S TASK TO SATISFY THE KING'S DESIRE TO STRENGTHEN FRANCE'S POSITION IN THE PACIFIC; TO RESTORE SOME DISCIPLINE ON BOARD THE WHALERS; AND FOLLOWING IN THE FOOTSTEPS OF PETIT-THOUARS, TO TAKE UP THE DEFENSE OF CATHOLICS, ALTERNATING SOFTNESS AND AGGRESSION IN HIS DEALINGS WITH FOREIGN MONARCHS.

Planche 90 par George Wolfgang Knorr, 1751.

# THE CIRCUMNAVIGATION CAMPAIGN OF THE *ARTÉMISE*

**1837–1840**

**At the age of fourty-five, eight years after his first world tour, Laplace embarked on another three-year circumnavigation, in course of which he was ready and willing to declare war on the king of Hawaii to defend France's trading interests.**

On returning from his voyage aboard the *Favorite* in 1832, Laplace needed to rest. His account of his voyage was published the following year, at which point he joined the Dépôt des Cartes. He then resumed duty in 1835, sailing to New York and Canada aboard the *Artémise* on a diplomatic mission for the French government. No sooner had he returned to Toulon than he was summoned by the Minister of the Navy. France was worried about her merchant ships, and more specifically her whalers, which were finding it hard to keep their crews at full strength during stopovers. Whaling was still dominated by the Americans and the English, and the French wanted a piece of the action. So Laplace embarked on a second around-the-world voyage.

When the *Artémise* left the port of Toulon on January 20, 1837, she headed east, calling at the same ports as the *Favorite*. This enabled Laplace to compare and judge how things had developed in the interim—the advance or decline of countries on the raod to independence and any changes in local populations are the influence of Europeans. After Pondicherry, he sailed up as far as Calcutta, where the dominance of the East India Company had sharply diminished and the British government was now calling the shots. He then headed for Sumatra, where a cholera epidemic cost him six men, followed by Sri Lanka and Bombay, where the frigate made quite an impression and enhanced the status of the French flag.

The *Artémise* headed for Persia and Arabia, dropping anchor in Muscat on March 2, 1838. The English plainly controlled the straits, prompting Laplace to write: "France can do no thing here for the moment, England is too strong." He then anchored off the town of Mokha, which was becoming famous for its coffee; and in May 1838, our navigator set sail for India.

In the strait of Malacca, Singapore, a spate of epidemics cost the lives of fifteen sailors before ever the *Artémise* arrived in Tourane [Da Nang] in August. Yet again, Laplace could not complete his diplomatic mission. With Cochinchina [Vietnam] still fearing British reaction, he was unable to obtain an audience with the Emperor and continued on to Manila and then China. As before, Laplace observed that French trade was virtually non-existent and that the British East India Company dominated trade with the Chinese. Opium smuggling was on the increase, and he noted in his notebooks the failings of the system that would lead, years later, to the Opium Wars. He also severely criticized French traders for export incompetence and dumping poor-quality products onto the market, while other Westerners sold their best products in China.

The *Artémise* then headed south, passing the Anambas Islands and entering the Indian Ocean via the Sunda Strait en route for Tasmania, where Laplace had been so well received on his first trip. In Sydney, by contrast, he found the Australians aloof and concerned about the French military presence in Tahiti and the Pacific, making his diplomatic mission there rather more complex. The next stop was Tahiti, where his arrival was by no means plain sailing. Approaching the island in April 1839, the frigate ran aground on a reef not mentioned on any of the charts and suffered extensive damage, not least the loss of her rudder. It is to the credit of her crew that they managed to enter the lagoon rudderless, with the keel ripped off, and the frigate taking on water through the breach. She was eventually rescued from her predicament by an American whaler that helped her reach Papeete.

It took two months of work; the frigate had to be hove down for repairs but was eventually refloated with the help of the Tahitians. In the meantime, Laplace could observe the island's development at his leisure. Despite the treaties signed with Queen Pōmare, the English Protestant missionaries were still very influential and the French Catholics were ill-treated. To remedy the situation, Laplace opted for a soft approach and invited the sovereign on board, where he had her sign an amendment to the Petit-Thouars Treaty providing for the free practice of the Catholic religion. He noted in his logbook that the population would be better treated by Catholic priests if Queen Pōmare were to cede her lands to France. The battle for Pacific ascendancy was thus as much a religious war between English Protestants and French Catholics as a conflict of visions between two countries with very different ideas of colonization.

Satisfied that he had done all he could, Laplace left Tahiti and headed north to Haiti, where he arrived on July 9, 1839 to find that a number of Catholics had been imprisoned. Appalled at such persecution, Laplace made it his business to put things straight. On July 10, he issued an ultimatum to King Kamehameha III to the effect that there would be war if His Highness did not sign a text guaranteeing the free practice of religion. An edict of toleration was signed that same evening, and Laplace obtained a plot of land and twenty thousand piasters to build a church. He asked a young Irish priest, Abbé Walsh, to celebrate a high mass for people of all beliefs, a move that won applause from the British and American consuls. A grand *Te Deum* played by the frigate's musicians brought the ceremony to a close, restoring dignity to France and the Catholic Church alike. And with that, the *Artémise* was off again, sailing by way of Peru and Chile, rounding Cape Horn and restocking in Brazil before heading back up the Atlantic.

On April 14, 1840 Laplace arrived in Lorient after a grueling but no less fruitful tour of the world. Promoted to Rear-Admiral the following year, he set sail again, heading the Antilles division from 1844–1847. He was promoted to maritime prefect and vice-admiral in 1853, and died in Brest on January 24, 1875 at the age of eighty-two.

► The *Artemise* in Muscat, Arabia, 1837.

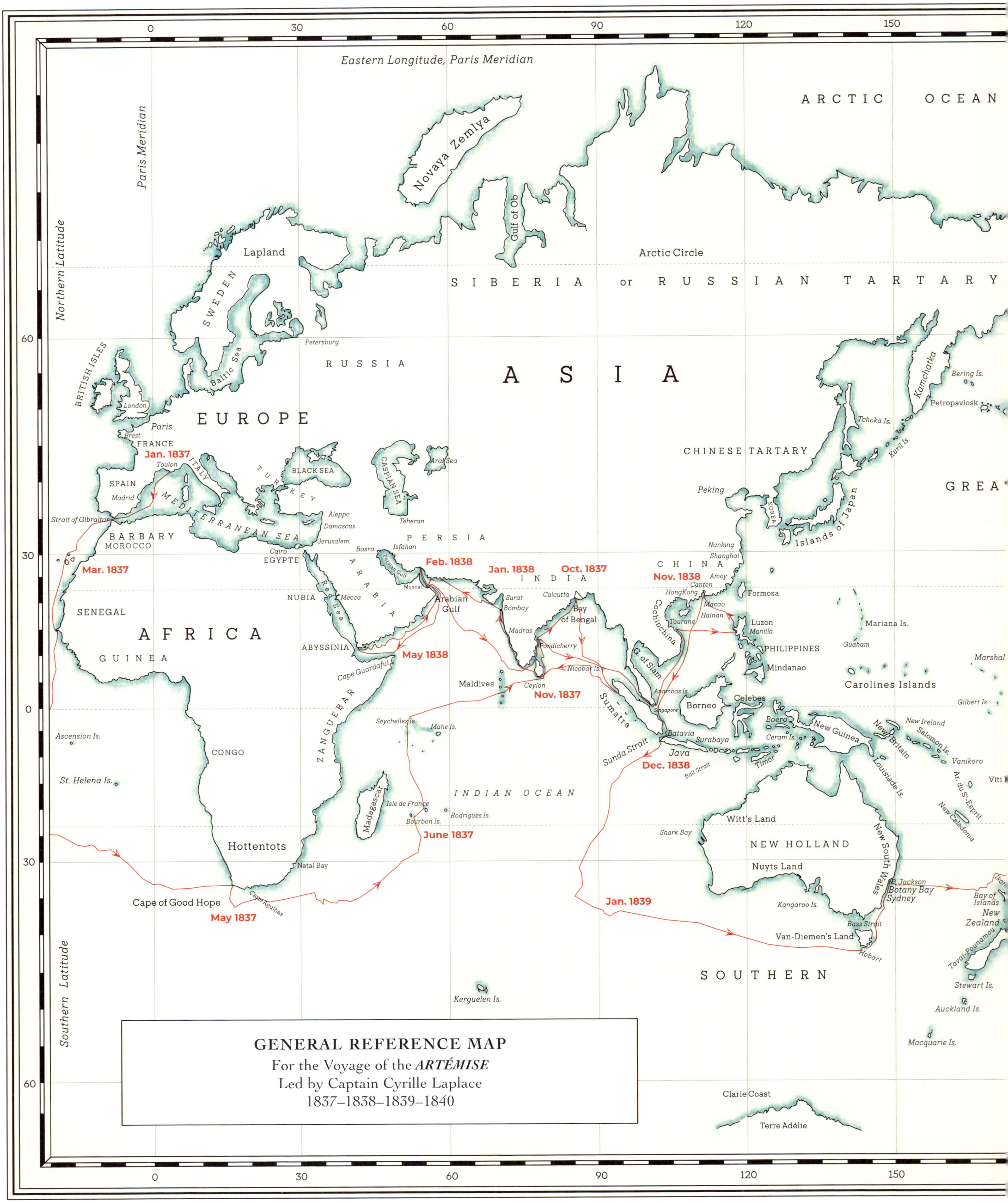


GENERAL REFERENCE MAP
For the Voyage of the *ARTÉMISE*
Led by Captain Cyrille Laplace
1837–1838–1839–1840

Western Longitude, Paris Meridian
Paris Meridian
Melville Is.
Lancaster Strait
Baffin Bay
Davis Strait
GREENLAND
Iceland
Bering Strait
Cook's River
Fort des Français
Lituya Bay
Kodiak Is.
Alaska
Sitka
Queen Charlotte's Is.
Vancouver Is.
Nootka
eutian Islands
Hudson Strait
Hudson Bay
Cape Farewell
LABRADOR
CANADA
NORTHERN
BRITISH ISLES
SWEDEN
Baltic Sea
London
St Lawrence River
Newfoundland
Apr. 1840
Paris
Brest
FRANCE
NORTH AMERICA
Quebec
Acadia
UNITED STATES
ATLANTIC
Toulon
ITALY
SPAIN
Madrid
ORTHERN OCEAN
San Francisco
New York
Washington
Azores Is.
MEDITERRANEAN SEA
Strait of Gibraltar
Monterey
Aug. 1839
Bermuda Is.
New Orleans
San Antonio
Madeira Is.
BARBARY
MOROCCO
California
MEXICO
Florida
Canaries Is.
Gulf of Mexico
OCEAN
Tropic of Cancer
Mazatlán
Lucayan Is.
Sandwich Is.
Hawaii Is.
San José
Mexico City
Santo Domingo
SENEGAL
Caribbean Sea
Jamaica
Acapulco
Guadeloupe
Martinique
Cape Verde
AFRICA
GUINEA
Panama
COLOMBIA
Cayenne
Equinoctial Line
GUYANA
Gallapagos Is.
Oct. 1839
Paita
SOUTH AMERICA
PERU
Ascension Is.
Marquesas de Mendoza
CONGO
Lima
avigators Is.
July 1839
St Helena Is.
Bahia
Pisco
BRAZIL
Maouna Is.
Tahiti
Arica
Friendly Is.
Society Is.
Tropic of Capricorn
Rio Janeiro
Traitors Is.
Mangareva
Pitcairn Is.
C. Frio
Feb. 1840
Easter Is.
Coquimbo
CHILE
St Catherine's Is.
SOUTH
pr. 1839
Valparaiso
Montevideo
Buenos Aires
Rio de la Plata
Cape of Good Hope
Concepción
ATLANTIC
atham Is.
OCEAN
PATAGONIA
Jan. 1840
Falkland Is.
OCEAN
Strait of Magellan
Tierra del Fuego
Is. de Los Estados
Cape Horn
Sandwich Is.
Antarctic Circle
150
120
90
60
30
0

## Instructions from the Minister of the Navy to Monsieur Laplace

The King, Sir, has declared his intention of sending a Royal Navy ship to visit the various trading posts in India. At my suggestion, His Majesty has kindly agreed that this mission should be given to the *Artémise*, with whose command you are entrusted.
On leaving Toulon you will head for the Strait of Gibraltar. It is desirable that you measure your pace in such a manner as to arrive at our factories in India by the month of June at the latest, whereupon you will commence the exploration campaign for which you have been engaged. For six months you will be at leisure to visit the different trading posts established by the English, remaining for as long as required to gather intelligence on such issues as may be relevant to our political and trade relations.
You will find in Bombay every means of re-provisioning your ship and you will make its port your point of departure for excursions along the shores of Persia and Arabia, as far as the mouth of the Red Sea.
The mission that is yours to fulfill is essentially to provide French commerce with all such services as fall within the scope of your duty, whether by visiting foreign trading posts in order to gather intelligence of use to the speculations of the ship owners and merchants in our ports, or by providing their ships with the protection that state vessels are everywhere required to provide.
On reaching the west coast of America in 1839, you will begin work of a different order.
In the ports of Chile and Peru, you will encounter ships stationed in the great Pacific Ocean whose mission is to pursue the same interests as those that you will have sought to further in India, in China and in New Holland. Your first communications with these ships will acquaint you with the situation in this part of America and Europe alike.

On departing from here, you will continue the work undertaken by the frigate *Vénus* in the previous year on the northwest coast of America and in the Galapagos and the Sandwich Islands.
That is to say that in visiting the ports of the American Continent as far as the entrance to the Gulf of California, you will identify areas where French whalers may ply their trade.
Examples abound of state vessels exerting a positive impact on whaling by intervening to settle disputes arising from a lack of discipline aboard the ship herself.
I recommend, Sir, that you apply yourself in this respect to assist such French whalers as may turn to you for help. Use your authority to restore order on ships where indiscipline prevails and give their captains the benefit of your counsel, offering them such material assistance as it is in your power to provide.
A most profound peace reigns between the European powers and our differences with the United States of America have been resolved.
There is therefore no reason to believe that you should prepare yourself for attack from any quarter whatsoever.

VICE-ADMIRAL AND MINISTER
OF THE NAVY AND COLONIES,
CLAUDE DU CAMPE DE ROSAMEL
PARIS, JANUARY 6, 1837

*Men are wrong not to consult their wives more often, for were they to do so we would see far fewer reversals of fortune and troubled families.*

JANUARY 1837
TOULON, FRANCE

On January 20, our preparations were all complete and I gave the order to cast off. It was time to put an end to that mental and physical agitation that takes hold of men on a warship that is about to depart on a long voyage.

Off the coast of Spain we encountered quite favorable winds that after a few mishaps carried us to the Strait of Gibraltar, which the *Artémise* passed through in just fourteen days.

That this was by no means my first rapid crossing of the Mediterranean would tend to belie the view widely held by sailors in the north of France that in the matter of victualing naval expeditions headed for the Atlantic, the Port of Toulon is less suitable than the ports of Brest, Cherbourg and Rochefort. From here, they argue, it is easy to reach the open sea thereby escaping the perils of skirting the coast in winter.

MARCH 1837
GORÉE, SENEGAL, FRENCH TERRITORY

I encountered these saintly persons in all our distant possessions and was filled with admiration for their devotion and even their heroism. I refer not to that vainglorious heroism that feeds off the applause of the multitude. Nay, what I have in mind is that love of one's neighbor that is one thousand times more heroic precisely because it brings no glory or reward, at least not in this world. It is this sublime self-abnegation that pleads so strongly in favor of the Catholic religion, there being no other religion to exemplify the kind of self-denial practiced by our nursing orders, much to the envy of Protestant nations.

These good sisters do not restrict their care to the relief of human suffering. They also spread learning and piety among the black and brown people of our colonies, partly by their gentle perseverance and partly by their admirable selflessness, the former winning the hearts of young girls and the latter winning the confidence of their families. Thus have they worked quite an astonishing improvement in the habits and behavior of the young Signares of Gorée, the while commanding the respect and obedience of the older women. At present, a great many of the young women can read and write French, sew and keep house with economy, even to the point of copying the Frenchwoman's manners and ways of dressing.

▲ Village of Bel Air, African coast.

*Do you not fear lest from the midst of this youth should issue a liberator of the Indians? The man they require to shake off the yoke of the foreigner:*

**JULY 1837**

**ALONG THE INDIAN COASTLINE**

*The East India Company,* ever concerned to assert its dominance and routinely thwarted in its desire to do good for its subjects, often sacrifices their interests in favor of its own and even those of the British Empire, for which trade and politics always take precedence over the rights of peoples. Little wonder, then, that it regards Hindustan more as a mine ripe for exploitation than as a populous country entrusted to its care.

It can therefore be said that the destinies of British Asia are currently presided over by two powers. One is local government which, the more it identifies with the country under its rule, the more it understands that given the extreme penury to which India is reduced, its only hope of rebuilding its finances is to encourage internal trade and agriculture, which necessarily means paying more attention to the well-being of its people. The other power is Great Britain herself, that ever-selfish force that is eternally ready to sacrifice the whole world on the altar of her subjects' material interests, whose commercial influence counts above all else. Not content to have deprived the Indians of their main livelihood through competition from her cotton mills, she stops at nothing to extend her monopolies to every aspect of their production, the while proclaiming her concern for their welfare and ignoring the protests of the East India Company.

**AUGUST 1837**

**PONDICHERRY, INDIA, FRENCH TERRITORY**

*It was with mixed feelings* that I beheld these young girls. Pleasure certainly but also some distress at the sight of their gracious and candid countenance already tinged with the sadness and despondency of the submissive wife: the woman under the yoke of a despotic and indifferent husband, forced into a lonely life with fatigue and disgust as her only companions. Such is the thankless lot of most Asian women, and it shows in their faces. I could see the surprise in their eyes as they looked almost enviously at our own lovely womenfolk, watched them disport themselves, gayly dancing to music that made them gayer and more coquettish still. Oh happy women! they seemed to be saying. What good fortune is theirs to belong to a nation where women are cherished and respected and know nothing of the suffering and humiliations heaped upon their counterparts in almost every other part of the world! It is certainly true that in France more than anywhere else, we men know how to appreciate those precious qualities that Nature has been pleased to bestow on the weaker sex: softness, devotion, intelligence and unerring tact.

**OCTOBER 1837**
CALCUTTA, BRITISH INDIA

*Then there was the irksome duty* of observing English customs to the letter; customs that bear little resemblance to our own but whose observance is mandatory on pain of appearing rude. Toasts, for instance, were a constant source of concern. Our English neighbors, regardless of class, have a gift for holding forth on every topic under the sun and most especially at grand repasts where only the stiff-upper-lipped British could tolerate the drivel spouted on all sides. That people should feel free to mouth such nonsense never ceases to amaze me; yet even in this, say their admirers, the British surpass we French. I think they are wrong in this case just as they are wrong in so many comparable cases. When it comes to the art of ridicule, there are quite as many gifted orators and witty spirits around the table in France as there are across the Channel. So let us not look to England for example but rather take pride in remaining what we are: a nation of mild-mannered, happy, witty and intelligent people whose country, despite all the ill that is said of her, attracts foreigners from all over the world in search of an agreeable way of life and lessons in good taste and urbanity.

**NOVEMBER 1837**
CEYLON

*For most Europeans,* the mere mention of "Ceylon" is enough to conjure up all the brilliance and magic of India in terms of riches and wonders alike. Even I felt an admiration mixed with curiosity.

The very day after our arrival, I set off to explore the outskirts of Colombo. No sooner was I beyond the drawbridges than I fell under the spell of the delightful prospect that unfurled before my eyes. On my right was the ocean, its frothy waves breaking at my feet; on my left were the placid waters of the lake, burnished by the rays of the rising sun. Along its banks and surrounded by lawns, parterres of flowers and gardens were scattered a multitude of country houses. Some were in the style of Florentine villas, others were authentic

*China is dependent on Great Britain for her opium, which is as necessary to the Chinese as strong liquor is to the Europeans.*

◄ Facing-page:
*Entrée d'une pagode à Pondichéry.*
(entrance of a pagoda in Pondichéry).

▲ Top, left:
*Bayadères de Pondichéry.*
(*Bayadères* of Pondichéry).

▲ Top, right:
*Passage de la rivière de Yanaon*
(crossing the river Yanaon).

► Pages 330-331:
*Porte méridionale de Moka, Arabie*
(southern entrance to Mecca, Arabia).

English cottages, complete with immense red roofs and verandas that offered protection from the sun and rain. Both styles of house were equally seductive, set amidst vast prairies and groves of ancient trees under which fashionable society arrives by carriage every evening to take the air or sometimes, depending on the season, to attend those fashionable horse racing events that the English have brought with them to Ceylon.

**DECEMBER 1837**
**GOA, BRITISH INDIA**

*What a bustling crowd* was there here! Human beings of all ages and both sexes; numberless buffaloes surrounding the carts that they had drawn from the interior; and lodged under those carts, the drivers waiting patiently until such time as they could take the path home freshly loaded with goods.

And not one of these myriad travelers was inactive: some tending to their teams, other setting off in caravans heavily laden with parcels whose labels were a sufficient indication of their European provenance. Others still were at the end of their long voyage and come to unload their cargo at the fine wharf opposite our mooring and around which crowded large boats busy transporting cotton bales taken from the great piles that clogged the shore, to the ships at anchor in the harbor. Not one quarrel or disagreement issued from this crowd of men of different origins and different religions. The Afghans, men of athletic build and independent disposition, rubbed shoulders with their northern neighbors the Tartars, pale-skinned and not unlike the Chinese in appearance but more roughly hewn. The Marathas, medium tall and hard featured, their forehead bearing the mark of the Brahmin, coexisted peacefully with the Persians, graceful, slightly built, with faces at once soft yet expressive. The Marathas even deigned to mix with those debased and servile Hindus from the southern regions of the peninsular, whose humbled brow testifies to the yoke of slavery that has oppressed their land for centuries past. Each of these men went about his noisy business with a calm and tranquility that was truly admirable.

Top:
*Palais du gouverneur à Pondichéry*
(palace of the Governor of Pondicherry).

Facing-page:
*Cour d'une pagode à Pondichéry*
(pagoda courtyard in Pondicherry).

**JANUARY 1838**
**BOMBAY [MUMBAI] BRITISH INDIA**

*Were they really to abandon* to the mercy of foreign negotiators an article without which commerce between Great Britain and China cannot exist? A measure, what is more, that for the sake of some trumped-up philanthropic goal would mean sacrificing the eighty-four million francs yielded by their monopoly on opium every year? Nay, such a step was beyond question and not least because the British explicitly forbad their agents from transporting opium to the shores of the Celestial Empire. So it was that in 1832 the commission decided to leave things as they were and the cultivation of opium in British India has increased apace ever since. In 1838 twenty-seven thousand chests of opium were shipped from Bombay and around seventeen thousand chests were shipped from Calcutta, a staggering level of exports that would remain undiminished throughout the Sino-British conflict.

It is therefore with Canton, which provides an important outlet for its principal commodities, that Bombay mainly does business. Its affairs have continued to flourish since 1833 even though by that time the East India Company had lost its China monopoly and no longer wished to undertake such dealings on its own behalf, satisfying itself with the levy of charges on private transactions. This prosperity, however, has come to a temporary halt due to the hostilities between England and the Celestial Empire hampering cotton sales in this part of Asia. The Chinese abandoned cotton growing some forty years ago following a so-called great famine that drove them to plant rice instead, If, as is expected, they now return to cotton, this could well deal a death blow to one of British India's main exports.

**MAY 1838**
SOMEWHERE IN THE INDIAN OCEAN

*For more than a year now* we have been scouring the coasts of Asia, yet there are still a great many maritime places on this vast continent that remain to be visited. Not one place of interest in Hindustan had been overlooked: Sumatra, Ceylon, the Persian Gulf and the Red Sea had all seen our flag fly over their shores and their ports. Wherever it went, the *Artémise* had acquitted herself with dignity as the protector of trade, the while providing the inhabitants of those foreign climes with an opportunity to admire one of the finest flowers of the French navy. But her task was very far from over. Indochina remained yet to explore, and that on the eve of the great events that were to take place there. Thither did I therefore direct the *Artémise*, but only after revisiting several places that I felt required a second inspection before forever abandoning the shores of the Indian Peninsular Region.

**JULY 1838**
PULAU-PINANG [PENANG], MALAYSIA

*Pulau-Pinang appeared* before us at daybreak in all her splendor. Her high mountains were cloaked from top to bottom in a lush greenery whose infinitely varied shades were picked out by the sun that occasionally broke through the otherwise immovable layer of cloud that shrouds the highlands at the time of the southwesterly monsoon. At their peak and along the flanks of the mountains extended vast primeval forests that I easily distinguished by their dark green shapes, together with fine plantations of symmetrically planted nutmeg trees whose tufted foliage contrasted agreeably with the reddish tint of the soil at their feet. Further down, stretching all the way to the sea, were emerald-green rice fields and immense prairies on which my eye rested with infinite delight. In vain however did I look for those villages and hamlets that lend such grace to our own delectable countryside.

Man's handiwork was everywhere to be seen but nowhere, with the exception of a few despicable huts scattered here and there amid the fields, was there any sign of habitation. In truth, these lands on which Nature has bestowed all her riches are home to a race of men more terrible and more ferocious than the tigers that infest the surrounding forests.

*The Malays' principal occupation is to scour the straits in search of luckless natives who they capture and reduce to slavery.*

Bottom and facing-page:
Lakeside villages in Malacca.

Pages 336-337:
*La frégate L'Artémise abattue en carène à Papaiti, Taïti* (the frigate Artémise on its side for hull maintenance, Papeete, Tahiti).

**NOVEMBER 1838**
CANTON [GUANGZHOU] CHINA

*Metal, wood and stone* are as lovingly worked here as in the most advanced countries of the Old World. What would the subjects of the Celestial Empire not achieve if they could only give free rein to their creative genius—shake off the yoke of a barbaric and benighted government that has kept them cowed for centuries and, for the best of them at least, take their rightful place alongside the most adroit artisans of London or Paris? For despite all the obstacles, when it comes to handiwork in the real sense of the word, the Chinese are masters of their trade. This is as true of those beautifully finished artifacts destined for European markets as it is of all their other works and constructions. How solidly built are the houses of the affluent Chinese! How beautifully executed is the stonemasonry yet as hard as granite withal! Each detail of these vast buildings stands as proof of the patience and artistry that went into their construction. I had not appreciated just how talented was the Chinese workman before visiting the magnificent residence of a rich merchant who, probably more to show-off than to please me, invited me on a detailed tour of his property.

But there where the Chinese have the better of the Europeans is in the matter of lower-class housing. Their healthy, decent and roomy housing was all the proof I needed of an advanced civilization.

**AUGUST 1838**
MALACCA, MALAYSIA

*What events led to the downfall* of this once fine and flourishing city, whose name conjures up such a flood of unforgettable memories? The same events as brought the indomitable land of Goa to its knees. The Portuguese made the same mistakes here as on the shores of India, with the same disastrous consequences. To Malacca as to Goa, the royal court in Lisbon dispatched monks and disloyal administrators knowing neither energy nor talent, in place of warriors and honest men with the courage and ability to defend Portugal's enviable possessions in Indochina against the relentless attack of the natives or the no-less dangerous invasions of the Dutch.

Here as on the Malabar coasts and wherever traces of the former greatness of Portugal were to be found, the same painful reflections would disturb my admiration for the exploits of those men of iron of whom I have just spoken. What remains today of their many triumphs and victories? Nothing but ruins and a profound aversion for Europeans that is kept alive in the Malays by the legacy of the dreadful calamities that laid waste to these shores from the moment that the Cape of Good Hope was discovered.

*Gone, except a tower at the top of which the English placed a beacon to show that Malacca had not entirely withered into nothingness.*

**SEPTEMBER 1838**
SINGAPORE

*If French merchants* were to spend less time wailing about being abandoned and more time applying themselves with the same diligence as our neighbors to seeking ways to make their business prosper, then our ministers would no doubt employ every means at their disposal to advance their business interests, knowing that such intelligence issued from the only genuinely reliable sources. At that point our shipowners could operate to good effect in distant countries and would quickly realize that Singapore offers them opportunities for success that they have so far let slip. Hitherto French goods have always been shipped to Canton, where shipowners stand to make few sales but pay dearly for the privilege in terms of import duties, with no prospect of taking on cargo for the return journey.

In Singapore, by contrast, they might exchange French goods for Malay or Chinese products much prized here in France, though success would of course depend on informed activity, starting with a detailed understanding of what the Malays and Chinese actually want.

**DECEMBER 1838**
CANTON, [GUANGZHOU] CHINA

*The English have destroyed* China's best soldiers for the sake of establishing European discipline, ravaged her finest port cities, profaned her temples and insulted her women and their daughters. Their latest objective if reports are to be believed, is no less heinous: to incite the Chinese population to revolt against their Tartar masters. In the eyes of the ever-suspicious government this is justification enough to fear that what it has long suspected may well come to pass: namely that the East India Company destines the Chinese for the same fate as the princes of Hindustan.

There is no doubt that the court of Pekin believes it made a huge concession for the sake of peace in openly entering into negotiations with a foreign power. Particularly since until very recently Pekin studiously avoided any such relationship for fear that it might shake the serfdom's belief in the emperor's impregnable supremacy, an opinion that very much suited Pekin's books. Hence the efforts made to keep the peasantry in ignorance of Great Britain's triumph, with some success to judge by the continuing reverence of the Chinese for their motherland and their undying contempt for Europeans. The only exceptions are the inhabitants of the port cities that bore the brunt of the war.

One can hardly blame the Chinese for such blindness considering that since time immemorial they have isolated themselves from the rest of the civilized world, the while rising to become a remarkably prosperous nation and especially so since the beginning of this century.

*Will France seize this opportunity to regain the rank she held right up to the end of the past century?*

**JULY 1839**
TAHITI

*With her makeshift rudder* and her hold full of water, the frigate maneuvered with difficulty, not helped by a weak and variable breeze. In the circumstances the pilot durst not engage such a large ship in the narrow strait leading to the mooring across the reef. We had to find a refuge before nightfall, our limping ship being likely to go under at any moment taking all her brave crew with her; but I could not take the even greater risk of sending her crashing into the reef. It was as I was struggling with this dilemma that Captain Abrill came to offer his services, an Englishman as loyal and honest as ever there was, whose energy and first-rate seamanship I would come to appreciate even more thereafter. Having settled in this region several years ago, he offered to guide the frigate through the coral banks toward a nearby mooring where she could hove to in perfect safety.

**NOVEMBER 1839**
SAN FRANCISCO, ALTA CALIFORNIA

*Practically speaking,* what does it matter to Old World countries that San Francisco is now the capital of a new Union State and on its way to becoming the center of an immense business movement? The next few years indeed might well see it become the trade and maybe the political metropolis of all those countries bordering the Pacific Ocean north of the Equator. I even think that those countries may congratulate themselves on seeing one of Great Britain's deadliest rivals erect a city on these distant shores with the potential to serve as a local counterweight to Sydney, whose influence over Polynesia, China, the archipelagos of Asia and fairly soon Japan increases by the day and directly contradicts our own interests. Let us hope therefore, for France's sake, that California's present capital fulfills the lofty destiny that apparently await her.

*England has offended against China's religious beliefs, national pride and material interests—in short, every thing she holds most dear.*

San Francisco—its docks crowded with ships from all over the globe, and its roads dotted with magnificent caravanserais rarely equaled by any hotel in the great cities of Europe or Asia.

**APRIL 1840**
**LORIENT, FRANCE**

*On the afternoon of April 14* the long-awaited moment finally arrived: driven by an onshore breeze, the *Artémise* negotiated the Port-Louis narrows and dropped anchor for the last time, having sailed through all the maritime regions of the known world.

It was the end of the line for this fine frigate, my faithful companion throughout a four-year tour of duty fraught with all the dangers of such a very long stint at sea. The journey had taken a heavy toll, leaving the *Artémise* too wasted to remain in service. She was decommissioned in the very place that had witnessed her construction and became a pontoon boat for use along the quays of the arsenal, serving in the fitting out of young ships destined, like the *Artémise* herself four years earlier, to protect our trade, honor our navy and enforce respect for France in the most remote lands on the planet.

Did the *Artémise* acquit herself nobly of this task under my command? I leave it to the reader to judge whether this account is proof of my success. If as I hope his opinion is favorable then there can be no sweeter reward for those many sleepless nights and hard labors at sea.

**Facing-page:**
Villagers in period costume, China, 1839.

**Top:**
The Canton factories, China, 1839.

**Pages 340-341:**
*L'Artémise.*

A LEGENDARY VOYAGE FOR AN EXCEPTIONAL MAN! THE SOUTH POLE WAS A REGION THAT REMAINED TO BE EXPLORED AND CLAIMED FOR THE BRITISH CROWN. TO GET THERE, KING LOUIS-PHILIPPE CALLED ON THE SERVICES OF SEASONED NAVIGATOR JULES SÉBASTIEN CÉSAR DUMONT D'URVILLE, AN INTELLIGENT IF SOMEWHAT WHIMSICAL MAN WHOSE IMPRESSIVE KNOWLEDGE OF THE OCEAN WAS MATCHED ONLY BY HIS SELF-ESTEEM.... BRAVING ALL THE DANGERS, THIS WAS A NAVIGATOR WHO WOULD INSPIRE HIS MEN AND LEAD THEM TO NEW DISCOVERIES, MOST NOTABLY TERRE ADÉLIE ON THE CONTINENT OF ANTARCTICA.

# Jules Sébastien César Dumont d'Urville

**1790–1842**

Plate 65, George Wolfgang Knorr, 1766.

# JOURNEY TO THE SOUTH POLE AND OCEANIA

### 1837–1840

**At age twenty-four, Dumont d'Urville saved the *Venus de Milo* from oblivion; at age thirty-five, he collected the remains of Lapérouse's expedition in Vanikoro; and at age forty-seven he discovered Antarctica, a grueling expedition to the two capes from which he would return having lost fourty-one of his 160 men, twenty-eight of them dead and another thirteen having deserted.**

Dumont d'Urville was born in Condé-sur-Noireau, Normandy, on May 23, 1790. Aged just six when his father died, he was entrusted to the care of his mother's brother, the Abbé de Croisille. He was schooled at the Collège in Bayeux, followed by the Lycée Impérial in Caen from 1804 onward, but despite being a gifted student he failed to gain admission to the École Polytechnique. He joined the navy at the age of seventeen and went to train in Brest, at a low point in the history of the French navy whose ships could barely leave port because of the English blockade. Our novice sailor nevertheless passed all his exams with flying colors, winning promotion but not the longed-for sailing commission. He tried his luck in Toulon, where his daily duties—standing armed watch on the city heights—proved equally boring. Rather more interesting was his encounter with local girl Adèle Pépin, who became his wife in 1815.

In 1820 Dumont d'Urville took to sea for the first time, sailing aboard the *Chevrette* on a mission to the Black Sea and the Greek islands. He was put in charge of botanical sampling and would take his insect box and butterfly net with him on stopovers. In Milos harbor, he heard of the recent discovery of a marble by a certain Olivier Voutier. Always fascinated by archaeology, our young sailor went to see for himself and immediately realized that what we now know as the *Venus de Milo* was far too unique to remain where it was. And with that, he obtained permission from the authorities to purchase the statue and bring it back to France. He was then assigned to the Dépôt des Cartes in Paris, less glamorous, but useful when preparing an round-the-world voyage. Particularly since this was where he met a certain Duperrey freshly returned from his voyage aboard the *Uranie*. Together, the two men suggested to King Charles X that they undertake an expedition pursuing Freycinet's work, but feeding themselves only on fish and coconuts—an altogether cheaper proposition that saw them depart France in August 1822 with the King's consent, sailing aboard the *Coquille*. Duperrey was given command, with the thirty-two year-old Dumont d'Urville as his second. On their return in 1825, all honors went to Duperrey. This enraged Dumont d'Urville who pointed out his many contributions to the expedition and finally received the much-desired recognition. Having been promoted and awarded a medal, he wrote up his findings, classified his herbariums and prepared a new exploration plan, this time headed for the New Guinea coast. King Charles X approved and gave him charge of the *Coquille*. But Dumont d'Urville had no intention of taking command of the same ship as Duperrey, so he had her renamed the *Astrolabe*, in honor of Lapérouse.

On April 25, 1826, Dumont d'Urville left Brest harbor for a new adventure in Oceania. With him sailed such exceptional sailors as Charles Hector Jacquinot (first mate), François-Edmond Pâris (future admiral), Joseph Paul Gaimard (ship's surgeon and naturalist), and Lesson, a pharmacist who had been with him on the *Coquille*. It was not an round-the-world voyage, and the *Astrolabe* did not round Cape Horn, but what an expedition it was! The frigate explored southern Oceania, circumnavigated Australia, Micronesia, the Indonesian islands and the Marianas, before rounding the Cape of Good Hope and returning to France. It was on this voyage that he heard that Dillon had found remains of the Lapérouse shipwreck. Hence d'Urville's decision to reroute and spend three weeks in Vanikoro in 1828, finding various marine relics that proved the location of the shipwreck and were brought back to Paris, most notably an anchor and a cannon.

On his return to France, d'Urville was promoted Captain but, on the downside, he failed to win nomination to the Académie des Sciences despite his expedition's undeniable contribution to science. He was furious and, as if to add insult to injury, he was not presented to the King as is traditional for navigators of his stature—one more wound that would never heal. Fortunately, however, the wheel always turns and when King Charles X was forced to abdicate and go into exile in 1830, Dumont d'Urville was chosen to accompany the royal family to England. He was finally able to speak with the King, who was now simply the Comte de Ponthieu.

The next five years marked the beginnings of the July monarchy and were less exciting. But Dumont d'Urville took it in his stride and focused on his writing, publishing his immensely popular *Voyage Pittoresque Autour du Monde* in 1834. In January 1837, he sent a letter to Rosamel, Minister of the Navy, suggesting a new expedition. The latter replied that King Louis-Philippe wished him to explore Antarctica, where there might be new lands to discover ... Dumont d'Urville had just lost his daughter to a cholera epidemic and was worn out, but he accepted the challenge all the same. After meeting the King, he fitted out two corvettes, the *Astrolabe* and the *Zélée*, and enlisted Jacquinot and Gaston de Roquemaurel as his seconds-in-command.

Dumont d'Urville and his men left Toulon on September 7 for one of the most difficult and deadly expeditions of the era. It was a real feat of endurance and a personal revenge for d'Urville. But at what a cost: twenty-five dead, at least thirteen deserters and fourteen sick left behind at ports of call. And just to make matters worse, Roquemaurel, with whom he didn't get on at all, saved the ship from disaster on at least three occasions. It was also on this expedition that d'Urville learned of the death of his son.

In Oceania, dysentery wreaked havoc and death followed death. On a rest stop in Tasmania, d'Urville told his travel-weary crew that he intended to head back south into the Antarctic ice. There could be no better way to motivate his men! So it was that in January 1840, a large land mass came into view, a new continent, perhaps the last continent, where Dumont d'Urville proudly planted the French flag and named it Terre Adélie in honor of his beloved wife. The return to France was as painful as the trip itself, and when the corvettes finally entered Toulon harbor on November 6, 1840, Dumont d'Urville was exhausted and sick. He spent a short while resting up; then newly decorated and promoted to rear-admiral, he set about writing up his adventures.

On May 8, 1842, he was in Versailles with Adèle and their son Jules for a party organized in honor of King Louis-Philippe. At 5:30 p.m., the family boarded France's latest attraction of the day: the train that ran from Versailles to Paris. Thirty minutes later, just as the train was entering Meudon at eighty kilometers an hour, the Matthew-Murray locomotive broke an axle and the train derailed. The carriages collided, a fire broke out ... and Dumont d'Urville and his family were killed in the first train accident in history, with more than 200 dead and 150 injured. He was honored with an exceptional ceremonial event, attended by the government, the Institut, the Académie and 15 admirals.

Bougainville was also in attendance, but the royal family were conspicuous by their absence. On May 11, 1842, on the orders of Minister Victor Guy Duperré, the Navy organized a search of the Dumont d'Urville family home, and all their documents were seized. The family would never have access to them for reasons that remain unknown to this day.

*L'Astrolabe faisant de l'eau sur un glaçon, 6 février, 1838, parages antarctiques* (the *Astrolabe* trapped in the ice, taking on water, 6 February 1838, Antarctica and its environs).

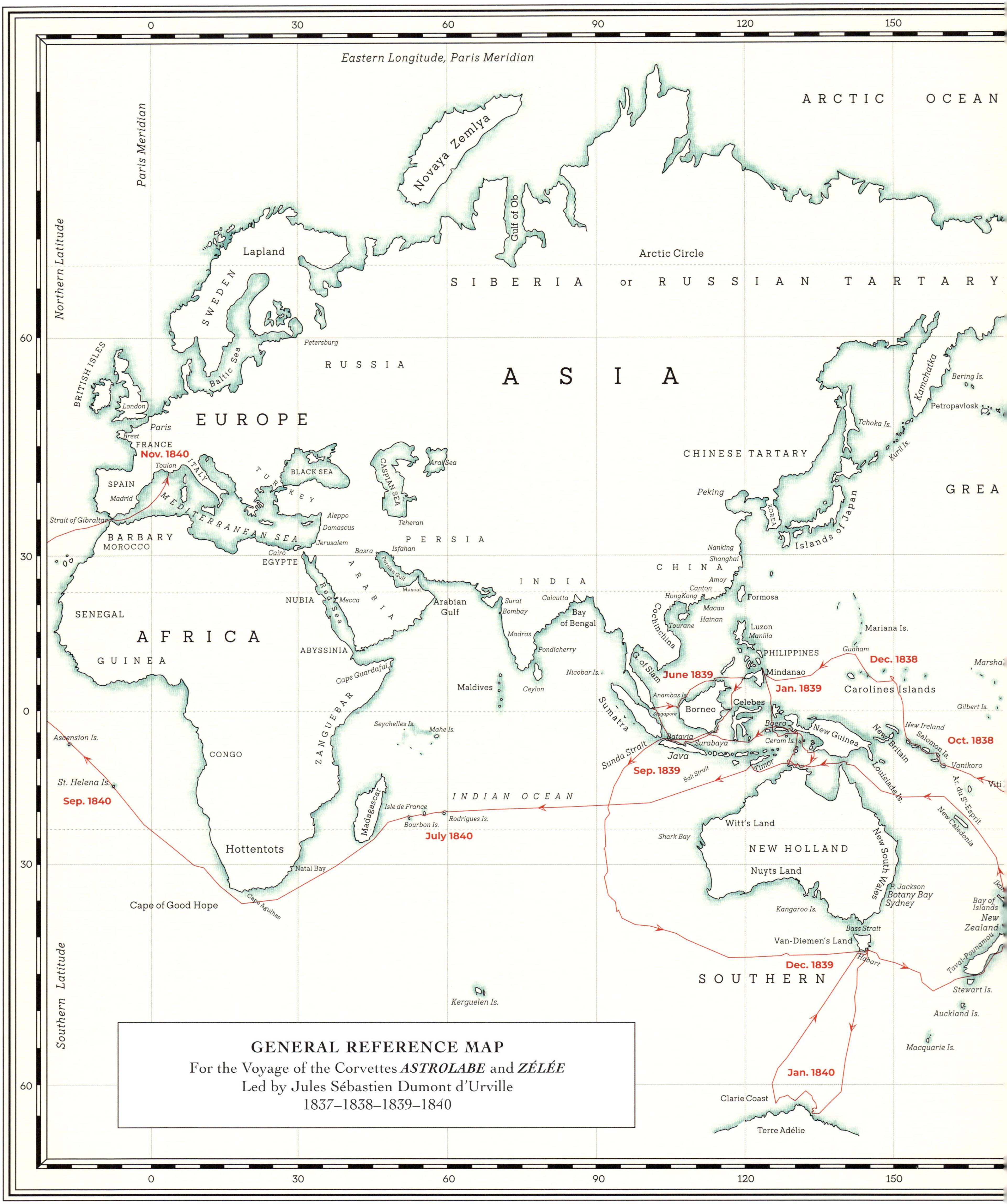


**GENERAL REFERENCE MAP**
For the Voyage of the Corvettes *ASTROLABE* and *ZÉLÉE*
Led by Jules Sébastien Dumont d'Urville
1837–1838–1839–1840

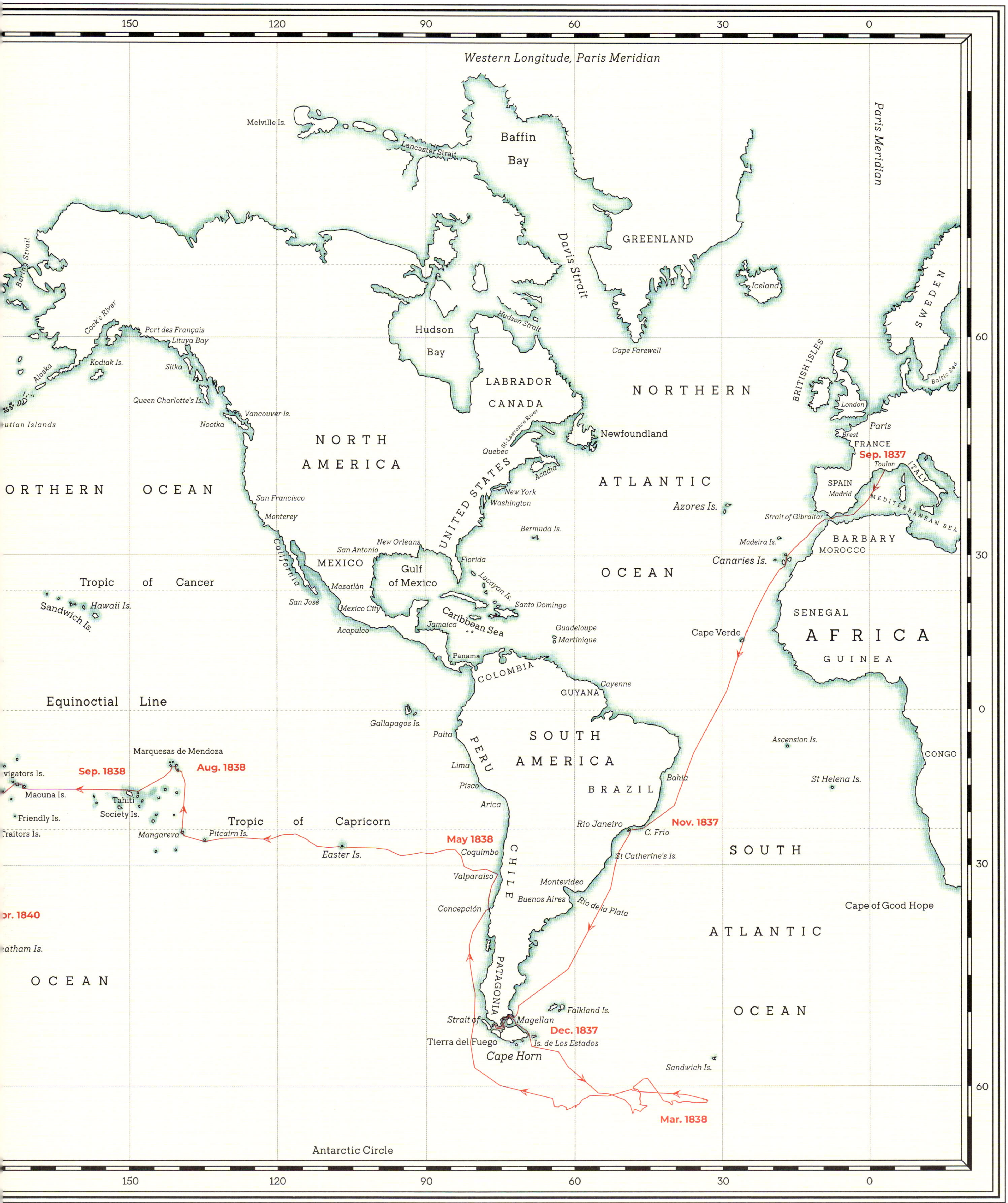

Western Longitude, Paris Meridian
Paris Meridian
150
120
90
60
30
0
60
30
0
30
60
Melville Is.
Lancaster Strait
Baffin Bay
Davis Strait
GREENLAND
Iceland
SWEDEN
Baltic Sea
Bering Strait
Cook's River
Port des Français
Lituya Bay
Kodiak Is.
Sitka
Alaska
Hudson Bay
Hudson Strait
Cape Farewell
BRITISH ISLES
London
Queen Charlotte's Is.
Vancouver Is.
Aleutian Islands
Nootka
LABRADOR
CANADA
NORTHERN ATLANTIC OCEAN
Newfoundland
St-Lawrence River
Quebec
NORTH AMERICA
Brest
Paris
FRANCE
Sep. 1837
Toulon
ITALY
SPAIN
Madrid
MEDITERRANEAN SEA
Acadia
UNITED STATES
New York
Washington
NORTHERN OCEAN
San Francisco
Monterey
Azores Is.
Strait of Gibraltar
Bermuda Is.
Madeira Is.
BARBARY
MOROCCO
New Orleans
San Antonio
California
MEXICO
Gulf of Mexico
Florida
Canaries Is.
Tropic of Cancer
Lucayan Is.
Mazatlàn
San José
Sandwich Is.
Hawaii Is.
Mexico City
Santo Domingo
SENEGAL
Caribbean Sea
Jamaica
Guadeloupe
Martinique
Acapulco
Cape Verde
AFRICA
GUINEA
Panama
COLOMBIA
Cayenne
GUYANA
Equinoctial Line
Gallapagos Is.
Paita
PERU
SOUTH AMERICA
Ascension Is.
CONGO
Marquesas de Mendoza
Navigators Is.
Sep. 1838
Aug. 1838
Lima
Bahia
BRAZIL
St Helena Is.
Pisco
Maouna Is.
Tahiti
Society Is.
Arica
Friendly Is.
Tropic of Capricorn
Rio Janeiro
Nov. 1837
Traitors Is.
Mangareva
Pitcairn Is.
C. Frio
May 1838
Coquimbo
CHILE
Easter Is.
St Catherine's Is.
SOUTH ATLANTIC OCEAN
Valparaiso
Montevideo
Buenos Aires
Rio de la Plata
Apr. 1840
Concepción
Cape of Good Hope
Chatham Is.
OCEAN
PATAGONIA
Falkland Is.
Strait of Magellan
Dec. 1837
Tierra del Fuego
Is. de Los Estados
Cape Horn
Sandwich Is.
Mar. 1838
Antarctic Circle

My orders were simply to betake myself to my home port of Toulon, where I arrived in the middle of the month of June in that fateful time when cholera was rampant in this part of Provence. One of its first victims had been a little girl idolized by her parents.
For more than eighteen months I enjoyed none of those numerous favors accorded by the Ministry and had to resign myself to those obscure duties that befall a ship-of-the-line captain in port.
In the month of January 1837 I wrote to the Minister of the Navy, Vice-Admiral Rosamel, to offer my services for a new campaign of exploration around the world.
Toward the end of February, I was informed that my project had been well received by the King but that His Majesty having learned of the presence of an American whaler close to the South Pole, desired that a French expedition be dispatched thither.
In consequence, I was offered command of two vessels and told that my campaign would begin with an advance toward the South Pole. The novelty, loftiness and even wonder of such an attempt could not fail to capture the attention of the public.
On May 19 the Minister of the Navy presented me to the King, who received me with great affability and talked to me at length of the efforts directed at the glaciers of the Southern Hemisphere, and the results obtained hitherto. His Majesty meanwhile paid detailed attention to a small map of the circumpolar regions that I had taken care to bring with me to facilitate his understanding of such explanations as I might proffer.
At midday on September 7, I gave the order to cast off, our corvettes already groaning under their weight and needing nothing further from the arsenal.

JULES DUMONT D'URVILLE

**SEPTEMBER 1837**
**TOULON, FRANCE**

*At midday precisely,* I bade my adieus to my wife and children; always a painful moment for me and doubly so on this third occasion, being older and free of those illusions that had been mine on my two previous trips. Nay, in 1837, I was old, weakened by a cruel malady and utterly disenchanted with my calling. So it was that I took my leave of every thing that was dear to me, forsaking my sole source of happiness to embark anew on an arduous and thankless endeavor, with no guarantee of any real reward in the offing. Such were the thoughts that assailed me as I kissed my Adèle for the last time, unable to hold back my tears and cursing my miserable destiny. But it was too late. I had made my bed, and now I must lie in it. With one last glance at the walls of my humble abode, I hastily beat my retreat and joined my ship.

Once aboard my strength of purpose returned and I directed my thoughts solely toward the mission that I had taken upon myself.

**JANUARY 1838**
**FIRST ICEFIELDS, SOUTH OF CAPE HORN**

*At three o'clock on the eighteenth,* the mist was so thick that it completely obscured a mountain of ice within a cable length of our position—an enemy that froze us in our tracks, more to be feared than the glaciers themselves.

The horizon having slightly cleared, I set an eastward course and we moved at a good speed of six knots. Eventually I noticed an ice island to the south some five or six miles distant that had not been spotted by the lookout—an oversight that cost him two stints on lookout duty for failure to warn us of an iceberg clearly visible from above the bridge. Such a harsh punishment, considering the cold, would serve as a salutary reminder to his shipmates to redouble their vigilance.

An hour later a second, longer ice block as high as the first but with steeper sides passed within a mile of our port side. In the afternoon a mixture of rain and sleet made the interior of the vessel unpleasantly damp.

Our calling becomes harder and more miserable by the moment. Quite apart from the ice, we must be constantly on the watch not to lose or crash into the *Zélée*.

At three forty-five on the morning of the twenty-second I am roused from my slumbers by the lookout shouting that our way is blocked. Moments later I am on the bridge and staring at an immense ice field formed from compacted layers of snow and ice that extend as far as the eye can see. The impassable sea ice lies before us, not more than two miles ahead—we are already there.

*No one would have guessed that this was the last line of defense for a crew facing the imminent loss of their ship.*

To have encountered sea ice so soon filled me with dismay, dashing all my hopes and raising the specter of the fatigues and dangers that lay before us, purely to demonstrate beyond any doubt the impossibility of progressing as far as might have been expected.

**FEBRUARY 1838**
**ON THE ICE PACK …**

*Heavy and unremitting snow* concealed the *Zélée* from view, despite her being just six hundred meters distant, and I could no longer see well enough to steer a course safely. We were presently situated in a sort of small interior basin barely two miles wide, surrounded by walls of ice on all sides except for the gap through which we had entered. I was initially for backing the sails or close hauling; but we were drifting so violently that I feared lest we be driven into the ice by the wind, powerless to resist.

We were meanwhile approaching an area of what appeared to be sea ice but visible beyond, about one mile hence, was an immense area of blue, open seawater with just a few blocks of ice here and there. It was the only glimmer of hope on an otherwise frozen horizon. Steeling my resolve but left with no other choice, I determined to force a way through this barrier using the hulls of the ships themselves.

With no more ado, I launched the *Astrolabe* forward under a press of sail, taking advantage of an opening in the ice and using the corvette's beakhead as a battering ram. She advanced over a distance of two or three times her own length then ground to a halt; at which point some of the men descended onto the ice with mooring ropes that they attached to the biggest blocks, leaving it to their shipmates on deck to pull with all their might and inch the boat forward. Other men meanwhile used pikes, clamps and pickaxes to clear a passage around the bow of the ship. It was a tough and exhausting job but our men acquitted themselves with a zeal, diligence and good humor that deserved the highest praise.

**Pages 350-351:**
*Chasse aux phoques, 6 février 1838, parages antarctiques* (seal hunting, 6 February 1838, AntarcticA and its environs).

**Top:**
*Les corvettes sortant de la banquise, 9 février 1838, parages antarctiques* (the refloating of the ice-trapped corvettes, 9 February 1838, Antarctica and its environs).

Right:
Araucanian woman, Concepción, Chile.

Facing-page, top:
Funeral scene, Nuku Hiva, Marquesas Islands.

Facing-page, bottom:
The schooner in Aukena, Mangareva, Gambier Islands.

Pages 356-357:
*Agression des naturels des îles Opoulou, Océanie* (under attack from the Opoulou Islanders, Oceania).

**MAY 1838**
**VALPARAISO, CHILE**

*My heart was pounding* when I received the packet: around eighty to one hundred letters, which I duly distributed to the crew albeit with mounting anxiety at the absence of any thing for me. In the end, however, there remained a note bearing almost illegible handwriting that I recognized to be that of my wife; I broke the seal with a sort of dread that proved all too well-justified. Our young son had passed away a month after my departure and his mother, barely recovered from her unspeakable grief, implored me to hasten to her side before it was too late. She said she could not wait until the end of my journey to put an end to her suffering and that her only remaining son was joining his prayers to her own.

Her letter hit me hard and I could scarcely hold back my tears, cursing the moment when I had agreed to this fateful voyage. However, I recovered my composure and somehow found the courage to steer the *Astrolabe* toward a mooring.

At around two o'clock the pinnaces of the *Ariane* and the two English vessels now also under my command arrived to tow our two corvettes into harbor. Free to grieve at last, I shut myself away in my bed chamber where I spent the rest of the day writing to my wife to express my deepest and heartfelt condolences. Fulfilling this painful duty was some consolation, not least because it left me more determined than ever to resist Adèle's entreaties and continue my campaign come what may, no matter what destiny might have in store for me.

**MAY 1838**
**VALPARAISO, CHILE**

*Two deaths,* half a dozen sick that we had been forced to leave behind, plus two others who had disembarked at their own insistence and most especially seven or eight who had cravenly deserted in Talcahuano: all of this had seriously reduced our crew numbers. The corvette *Ariane* approaching the end of her tour of duty and being due to return to France within three or four months, I therefore impressed upon her captain, Duhaut-Cilly, that he would be doing our mission a great service if he might spare us a few seamen of cheerful disposition.

His associate and a third gentleman were announced as we were finishing breakfast. I greeted them as my distinguished guests and invited them to examine the expedition drawings. They showed great interest in our work, Monsieur Duhaut-Cilly especially admiring the maps we had drawn up as he had hit her to believed our campaign to have been entirely fruitless. At this, I read them the detailed report that I was submitting to the Ministry of the Navy, which they attended to in total silence, apparently much impressed. On the morning of the twenty-eighth, Monsieur Duhaut-Cilly had the whistle blown to invite those members of the *Ariane's* crew interested in joining our circumnavigation to present themselves. Eight of them immediately stepped forward, of whom six were retained by M. Jacquinot, three of them being assigned to the *Zélée* and three to the *Astrolabe*. Thanks to these reinforcements the *Zélée* now has her full complement of men and I am now only missing three crew members.

*M. Duhaut-Cilly said he could not understand how it was that our mission had been so disparaged and the facts so grossly misrepresented.*

**SEPTEMBER 1838**
TAHITI

*The closer we draw* to this delectable island, the more we admire her beautiful beaches and lush hillsides. Point Venus with its tufts of coconut palms projecting out to sea shows me where to drop anchor; this though I know that Matavai Bay has lost its former privileges and that Papeete is now the main port of call for navigators. Papeete is indeed the seat of the island's main establishments, and its harbor is vastly more secure than that of Matavai Bay. But I give preference to Point Venus for the simple reason that our main purpose in visiting Tahiti is to adjust our chronometers in a place with a definite geographic location, and Point Venus is widely reputed to have the best-determined longitude in the whole of Oceania.

*Before they were conquered by the Dutch in 1621, the Banda Islands were populated by a warlike people…*

A canoe manned by natives draws up alongside us. I tell the men that I have no need of pilots, giving them to understand that I am perfectly acquainted with the routes into their harbors and have no intention of paying the dues charged by the missionaries for their services.

I breakfast with Captain du Petit-Thouars, and it is with no little surprise that I learn that the *Voyage Pittoresque Autour du Monde* had arrived in Tahiti.

**Top:**
*Rade de Banda, îles Moluques* (Banda harbor, Moluccas).

**Above:**
*Un intérieur dans le village de Mystern-Cornlis près de Batavia, île Java* (interior scene in the village of Mystern-Cornlis, near Batavia, Java).

**Facing-page:**
*Rade et village de Warrou, île Ceram* (Harbor and village of Warrou, Seram Island).

**Pages 360-361:**
Visit to the Sultan of Solo, Indonesia.

**FEBRUARY 1839**
**BANDA ISLAND [BANDA NEIRA], DUTCH EAST INDIES [INDONESIA]**

*The Banda Islands are the jewel* in the crown of the Dutch East Indies, long renowned for a spice production that belies their small size. The quality of their spices, particularly their nutmeg, is second to none. The Dutch have always zealously guarded their spice monopoly, from which they continue to profit handsomely despite the English polities on the Malay Peninsula now competing for customers with the Dutch in the Moluccas. Nowhere has the nutmeg tree been more intensely cultivated than in the Banda Islands. The colonial government in Batavia that was for many years Europe's sole source of spice was adamant that nutmeg production should remain the exclusive prerogative of these little islands.

The Dutch conquest culminated in a series of wars that saw all these primitives killed or driven out, preferring self-imposed exile to enslavement on the very lands of which they were dispossessed. The Dutch East Indies company then hastened to divide the conquered territory between its partisans, retaining for itself the exclusive right to buy and export the growers' production, at the price set by said company.

**APRIL 1839**
**SERAM, DUTCH EAST INDIES, [INDONESIA]**

*For more than half an hour* I continued along a path that seemed to lead to the mountain, until I reached a hut that looked to be usually occupied. I found it empty, doubtless because the occupants had been scared away by my arrival and were hiding in the woods watching my every move. I noticed some straw mats that were quite well woven, a few China cups and several pieces of Malay fabric. I retraced my steps to take a path that led off from the one I had just followed, warding off the insects that are many and various in these parts. The butterflies are especially striking, not only for their abundance but for their rich and contrasting colors and enormous sizes. Nothing is more majestic than the perfumed forests of New Guinea. Relatively little grows at the base of the trees; their dense leaf canopy prevents the sun from reaching the ground, keeping their roots cool and retaining that soil moisture essential for good growth. However, the entire forest is a tangle of creepers that make walking very difficult, many of them laden with fruit of deceptive appearance.

*...whose love of independence and valiance were their downfall.*

▼ Bottom:
*Rade et ville de Singapour* (harbor and town of Singapore).

► Facing-page, top:
*Observatoire de la baie Raffles à Singapore* (observatory in Raffles Bay, Singapore).

► Facing-page, bottom:
Two birds observed by the naturalists.

► Pages 364-365:
Fish and birds brought back to France by the naturalists.

**MAY 1839**
MAKASSAR, DUTCH EAST INDIES, [INDONESIA]

*"The kris with straight,* curved blade or wavy blade adorns the belt of all the inhabitants," writes Monsieur Roquemaurel. "Though excellent for an assassin, these weapons are of no use to the honest man who refuses to attack his enemy from behind. The hilt is so badly conceived as to offer no protection to the hand, retaining the blade so loosely that it is ever at risk of separating. Be that as it may, the very existence of this weapon suggests a certain understanding of ironwork on the part of the makers. The blades are thirty-one to thirty-nine inches long (eight to ten decimeters) and made of wrought iron forged from layers of pattern-welded steel. The result is a Damascus-like finish of such finesse that there can be no denying the skills of the artisan. Kris daggers are the most widely produced items in Makassar, and so common that the natives will usually part with theirs for ten or fifteen francs, though one hears of some who demand a much higher price, not so much because of the object itself as for the services it has rendered to its owner."

**JULY 1839**
SINGAPORE

*The plantations* here are often ravaged by tigers that swim across from the mainland and are said to be present in abundance in Singapore. A few days before our arrival, a Malay had been devoured by one of these ferocious animals, which the locals only seek and destroy when they prey upon a family member. If Singapore were only occupied by Malays, tigers would still be laying waste to the island; but fortunately the Chinese have other ideas. Self-interest prevailing over all else, the Chinese go tiger hunting. Tiger flesh is much appreciated in the Celestial Empire and said to possess wondrous properties that make it a favorite with gourmets. The skin is considered a precious ornament and the fat, entrails and especially the brain play a major role in the Chinese pharmacopeia. One hears of a Chinese man who made his fortune with just one of these animals, which having destroyed he then sold off piece by piece, earning himself the princely sum of 350 piasters (nearly 2,000 francs). This peculiar taste of the Chinese will very probably save the English from having to offer a reward for killing tigers—just as land clearance will probably put an end to the tigers' incursions and save the peaceful Singaporeans from having to drive them out.

*People in China believe that eating tiger flesh gives them the muscular strength and courage of the animal itself ...*

*... hence the meat being preserved and fetching a high price on the Canton market.*

*Bevalet pinx.* *Schmelz sc.*

1 MÉROU MILLE-ÉTOILES. (NOUV.-GUINÉE)

2 MÉROU MOUCHETÉ. (NOUV.-IRLANDE)

3 MÉROU GAIMARD (NOUVELLE-GUINÉE)

4 MÉROU BOELANG. (DÉTROIT DE LA SONDE)

J. Tastu édit. Dumenil imp.

Peint par A. Prévost. Dirigé par Borromée. Gravé par A. Duménil.

1. CACATOIS DE DUCORPS. (Nob.) Mâle. 2. CACATOIS A HUPPE ORANGÉE. (Nob.)

Gide Editeur. Impr. de Bougeard

**NOVEMBER 1839**
**AT SEA, SAILING FROM SUMATRA TO TASMANIA**

### *Ship's surgeon*

Monsieur Leguillou arrived aboard the pinnace of the *Zélée* to report in detail on the health of her crew: two men were dead, two others were gravely ill and twelve were bed-ridden; nobody could say how the disease would end.

I straightway resolved to go aboard the *Zélée* and inspect the men for myself. I found things to be no worse and if any thing better than aboard the *Astrolabe*, the *Zélée* counting fifty-seven able bodied men fit for duty, compared to the *Astrolabe's* fifty-six. I ended my visit with a few words of comfort for the officers, chiefly in the hope of reviving their courage and enthusiasm.

The winds continuing to blow from the east, it was our best course to pursue our route southward until such time as we would encounter those westerlies whose strong swells we had been feeling for so long. Though the sickness appeared at a standstill, several of the men were dangerously unwell. The *Astrolabe's* Roquemaurel, Demas, Dumoulin and Desgraz were particularly affected, the first two being released from duty without delay.

November 23 was one of the worst days of the entire trip. In the morning, the *Zélée* lost two of her finest sailors: De Lorme, who had borne his suffering with great resignation for fifty days, and the courageous and dutiful Fabry, one of the ship's best-loved men. As one man after another succumbed to his afflictions, the corvette left a trail of corpses in her wake until our arrival in Hobart-Town on December 12.

**DECEMBER 1839**
**HOBART, TASMANIA, NEW HOLLAND**

### *I welcomed aboard*

Sir John Franklin and his family, whose kindness toward us was unfailing throughout our stay in Hobart. The Governor took pleasure in reviewing the mission's achievements thus far, pausing only to express his satisfaction with our efforts. He was accorded a seventeen-gun salute as he disembarked from the *Astrolabe*. I accompanied him ashore to take him to the observatory of the expedition hydrographer, Monsieur Dumoulin, whose instruments were naturally of interest to a fellow explorer like Sir John who had contributed so much to science himself. Hence the many questions he put to Monsieur Dumoulin as to the construction and use of his instruments.

Crossing the Antarctic Circle aboard the *Astrolabe*, 19 January, 1840.

Our preparations for departure being almost complete, I was anxious to set sail without delay and it was decided to cast off on the morrow. We spent our last day in port settling our accounts and bidding our adieus. In the evening we attended a ball at the Governor's residence—a sparkling affair held by Lady Franklin in honor of *les officiers français* but also the occasion to celebrate the New Year of 1840 with much merriment and rejoicing (as I said earlier, the English of Hobart were a day ahead of us in calendar time.)

### JANUARY 1840
### IN ANTARCTICA

*We had reached* the sixty-fourth parallel south. The weather was mild and humid and there was every hope that we would shortly pass beyond the seventieth parallel. At midnight however, we suddenly found ourselves surrounded by five huge blocks of ice shaped like tables.

At six o'clock in the morning there were six ice islands floating around us. At eight o'clock, there were ten times that many. They were all identically shaped, flat with sheer sides, and generally bigger than those we had already encountered. They varied in height from thirty to forty meters, with a width that in many cases exceeded one thousand meters. They were all identical-looking, resembling those we had seen close to land on our first polar expedition.

We were so hemmed in by these floating masses that I feared lest our corvettes should at any moment crash into them. But squeezing past their base was really the only way to gauge their height. It was an awesome sight—an impression somewhat akin to wandering through a city of giants. At the feet of these frozen monoliths were immense caves carved out by the sea, into which the waves hurled themselves with a roar as if to defy the sun that darted its slanting rays onto the crystalline walls towering above us. The contrast of light and shadow was spellbinding. From the top of these mountains myriad streams cascaded into the sea, fed by what seemed to be actively melting snow. Eventually we came to a vast basin formed by the land on one side and on the other side by the chain of floating islands we had just passed. At midday we were within no more than three or four miles of a fresh discovery. Peering through our binoculars, we scoured the horizon in search of this mysterious land, whose existence no longer seemed in any doubt.

### JANUARY 1840
### TERRE ADÉLIE [ADÉLIE LAND]

*At last, just when* we were beginning to lose hope, we sighted several tiny islands whose snowless strands exhibited that much desired blackish soil. Moments later we watched the full-rigged pinnace detach itself from the *Astrolabe* and make for the shore carrying an officer and two naturalists.

Here is the account logged by Monsieur Dubouzet: "It was nearly nine o'clock when to our great joy we made land on the western part of the easternmost and highest island. The pinnace of the *Astrolabe* had arrived shortly before us and the men on board had already climbed to the top of this steep-sided rock. Following the long-established English tradition, we claimed possession of the island in the name of France, together with the neighboring hillside that the ice prevented us from approaching. Our enthusiasm and joy at this point knew no bounds. To think that this peaceful conquest had just added a new province to France! For while such acts of possession are so overused as to have become worthless, in this case at least we felt entirely justified in making our claim on behalf of France. We were dispossessing nobody and clearly within our rights. From this point forward, we considered ourselves to be on French soil, in a land that, if nothing else, would never wage war against our homeland."

This excursion left no doubt as to the reality of our discovery, which it now only remained for me to make known as widely as possible. While doubt had persisted, I had resisted giving a name to this new land but on the return of our pinnaces I named it Terre Adélie. The most prominent cape that we had sighted in the morning, received the name of Cap de la Découverte [Cape Discovery]. The point near the spot where our vessels had made land and collected the geological samples was named Pointe Géologie [Point Geology].

### SEPTEMBER 1840
### ST HELENA ISLAND

*We learned* that any day now a division under the command of Prince de Joinville would be arriving to collect the remains of the Emperor who was buried on St Helena Island. The English had lately sent orders aimed at speeding Napoleon's exhumation, such that it was just a matter of days before he could finally depart this island—a place so often visited by travelers with a fascination for great events and the scenes where they took place.

Facing-pages:
*Reconnaissance de la terre Adélie, 20 Janvier, 1840* (reconnoitering Terre Adélie, 20 January, 1840).

Pages 370-371:
*Débarquement sur la terre Adélie, 21 janvier, 1840* (landing on Terre Adélie, 21 January, 1840).

*I straightway sent one of our sailors to raise the tricolor flag, on this land unseen and untrodden by any human being before us.*

I was aware that our sailors wished to visit the emperor's last residence and tomb, an understandable request that I hastened to anticipate by granting the crews leave ... to go ashore. So it was that on the afternoon of the eighth, despite the unsettled weather, the sailors of the *Astrolabe* and the *Zélée* could be seen making their way from the town across the surrounding mountainsides toward Longwood House. It is to their credit that they did not take advantage of their special leave to visit the town's great many cabarets, conducting themselves in an orderly fashion throughout their time ashore and giving me no reason whatsoever to issue reprimands.

I was shown the funeral carriage that was to transport Napoleon's remains to the shore, where his coffin would be placed in the charge of Prince de Joinville.

**NOVEMBER 1840**
**TOULON, FRANCE**

*On November 6 at ten o'clock* at night we dropped anchor for the last time in Toulon harbor, in the midst of a squadron under the command of Admiral Julien Pierre Anne Lalande. It was thirty-eight months to the day since our two corvettes had departed the port of Toulon to undertake their exploration campaign.

At six o'clock on the morning of the morrow we were cleared to enter. To our delight, Monsieur Lafond, who had remained a serious concern since his departure in Batavia, came rushing toward the *Astrolabe*. He took charge of all the on-board formalities, leaving us free to go ashore, to be reunited with our families and find out every thing that had happened in our absence. For my part, it was high time I was home. My sufferings were much worse and it was only at the close of two months' constant treatment that I could finally repair to Paris as ordered by the Ministry of the Navy. I immediately set about compiling the material gathered in the course of the campaign.

GASTON DE ROQUEMAUREL WAS THE LAST
OF THE MOHICANS AND HIS SHIP, THE *CAPRICIEUSE*,
WAS THE LAST FRENCH THREE-MASTER
TO CIRCUMNAVIGATE THE GLOBE.
IN HER WAKE CAME STEAM POWER TO POLLUTE
THE OCEANS. ROQUEMAUREL, FOR ALL HIS NOSTALGIC
LONGING, COULD BE UNDER NO ILLUSIONS
ABOUT WHAT WAS REALLY HAPPENING IN THE WORLD:
THOSE ONCE "NOBLE SAVAGES" OF OCEANIA
HAD ALL BEEN CORRUPTED BY THE WEST,
AND CONFLICT WITH CHINA WAS INEVITABLE.
CONTINUING THE WORK OF LAPÉROUSE,
HE MAPPED THE COASTS OF JAPAN AND KAMCHATKA
AND RETURNED HOME JUST AS FRANCE WAS
ABOUT TO DECLARE WAR ON RUSSIA.

# Gaston de Roquemaurel

**1804–1878**

Plate 30, George Wolfgang Knorr, 1766.

# VOYAGE OF THE *CAPRICIEUSE*

**1850–1854**

**Roquemaurel was fifty when he returned to Toulon from his expedition. It was the end of the great scientific explorations and despite his eminent talents he was forgotten, or almost. Fortunately, there still remain traces of his passage through Oceania, China and the Korean Sea ...**

Roquemaurel was born in Toulouse on September 20, 1804, in the year of Napoleon Bonaparte's coronation. His father died when he was still very young, and his education was entrusted to a tutor. His uncle, Baron de Puymaurin, a minister under King Louis XVIII, then took him in hand and enrolled him at the Collège in Auch, where he quickly made up for lost time thanks to his intellectual abilities, particularly his aptitude for science. From there he entered the Collège Royal in Toulouse, followed by the École Polytechnique in 1823. He already knew that he wanted to be a sailor but France at the time had no naval school as such, just a naval college in Angoulême. Fortunately, six places in the navy were reserved for graduate students of the Ecole Polytechnique, one of whom was Roquemaurel when he qualified from the school in 1825. He was all set to fulfil his dream and pursue a career as a naval officer.

For the first twelve years of his service, his horizons were strictly Mediterranean. In the course of his training he became fascinated by Freycinet, Duperrey and Bougainville—those men who had set off around the world and returned with glowing reputations. His own debut came in 1830 on the Algiers expedition, under the command of Captain Jacquinot, future commander of the *Zélée* on Dumont d'Urville's expedition. Promoted to lieutenant in 1834, Roquemaurel continued to sail the Mediterranean, and in 1837 tried his luck with Laplace to join the *Artémise*, but was unsuccessful. Jacquinot then introduced him to Dumont d'Urville, and Roquemaurel became his first mate on the *Astrolabe*, setting off to discover Antarctica.

Living together on the corvette was no easy task for two men with such radically different personalities. On the one hand, the cold and austere Roquemaurel, thirty-three years old and just setting out on his career. And on the other hand, the altogether sunnier Dumont d'Urville, forty-eight and at the height of his career despite describing himself as old and ill. In his account of the voyage, Dumont d'Urville says very little about his second-in-command, unlike ship's surgeon Élie Le Guillou who fortunately for us says rather more. Yet it was Roquemaurel's skill that saved the *Astrolabe* from breaking up on three occasions when Captain Dumont d'Urville, weakened by illness and gout attacks, was no longer in shape to command. When passing through the Polar ice for the first time, it was Roquemaurel who prevented the two ships from becoming stranded. The second occasion was on the approach to Terre Adélie when he saved the corvette from being driven into the pack ice and icebergs by strong winds. The third time was when he refloated the *Astrolabe* in the Torres Strait, between Australia and New Guinea. Dumont d'Urville congratulated him, of course, but never publicly acknowledged his exploits. The officers, on the other hand, wrote to the Minister of the Navy, declaring that their awards would be worthless in their eyes if the most deserving among them were not treated as his services demanded. And with that Roquemaurel was belatedly awarded the Légion d'Honneur on his return to France in 1841. He was also promoted to frigate captain.

But decorations were the last thing on Roquemaurel's mind. What he wanted, to quote the man himself, was, "to see to it that through my efforts I am entrusted with the command of a long-haul mission. "A request that went unheeded. By 1843, interest in around-the-world exploration had waned. What could there possibly be left to discover? Roquemaurel would have to wait another seven years before achieving his aims. Thanks to a meeting with François Arago, one of his former teachers at the École Polytechnique and the new Minister of the Navy following the revolution of 1848 that established the Second Republic, he joined the Ministry as Chief of Staff. But he soon became bored and, though appointed Captain, he resigned in 1849. A year later, his luck changed. Rear-Admiral Joseph-Romain Desfossés, the new Minister of the Navy, obtained the permission and budget for a new round-the-world expedition. Roquemaurel was the man for the job. He had the experience, he had the drive and he had the ambition. And on May 28, 1850, at the age of forty-five, he departed Toulon as the newly appointed captain of the corvette *Capricieuse*.

For four years, he travelled the world under the colors of France in South America, Oceania and China, where he stayed for almost two years, then along the coast of Korea and Japan. By the time of his return on March 15, 1854, he had become disenchanted with the native way of life and deplored the corrupting influence of Western civilization. But he had accomplished his mission perfectly, having been able to complete the maps drawn up by Lapérouse for the coasts of Japan, Korea and Kamchatka. In terms of hydrography, ethnology, astronomy, science and geography therefore, his mission was a great success. His collections, which he donated to the city of Toulouse (France) formed the basis of the Museum's anthropology department, and are certainly the largest oceanic ethnographic collection since the destruction of Dumont d'Urville's collections in Caen in 1944. Roquemaurel had most certainly distinguished himself as a French sailor yet his work earned him no honor, glory or promotion. To be fair, times had changed in his absence. French political life had been through a rocky patch; there had been five different Ministers of the Navy one after the other; and in 1852 the Empire had made way for the Republic. Now France and England were preparing to fight in the Crimea; and in the press, all the talk was of war with Russia. The world of navigation was likewise turned on its head as coal-fired engines arrived to replace sails ...

In the years that followed his return, Roquemaurel found none of his missions interesting and left the navy in 1862 to retire to Toulouse, where he died on April 1, 1878, six years after Jules Verne published *Around the World in Eighty Days*. Although he wrote a richly descriptive journal aboard the *Astrolabe*, Roquemaurel never published an account of his voyage, nor did he produce a historical or geographical atlas. He did however leave some very moving letters, together with a few reports and anecdotes. But we have no day-by-day account of his voyage aboard the *Capricieuse*. To illustrate his adventures and pay him the tribute he deserves, we have therefore drawn on the following sources: the in-depth work of Jean-Philippe Zanco, historian and biographer of Gaston de Roquemaurel, based on documents handed over by members of his family, in particular by Jacques de Roquemaurel; also the correspondence of his young aide de camp, the future vice-admiral Abel Bergasse Du Petit-Thouars, published in 1906; and of course, Roquemaurel's own logbook.

► L'Astrolabe *arrêtée par un glaçon avant sa sortie de la banquise, 9 Février, 1838* (the ice-trapped *Astrolabe* before its refloating on February 9, 1838).

**GENERAL REFERENCE MAP**
For the Voyage of the *CAPRICIEUSE*
Led by Captain Gaston de Roquemaurel
1850–1851–1852–1853–1854

150
120
90
60
30
0
Western Longitude, Paris Meridian
Paris Meridian
Melville Is.
Lancaster Strait
Baffin
Bay
Davis Strait
GREENLAND
Iceland
Bering Strait
Cook's River
Port des Français
Lituya Bay
Kodiak Is.
Alaska
Sitka
Queen Charlotte's Is.
Vancouver Is.
Nootka
eutian Islands
Hudson Strait
Hudson
Bay
LABRADOR
CANADA
Cape Farewell
NORTHERN
BRITISH ISLES
SWEDEN
Baltic Sea
London
Paris
Brest
FRANCE
May 1850
Toulon
ITALY
SPAIN
Madrid
MEDITERRANEAN SEA
Strait of Gibraltar
BARBARY
MOROCCO
Madeira Is.
Canaries Is.
Newfoundland
St-Lawrence River
Quebec
Acadia
NORTH
AMERICA
UNITED STATES
New York
Washington
ORTHERN
OCEAN
ATLANTIC
Azores Is.
Bermuda Is.
OCEAN
San Francisco
Monterey
California
San Antonio
New Orleans
MEXICO
Gulf
of Mexico
Florida
Lucayan Is.
Mazatlán
San José
Mexico City
Acapulco
Santo Domingo
Caribbean Sea
Jamaica
Guadeloupe
Martinique
Tropic of Cancer
Sandwich Is.
Hawaii Is.
Cape Verde
SENEGAL
AFRICA
GUINEA
Panama
COLOMBIA
Cayenne
GUYANA
Equinoctial Line
Gallapagos Is.
Paita
PERU
Lima
Pisco
Arica
SOUTH
AMERICA
BRAZIL
Bahia
Ascension Is.
CONGO
St Helena Is.
Marquesas de Mendoza
Dec. 1850
avigators Is.
Maouna Is.
Tahiti
Society Is.
Friendly Is.
Traitors Is.
Mangareva
Pitcairn Is.
Tropic of Capricorn
Easter Is.
Rio Janeiro
C. Frio
St Catherine's Is.
SOUTH
Coquimbo
CHILE
Valparaiso
Oct. 1850
Concepción
Montevideo
Buenos Aires
Rio de la Plata
July 1850
Cape of Good Hope
ATLANTIC
atham Is.
OCEAN
PATAGONIA
Falkland Is.
OCEAN
Strait of Magellan
Tierra del Fuego
Is. de Los Estados
Cape Horn
Sandwich Is.
Antarctic Circle

**In 1854, the *Capricieuse*, a 141 foot [43 meter], 32 canon sailing corvette, dropped anchor in Toulon harbor following a three-year campaign that had taken it across the Pacific and into the unknown reaches of the China Sea, skirting a hostile Japan, sailing in the tracks of the great French explorers of the eighteenth and nineteenth centuries. "Officially," however, France's great voyages of exploration in the age of sail had ended with the second of Dumont d'Urville's expeditions. The journey of the *Capricieuse* is no less the last circumnavigation of the globe by a French sailing ship on a hydrographic and scientific expedition, the swan song of that great tradition of the Pacific exploration that had begun with Lapérouse and nominally ended with Dumont d'Urville. Having cast off in July 1849, the *Capricieuse* was already obsolete on her return from the Far East. After a second and final journey to Canada (that would make her famous across the Atlantic) she was decommissioned in 1861. In the three years that she had been away in the Far East, the French Republic had given way to the Empire and war had given way to peace. And with so much else going on, everybody had forgotten about the *Capricieuse*. Her return to the port of Toulon took place amid general indifference and the puffing smokestacks of steamers making ready to cast off for the Crimea. The fall into obscurity of the *Capricieuse* and her captain (and project initiator) must therefore be seen in the particular historical, technological and political context in which it occurred.**

*THE FORGOTTEN FINAL VOYAGE OF THE AGE OF SAIL*
JEAN-PHILIPPE ZANCO,
DOCTOR OF LAW AND HISTORIAN

**MAY 1850**
**TOULON, FRANCE**

*Gaston de Roquemaurel* had three objectives in mind when he departed Toulon on May 28, 1850 aboard the *Capricieuse*, even though his primary mission was to take command of the naval base in the China Seas. Firstly, a geographic and hydrographic objective: "Having conducted a general reconnaissance under sail, the safety of navigation required more detailed work to allow sailors to approach the shore and cross the barrier reefs surrounding most of the islands of Oceania, without coming to harm." Secondly, an ethnographic and naturalistic objective: "The natives in these lands have barely glimpsed the first dawning of civilization that is happening to the east of Oceania. The study of primitive habits may well add an interesting page to the history of mankind." Thirdly a scientific objective: "The determination of a purely celestial departure point would have the great advantage of providing geographers with new milestones in this vast sea where the calculation of longitude remains subject to all the whims of timepieces."

**JULY 1850**
**CAPE HORN**

*We could clearly see five natives* grouped around a fire, seemingly too busy fanning its flames to pay much attention to our maneuvering; now and then one of them would stand up and walk toward us, gesticulating as he went, or simply remain standing and look out to sea at the ships. The men appeared to be dressed in red skins similar in color to the coat of the guanaco. They wore this garment over their shoulders in the manner of a cloak, unfitted and unfastened. One of them had on a kind of knee-length jupon or baggy breeches of a pale color.

Hearing weeping, we entered a tent where we found an old woman wrapped in skins and obviously unwell. Her companions signaled us to approach on tiptoe, whereupon they uncovered her and gave us to understand that she was dying. This they did by blowing on their fingers then raising them above their heads, as if to express the old woman's dying gasp or her soul taking flight to Heaven. Just then, however, the patient sat up and complained about feeling a weight on her chest and her airways being blocked. Lifting an emaciated arm, she indicated with the other hand that bloodletting was the only remedy for her ills, an embarrassing request that we were forced to refuse having brought with us only knives and cheap scissors to trade. But after taking the old woman's pulse and inspecting her tongue we did at least manage to convey that there was still room for hope. Some hard tack distributed to the whole family was devoured in a trice, considerably raising the spirits in this Kawashkar tent.

*On the coast of Patagonia, we enjoyed four or five days of friendly intercourse with the natives who we used to regard as fierce giants.*

In accordance with our principles we visited the Patagonians in their tents, and we shared their meals of shellfish, roots and sometimes pieces of guanaco and ostrich when the hunting had been fruitful. We warmed ourselves by their fires and we slept on the same animal skins, in a confused jumble of women, children and snarling dogs.

**NOVEMBER 1850**
**GAMBIER ISLANDS, AKA-MAROU [AKAMARU], FRENCH POLYNESIA**

*Unable to land* because they had lost their canoes, the natives were burning with impatience to return home. But so gentle is their conduct that we had nothing but praise for these good people. Their guileless, naïve countenance and innocent jollity enchanted every one of us. What is more, having joined our religion and come to respect and cherish the name of France, they most certainly could no longer be called uncivilized. One of them indeed was even kind enough to allow us to make a mold of his head, a devotion to science that earned him a handkerchief as a reward.

Each of the natives received a sailor's ration, which they appeared to appreciate quite as much as their own, exclusively vegetable, diet, but which they only brought to their mouths having crossed themselves, with all the zeal typical of the new convert. Having seen no religious practices observed on board, they likely doubted that we belonged to the same family as had given them the missionaries and repeatedly asked us if we were Christian. When we replied in the affirmative, our tan friends seemed overjoyed, saying that they too were Christian and showing us the medallions and amulets hanging around their necks or reciting the paternoster in very garbled Latin.

Under the direction of three French missionaries, these natives have been transformed from the fierce cannibals that they once were into good and hospitable Catholics. But it will not be long before the fundamental innocence of these childlike people becomes sullied through impure association with all those nations that sperm-whale hunting and pearling attract to these shores.

Top: *Campement de Patagons au havre Peckett, Détroit de Magellan* (Patagonian camp, Peckett Bay, Strait of Magellan).

Pages 382-383: *Environs de Samboangan, Île Mindanao* (outskirts of Zamboanga, Mindanao Island).

*Women are fined when they are caught in such infamous trade but the income from those fines is good revenue for the missionaries.*

**JANUARY 1851**
MARQUESAS ISLANDS, FRENCH POLYNESIA

*These Oceanian beauties* are very far from attaching any idea of modesty or chastity to their dealings with men. I even doubt that these words have any equivalent in the language of the islanders. As mistresses of their own bodies, it is of no consequence to these girls whether they unite themselves with a man through love, need or self-interest. They can offer themselves to whoever takes their fancy whenever and wherever they wish without any one being entitled to take offence. Only when a girl sells her sexual favors to a stranger can her parents intervene—to ask for their share of the profits. Such are the habits of this land of damnation where women are only esteemed for the paltry revenue they fetch as girls and the passions they arouse and quell by turn as women. Hence their behaving as if they had been created and put into this world purely for the purpose of sin, with none of that delicacy shown by our European womenfolk. Far from merely making eyes at a man, these women get straight to the point, often presenting themselves to you bare-bosomed as if to say: "Here I am! Take me! I am young and nubile, feel this breast and let me show you what I can do!" And so saying they proceed to move and thrust in such a way as to convince you of their being perfectly schooled in such matters.

**FEBRUARY 1851**
TAHITI, FRENCH POLYNESIA

*We once again crossed* the Archipel Dangereux [Tuamotu] and dropped anchor off the island of Tahiti, which is no longer that queen of Oceania described in such glowing terms by Bougainville and Cook. Its population is considerably reduced and so indoctrinated by the Anglican missionaries that these once innocent people are now devious, mercantile, mendacious and false. The same licentiousness prevails here as in Nuku Hiva, with hypocrisy thrown in for good measure.

Tahiti is no longer what she was in the days of your Wallis, Bougainville and Cook. To see these ever-lush shores watered by cool clear streams or these deep ravines covered in dense forest all the way to the top, you might think you were back in the Tahiti of yore. But one look at these dirty, ragged people is enough to disabuse any traveler. Gone is their softness, naivety and inborn virtue. In their place, we find only cunning, drunkenness and prostitution. Gone are all those little crafts handed down by their ancestors that kept them in gainful activity. The building and maintenance of large war canoes, like the construction of *morais* (funeral monuments), and those thousand little chores they used to perform for their chiefs, all are forgotten today. These days Tahitian naval art seems to have regressed, consisting solely of small dugout canoes made from hollowed-out tree trunks. The bark cloth less prized by the natives than our European fabrics now provides work for very few hands, and the *morais* have made way for modest Protestant graves. Once again, the missionaries have failed in their duty and no thing they have done over the past years can possibly atone for their having usurped and then exploited temporal power to serve their own mean and selfish ends.

**MARCH 1851**
**CAROLINA ISLANDS, MICRONESIA**

*The main island here occupies* the easternmost part of the archipelago, while the western part is composed of a continuous reef system. A fine row of large huts makes up the village that overlooks the sea on the western side of the main island. Needless to say, all of these islands are covered in fruit trees. The natives sailed out to sea for nearly three miles to greet us, and far from showing any fear as they came alongside they showed every sign of being delighted to see us. They babbled endlessly, roared with laughter and in short seemed quite overjoyed by our presence. These good people had sailed out over nearly one league to bring us a few coconuts and a small quantity of fish, battling a cold head wind and heavy seas all the way. Several of them indeed had brought nothing to trade and contented themselves with flitting around the corvettes to take a good look.

These were men of the brown-skinned or tan type. From what I could see, they were of medium size and well-proportioned, though many of them appeared to have rather feminine bodies. Their faces, albeit somewhat flat, are not devoid of charm. Their noses are broad and not very prominent, their foreheads are well-developed but receding and their teeth are the finest you could wish to see, so they have no use for betel leaves, though I did notice several chewing a plant unknown to me. Their countenance is soft and full of grace.

Facing-page:
*Maison de l'évêque à Ao-Kena, Îles Manga Reva* (house of the bishop of Aukena, Mangareva Islands).

Bottom:
Native huts, Houka Hiva, Marquesas Islands.

▸ **Right:**
Members of the Royal Family, Papeete, 1850.

▾ **Bottom:**
Mangareva islanders, 1850.

▸ **Facing-page:**
Seashells collected for the Natural History Museum in Paris.

*In Canton, strangers are murdered in the two streets reserved for trade and in Hong Kong no-one is safe within a mile of the city.*

Gaston de Roquemaurel

*Though France has spent vastly in China, she engages in no trade there as our interests in that country ...*

**APRIL 1851**

**HONG KONG, CHINA, BRITISH TERRITORY**

*It is impossible to describe* our astonishment on sighting this city, which dates from barely ten years ago yet already covers as great an expanse as Valparaiso, which it resembles in every respect. There is no continuous place for mooring, only magnificent wharfs here and there. A wide avenue follows the shoreline, on the same side as the goods warehouses, with the city rising up like an amphitheater on the other side.

War is bound to break out between the English and the Chinese when the treaty expires next year; and if truth be told, China's position is untenable. Her trade has ground to a halt and the new emperor has just enacted a decree forbidding the smoking of opium on pain of severe punishment. I do not doubt that what the commentators say is true and that when the treaty ends all the ports will be wholly closed. The English are preparing themselves for what lies ahead and enduring countless humiliations for the sake of holding on until then. One sees them being thrashed by the Chinese in the streets of Hong Kong, which are infested with thieves ready to fleece you at any

*... are strictly political and moral and quite distinct from those of other nations.*

moment. Two of our own officers were among their victims. I believe that the establishment of a French naval base has never been more important than now, when the Americans are reinforcing theirs and hope to obtain a concession granting them a berth in Hong Kong for their steamships.

## FEBRUARY 1852
### SEMARANG, DUTCH EAST INDIES, INDONESIA

*The town of Semarang,* like Batavia, [Jakarta] is crossed by a partly canalized river whose mouth is clogged with sludge and no more than two feet deep at low tide. Here, as elsewhere on the Javanese coast, there is but one high tide and one low tide within each twenty-four hours.

The countryside around Semarang is neatly planted and produces an abundance of rice and cotton, together with some sugar cane and coffee. This part of the island was conquered by the Dutch in 1708, since when they have enjoyed the possession of all this flat country on the northern coast of Java, from the Sunda Strait to the Bali Strait.

The officers of the Semarang garrison welcomed us most cordially and plied us with all the pleasures that this charming residence can provide. Local society threw open its doors to all the expedition officers and our entire stay indeed was one long succession of horseback and carriage rides, dinners, soirees, and bucolic delights. The port captain, Van de Velde, the captain of the infantry, Van Exter, and Lieutenant Boon, who had already given us such a warm reception in Ternate, were attentive to our every need and could not have been more friendly.

It was here that we encountered Monsieur Boll who had left the Amboine residence for that of Semarang and who celebrated our arrival by treating us to a broiled nautilus: a very rare mollusk whose shell is all that France has ever seen. But even more precious to us was to rediscover that bonhomie and generous hospitality of the Dutch, whose kindliness in Semarang was in sharp contrast to the chilly reception they gave us Batavia.

**Facing-page, top:**
*Village d'Umata, Île Gouacham*
(Umatac village, Guam island).

**Facing-page, bottom:**
*Aotahi, chinois habitant Makassar, Îles Célèbes*
(Aotahi, a Chinese inhabitant of Makassar, Celebes Islands).

**Top:**
*Site sur les bords du canal de Samarang, Île Java*
(site on the borders of the canal of Semarang, island of Java).

**Pages 390-391:**
*Messe célébrée par Monseigneur l'évêque de Nicolopolis à Manga-Réva* (mass celebrated by My Lord Bishop of Nicolopolis, Mangareva).

**JUNE 1852**
**SHANGHAI, CHINA**

*The Jesuits here operate* a very successful establishment, having some sixty young Chinese men under their tutelage who all look as interesting as the rest of the Chinese do not. Several have already completed their philosophy studies and are now embarking on theology. Most of them are destined for a secular life and will sit the bachelor's degree examination, which is extremely hard in China. I have no doubt that they will pass, for they are full of good will and their teachers are first-rate. What a splendid way for these Jesuits to make new converts! They no longer have room to accommodate new applicants so their first concern on completing their other works will be to expand their premises to provide room for more students. They quickly learn to speak Latin among themselves, a common language that all Europeans can understand and to which they apply themselves with great success.

The Jesuits have a great mix of people in their congregation: masons, architects, engineers, astronomers and many more. They are currently building a beautiful church and cathedral under the supervision of the brother who drew up the plans and then became a master mason himself to train the Chinese. The churches are decorated by a brother who crafts impressive wooden sculptures that would draw admiration anywhere in the world.

**JULY 1852**
**SEA OF KOREA**

*We were en route* for the Sea of Japan to follow in the tracks of Lapérouse and review his work in the Strait of Tartary. We passed within sight of Jeju, South Korea, one thousand nine hundred meters high [six thousand, two hundred feet], with no mooring. On the morrow we sighted the very elevated and jagged eastern shoreline of Korea that drops down in the north to form Cap Clonard, as reconnoitered and perfectly identified by Lapérouse.

The weather remained glorious all day, allowing us to distinguish several groups of scattered huts and two or three boats, together with mountains clad in dark green vegetation that extended all the way to the top to most delightful effect. The corvette was meanwhile surrounded by a great number of whales that sent up huge shoals of big fish. The horizon was shrouded in the plumes of mist spouted by countless cetaceans; ahead was the unknown coast of Korea, its high peaks swathed in cloud; before us those unchartered waters said to be so dangerous. Every thing about this place was so grandiose and so attractive that every one remained on bridge until nightfall.

I had never felt such hope and enthusiasm. We were finally about to see a totally new country and no doubt render the justice due to our illustrious compatriot. There was even a chance of following him beyond to make a discovery of our own. On July 28 at five o'clock in the morning, we found ourselves within sight of the Ile Dagelet, which had been erased from our maps on the recommendation of the Englishman Broughton who claimed it lay farther away. You can imagine how pleased we were to discover it precisely where the French expedition had placed it.

Having thus identified two of Lapérouse's discoveries, located to the nearest inch, the captain decided to return to the coast of Korea to review the somewhat questionable findings of the aforesaid Broughton, the only man ever to have made a survey of the stretch of coastline between Cap Clonard and the mouth of the strait of Tartary.

Top:
*Mandarin of Quelpart, Corea* (Mandarin of Quelpart, Korea).

Facing-page:
*Rade de Singapour prise de la maison du gouverneur* (Singapore harbor seen from the Governor's residence).

*Men must summon all their ingenuity to rise to the occasion; the more the obstacles, the greater the glory in overcoming them.*

**AUGUST 1852**
**SEA OF KOREA**

*At first glance,* this land has little to offer navigators in the way of supplies. Apart from an abundance of water, wood is scarce, vegetables are few and lack variety, and what fishing and hunting exists is hardly worth the effort. Such however were our only sources of victualing for the duration of our stay. But what a beautiful country this is! The mountains are cloaked in lush greenery all the way to the crest and where ever we looked we found signs of vegetation not unlike that of our own climes. I do not doubt that in exchange for appropriate goods, the Tartars would have provided us with victuals in abundance. For pretty soon the *Capricieuse* was entirely surrounded by their canoes, some fifty in all made from hollowed-out tree trunks and propelled by two or four oars, the Tartars meanwhile observing our every maneuver with a wide-eyed astonishment such as I had only ever noticed among the childlike people of Oceania.

It was not long before we had become acquainted with all of the Tartar tribes scattered around the gulf, whose only source of livelihood appears to be wrack gathering.

**APRIL 1853**
**CANTON [GUANGZHOU], CHINA**

*The American consul in Canton* is Mr. Forbes, a rich merchant who also commands the French Vice Consulate. Meaning no offense and wishing to display our flag, he had it flown alongside his own on a yardarm slung from the mast. In the meantime, the minister-resident of Alphonse de Bourboulon on his way to Canton saw our flag in this position and told Mr. Forbes that he preferred not to see it at all than to see it take second place to another. At which our vice-consul, ever a real gentleman, had our flag raised to the tip of the mast and his own flag placed below it as a gesture of his good will.
Had this been any where other than China, the matter would have rested there. As it was the English told the Chinese, who are ever eager to meddle in what goes on between Europeans, that the addition of our flag on the American mast proved that France had just been annexed by the United States. The English had already told the Chinese that France was a secondary power, whose flag stood for no thing as her ships belonged to the English and only sailed at England's discretion. On being informed of this by the misionaries, Mr. Forbes straightway wrote to us saying that for the sake of France's honor we should fly our flag over the factories. But no

sooner was this done than all the inhabitants of Canton (including foreigners and Europeans) rose up against the French, bullying us to think twice before installing our flagpole as they themselves would just as quickly take it down again. Events were taking such a serious turn that the *Capricieuse* sailed back up to Wampoo [Whampoa], where between two hundred and fifty and three hundred employees immediately rushed out to plant themselves on the site of our future flagpole, saying we would have to use force to remove them. At this, fifteen unarmed men were ordered to dispel the mob by peaceful means; but eventually it was the threat of force that did the trick.

From the moment that we had won the game, opinion swung the other way so quickly that now all the papers are full of praise for the French officers' firm handling of the situation, mixed with sallies against the ridiculous conduct of the Cantonese. The impact on the Chinese surpassed any thing we could have expected. Throughout the five or six days that it took to plant our flagpoles, the whole of Canton filed past the factories sneering that the English had "lost face", which is the ultimate expression of contempt in China. In the streets we were escorted by cheering Chinese, with merchants even tendering their goods (crepe &.c) as gifts.

**MARCH 1854**
**TOULON, FRANCE**

*On March 1854*

the *Capricieuse* entered Toulon harbor after a four-year campaign crowned with success despite its many difficulties. Nobody was celebrating. On the twenty-seventh of the month France and England declared war on Russia, and Toulon prepared for a new type of war at sea, waged by ships with engines. The return of the *Capricieuse* passed unnoticed, despite its three masts, sails, exploits and an extraordinary collection of objects brought back from the South Seas. Describing their time aboard the ship, Abel Bergasse Dupetit-Thouars, adopted son of Abel Aubert Dupetit-Thouars, and orderly officer to Roquemaurel, would later write: "It was a harsh campaign. News from France was rare and sometimes worrying, victualing was difficult, disease was rampant and we were operating in dangerous and uncharted territories. Every thing combined to make our captain's task even more burdensome, adding to his already weary state after his previous tours of duty, and exacerbating his asthma, a cruel infirmity that gave him not a moment's respite. Yet through it all his resolve remained unshakable. By the time he returned to Toulon his crew was in tatters and his ship was already outdated. He arrived to find Admiral Bruat's squadron making ready to depart for the Crimea, with a fleet composed entirely of steam ships. Henceforth men were replaced by machines; it was the start of a new era. The future had arrived and the Capricieuse was already a thing of the past."

*Prise de possession de la terre Adélie, le 21 janvier 1840* (claiming possession of Terre Adélie for France, January 21, 1840).

## BIBLIOGRAPHIES

## HISTORICAL WORKS CITED, WITH PHOTOGRAPHIC CREDITS


- **Anson, George.** *Voyage autour du monde, fait dans les années 1740–1744, de George Anson,*
**Amsterdam: Arkstee & Merkus, 1749.**
**P. 67**

- **Arago, Jacques.** *Souvenirs d'un aveugle, Voyage autour du monde, par M. Jacques Arago, Chasses. Drame.*
**Paris: H.L. Delloye, 1840.**

- **Augustin, Zarate de,** *Histoire de la découverte et de la conquête du Pérou traduite de l'espagnol d'Augustin De Zarate par S. D. C.*
**Paris: Jean Geoffroy Nyon, 1716.**
**P. 22**

- **Billings, Joseph, Commodore.** *Voyage fait par ordre de l'impératrice de Russie Catherine II, dans le nord de la Russie asiatique, dans la mer glaciale, dans la mer d'Anadyr, et sur les côtes de l'Amérique, depuis 1785 jusqu'en 1794, par le commodore Billings; rédigé par M. Sauer … et traduit de l'anglais … par J. Castéra …*
**3 vols. and atlas. Paris: F. Buisson, 1802.**

- **Bougainville, Hyacinthe de.** *Journal de la navigation autour du globe de la frégate La Thétis et de la corvette L'Espérance pendant les années 1824, 1825 et 1826, publié sous les auspices du département de la marine par M. le baron de Bougainville, capitaine de vaisseau, chevalier de Saint-Louis et de l'ordre des Cincinnati, combattant de la Légion d'honneur, chef de l'expédition.*
**2 vols. and atlas. Paris: Arthus Bertrand, 1837.**
**P. 205, 209-221**

- **Bougainville, Louis de.** *Voyage autour du monde, par la frégate du roi la Boudeuse, et la flûte l'Étoile; Louis de Bougainville, en 1766, 1767, 1768 et 1769.*
**Paris: Saillant & Nyon, 1771.**
**P. 41, 45-47, 53-55**

- **Choris, Louis.** *Vue et paysages des régions équinoxiales, recueillis dans un voyage autour du monde, par Louis Choris, avec introduction et texte explicative.*
**Paris: Paul Renouard, 1826.**
**P. 17, 21, 24, 227, 232–233, 241**

- **Dillon, Peter.** *Voyage aux îles de la mer du sud, en 1827 et 1828, et relation de la découverte du sort de La Pérouse, dédié au roi, par le capitaine Peter Dillon. 2 vols.*
**Paris: Pillet Ainé, 1830.**

- **Duhaut-Cilly, Auguste.** *Voyage autour du monde, principalement à la Californie et aux Îles Sandwich, pendant les années 1826, 1827, 1828, et 1829; par A. Duhaut-Cilly, capitaine au long–cours, chevalier de la Légion d'honneur, membre de l'Académie d'industrie manufacturière, agricole et commerciale de Paris. 2 vols.*
**Paris and Saint–Servan: Arthus-Bertrand and D. Lemarchand, 1834 (vol 1) and 1835 (vol 2).**
**P. 236-238**

- **Dumont d'Urville,** *Jules Sebastien César. Voyage au pôle Sud et dans l'Océanie sur les corvettes l'Astrolabe et la Zélée, exécuté par ordre du roi pendant les années 1837-1838-1839-1840, sous le commandement de M. J. Dumont d'Urville, capitaine de vaisseau, publié par ordonnance de sa majesté …*
**23 vols. and 7 atlases. Paris: Gide, 1841-1854.**
**P. 347, 350–351, 353–367, 369–371, 377, 381, 382–383, 384-388 (top), 389–391, 393–395**

- **Dumont d'Urville,** *Jules Sebastien César. Voyage pittoresque autour du monde, résumé des voyages de découvertes … par Jules Dumont D'Urville.*
**2 vols. Paris: Tenre, 1834–1853.**
**P. 25, 142-145, 149**

- **Duperrey, Louis Isidore.** *Voyage autour du Monde, exécuté par ordre du Roi, sur la corvette de Sa Majesté, la Coquille, pendant les années 1822, 1823, 1824 et 1825 … par M. L. I. Duperrey …*
**11 vols. and atlas. Paris: Arthus Bertrand, 1826–1830.**
**P. 183, 187-199**

- **Forrest, Thomas.** *Voyage to New Guinea, and the Moluccas … illustrated with thirty copperplates. Performed in the Tartar galley, belonging to the honourable East India Company, during the years 1774, 1775 & 1776, by Captain Thomas Forrest.*
**London: G. Scott, 1779.**
**P. 48**

- **Freycinet, Louis de.** *Voyage autour du monde, exécuté sur les corvettes de S. M. l'Uranie et la Physicienne pendant les années 1817, 1818, 1819, 1820 …*
**12 vols, 4 atlases and one historical atlas. Paris: Pillet Ainé, Paris, 1824–1829.**
**P. 157, 162–170 (left), 172–173, 175-177**

- **Freycinet, Rose de Saulces de.** *Campagne de l'Uranie (1817–1820). Journal de madame Rose de Saulces de Freycinet, d'après le manuscrit original accompagné de notes par Charles Duplomb.*
**Paris: Société d'Editions Géographiques, Maritimes et Coloniales, 1927.**
**P. 161, 170–171, 174**

- **Gervaise, Nicolas.** *Description historique du royaume de Macacar divisée en trois livres.*
**Paris: Hilaire Foucault, 1688.**

- **Guignes, Joseph de.** *Voyage à Peking, Manille et l'île de France faits par intervalle des années 1784 à 1801, par M. de Guignes.*
**3 vols and atlas. Paris: Imprimerie Impériale, 1808.**
**P. 28, 240**

- **Knorr, George Wolfgang.** *Délices physiques choisies, ou Choix de tout ce que les trois règnes de la nature renferment de plus digne des recherches d'un amateur curieux pour en former un cabinet, par George Wolfgang Knorr.*
**2 vols. Nuremberg: 1766-1767.**
**P. 8, 12, 32, 56, 78, 108, 130, 152, 178, 200, 222, 242, 268, 292, 318, 342, 372**

- **La Barbinais, Le Gentil de.** *Nouveau voyage autour du monde. Par L. G. de la Barbinais. Enrichi de plusieurs plans, vues et perspectives des principales villes et ports du Pérou, Chili, Brésil et de la Chine. Avec une description de l'Empire de la Chine beaucoup plus ample et plus circonstanciée que celles qui ont parues jusqu'à présent, où il est traité des mœurs, religion, politique, éducation et commerce des peuples de cet empire.*
**3 vols. Briasson, Paris, 1728.**
**P. 30**

- **La Billardière, Jacques Julien Houtou de.** *Relation du voyage à la recherche de Lapérouse fait par ordre de l'assemblée constituante, pendant les années 1791, 1792 et pendant la 1ère, et la 2de année de la République Française.*
**2 vols. and atlas. Paris: H.J Jansen, Year VIII, 1799.**
**P. 49**

- **Laplace, Cyrille, Pierre, Théodore.** *Campagne de Circumnavigation de la frégate L'Artémise, pendant les années 1837, 1838, 1839, 1840, sous le commandement de M. Laplace, capitaine de vaisseau, publié par ordre du roi, sous les auspices du Ministre de la Marine.*
**6 vols. Paris: Arthus Bertrand, 1841 (vol. 1), 1842 (vol. 2), 1844 (vol. 3), 1848 (vol. 4), 1853 (vol. 5) and 1854 (vol. 6).**

- **Laplace, Cyrille, Pierre, Théodore.** *Voyage autour du monde par les mers de l'Inde et de la Chine, exécuté sur la corvette de l'État la Favorite pendant les années 1830, 1831 et 1832 sous le commandement de M. Laplace, capitaine de frégate; publié par ordre de M. le vice-amiral comte de Rigny, ministre de la Marine et des Colonies.*
**4 vols. and historical atlas. Paris: Imprimerie Royale, 1833 (vols. 1 & 2), 1835 (vols. 3 & 4) and 1835 (historical atlas).**
**P. 68–69, 75, 247, 250–251, 253–267, 327-329, 332-335, 338-339**

- **Le Guillou, Elie.** *Voyage autour du monde de l'Astrolabe et de la Zélée sous les ordres du contre-amiral Dumont-D'Urville pendant les années 1837, 38, 39 et 40, par Elie Le Guillo.*
**Paris: Berquet et Pétion, 1844.**

- **Leguat François.** *Voyage et avantures de François Leguat, & de ses Compagnons, en deux isles désertes des Indes orientales.*
**2 vols. London: David Mortier, 1708.**
**P. 31**

- **Lesseps, Jean-Baptiste Barthelemy de.** *Journal Historique du voyage de M. de Lesseps, Consul de France, employé dans l'expédition de M. le comte de la Pérouse, en qualité d'interprète du roi …*
**2 vols. Paris: Imprimerie Royale, 1790.**

- **Lesson, René, Primevere.** *Voyage autour du monde entrepris par ordre du gouvernement sur La Corvette La Coquille; par P. Lesson, membre correspondant de l'institut.*
**2 vols. Paris: P. Pourrat Frères, 1839.**
**P. 44, 107**

- **Malpière, D. Bazin de.** *La Chine et les chinois, mœurs, usages, peines et châtiments, fêtes, cérémonies religieuses - Costumes civils et militaires - Arts et métiers - Architecture, monuments, maisons, intérieurs, vues et paysages, voitures et vaisseaux ... avec des notices explicatives par D. B. de Malpière ...*
**4 vols. Paris: J. Caboche Demerville et Cie, 1848.**

- **Marchand, Etienne.** *Voyage autour du monde, pendant les années 1790, 1791, et 1792, par Étienne Marchand, précédé d'une introduction historique; auquel on a joint des recherches sur les terres australes de Drake, et un examen critique du voyage de Roggeween; avec cartes et figures : par C. P. Claret Fleurieu, de l'Institut national des sciences et des arts, et du Bureau des longitudes.*
**4 vols. Paris: Imprimerie de la République, Year VI (1797–1798).**
**P. 113, 120-121, 124-126, 129**

- **Marryat, Franck.** *Borneo and the Indian archipelago with drawings of costumes and scenery, Franck Marryat (1826–1855).*
**London: Longman, Brown, Green, 1848.**
**P. 73, 388 (bottom), 392**

- **Marsden, William.** *The History of Sumatra, containing an account of the government, laws, customs, and manners of the native inhabitants, with a description of the natural productions, and a relation of the ancient political state of that island.*
**2 vols. and atlas (27 plates). London: Longman, Hurst, Rees Orme and Brown, 1811.**
**P. 29, 61**

- **Mason, George, Henry.** *The Costume of China, illustrated by sixty engravings: with explanations in English and French.*
**London: William Miller, 1804.**
**P. 239**

- **Milet de Mureau, Louis, Antoine.** *Voyage de La Pérouse autour du monde, publié conformément au décret du 22 avril 1791, et rédigé par M. L. A. Milet-Mureau, général de brigade dans le corps du génie, directeur des Fortifications, ex-constituant, membre de plusieurs sociétés littéraires de Paris.*
**4 vols. and atlas. Paris: Imprimerie de la République, 1797.**
**P. 86–87, 89-106**

- **Pagès, Pierre-Marie-François, Vicomte de.** *Voyages autour du monde, et vers les deux pôles, par terre et par mer, pendant les années 1767, 1768, 1769, 1770, 1771, 1773, 1774 et 1776. Par M. de Pagès, capitaine des vaisseaux du roi, chevalier de l'ordre royal et militaire de Saint-Louis, correspondant de l'Académie des sciences de Paris.*
**2 vols. Paris: Moutard, 1782.**
**P. 64–65, 74, 76–77**

- **Paris, François-Edmond.** *Album pittoresque d'un voyage autour du monde exécuté par ordre du gouvernement français, gravures en taille-douce entièrement dessinées par l'amiral Paris, textes par M. Casimir Henricy, publiciste, ancien marin, ayant fait partie de l'expédition.*
**Paris: Noblet, 1842.**
**P. 323, 330–331, 336–337, 340–341**

- **Pernetty, Dom.** *Histoire d'un voyage aux isles Malouines, fait en 1763 & 1764; avec des observations sur le détroit de Magellan et sur les Patagons, par Dom Pernetty.*
**Paris: Saillant & Nyon, 1770.**
**P. 37, 42–43**

- **Péron, François.** *Voyage de découvertes aux terres australes exécuté par ordre de Sa Majesté l'empereur et roi, sur les corvettes Le Géographe, le Naturaliste, et la Goëlette Le Casuarina, pendant les années 1800, 1801, 1802, 1803 et 1804; rédigé par M. F. Péron.*
**2 vols. Paris: Imprimerie Impériale, 1807.**

- **Petit-Thouars, Abel du.** *Le Vice-amiral Bergasse du Petit-Thouars d'après ses notes et sa correspondance 1832–1890, préface du contre-amiral Dupont.*
**Paris: Perrin et Cie, 1906.**

- **Petit-Thouars, Abel du.** *Voyage Autour de Monde sur la Frégate la Vénus, pendant les années 1836–1839, publié par ordre du roi, sous les auspices du ministre de la marine, par Abel du Petit-Thouars, capitaine de vaisseau, commandeur de la Légion d'honneur.*
**11 vols. and atlas. Paris: Gide, 1840–1855.**
**P. 297, 301–317**

- **Raffles, Sir Stamford.** *The History of Java by Thomas Stamford Raffles.*
**2 vols. London: Black, Parbury and Allen, 1817.**

- **Reuilly, Jean, Baron de.** *Voyage en Crimée et sur les bords de la mer noire pendant l'année 1803 .... , par J. Reuilly.*
**Paris: Bossange, Masson et Besson, 1806.**

- **Roquefeuil, Camille-Joseph de.** *Journal d'un voyage autour du monde, pendant les années 1816, 1817, 1818 et 1819, par M. Camille de Roquefeuil, lieutenant de vaisseau, chevalier de Saint-Louis et de la Légion d'honneur, commandant le navire le Bordelais, armé par M. Balguerie Junior, de Bordeaux.*
**2 vols. Paris: Lesage-Gide-Ponthieu, 1823.**
**P. 135**

- **Rossel, Edouard de.** *Voyage de Dentrecasteaux envoyé à la recherche de La Pérouse publié par ordre de sa majesté l'empereur et le roi, sous le ministère de S. E. le vice-amiral Decrès, comte de l'empire. Rédigé par M. de Rossel, ancien capitaine de vaisseau.*
**1 vol. Paris: Imprimerie Impériale, 1808.**

- **Shipley, Conway.** *Sketches in the Pacific, the South Sea Islands, drawn from nature and on stone by Conway Shipley Esq., Lieut. R.N.*
**London: T. McLean, 1851.**
**P. 117-119, 122–123**

- **Sonnerat, M.** *Voyage à la Nouvelle Guinée ... par M. Sonnerat.*
**Paris: Ruault, 1776.**
**P. 50-52, 70-71**

- **Staunton, George Leonard.** *An Authentic Account of an Embassy from the King of Great Britain to the Emperor of China ... taken chiefly from the papers of His Excellency the Earl of Macartney ...*
**2 vols. and atlas. London: W. Bulmer and Co. for G. Nicol, London, 1797.**
**P. 127-128**

- **Thorn, William.** *Memoir of the Conquest of Java; With the Subsequent Operations of the British Forces in the Oriental Archipelago.*
**London: Egerton Military Library, Whitehall, 1815.**

- **Tombe, Charles, François.** *Voyage aux Indes orientales pendant les années 1802, 1803, 1804, 1805 et 1806.*
**Paris: Arthus Bertrand, 1811.**
**p. 72**

- **Vaillant, Auguste, Nicolas.** *Voyage autour du monde exécuté pendant les années 1836 et 1837 sur la corvette la Bonite commandée par M. Vaillant capitaine de vaisseau publié par ordre du roi sous les auspices du département de la Marine, relation du voyage par A. de La Salle.*
**11 vols, 2 atlases. Paris: Arthus Bertrand, 1841–1852.**
**P. 23, 146–148, 273, 279–291**

- **Van de Velde, Charles William Meredith.** *Vues de Java, Sumatra, Bornéo, Célèbes et autres possessions hollandaises dans les grandes Indes, dessinées d'après nature par C.W.M. Van de Velde; officier de la marine royale dédiées à son altesse royale le Prince Henri des Pays-Bas.*
**Amsterdam: Franz Buffa & Zonen, 1846.**
**P. 150–151**

- *Vancouver, George. A Voyage of discovery to the North Pacific Ocean, and around the world ... By George Vancouver.*
**3 vols and atlas. London: G.G.&J. Robinson and Edwards, 1798.**
**P. 140–141, 234–235**

- **Wilson, William.** *A Missionary Voyage to the Southern Pacific Ocean, performed in the Years 1796, 1797, 1798 in the Ship Duff, Commanded by Captain James Wilson. Compiled from Journals of the Officers and the Missionaries; and Illustrated with Maps, Charts, and Views, Drawn by Mr. William Wilson, and Engraved by the Most Eminent Artists ...*
**London: S. Gosnell for T. Chapman, 1799.**
**P. 26–27, 231**

# FEATURED EXTRACTS
(IN ORDER OF APPEARENCE)


### Le Gentil de La Barbinais

*La Barbinais, Le Gentil de. Nouveau voyage autour du monde. Par L. G. de la Barbinais. Enrichi de plusieurs plans, vues et perspectives des principales villes et ports du Pérou, Chili, Brésil et de la Chine. Avec une description de l'Empire de la Chine beaucoup plus ample et plus circonstanciée que celles qui ont parues jusqu'à présent, où il est traité des mœurs, religion, politique, éducation et commerce des peuples de cet empire.*
**3 vols. Paris: Briasson, Paris, 1728.**

### Louis-Antoine de Bougainville

*Bougainville, Louis Antoine de. Voyage autour du monde, par la frégate du roi la Boudeuse, et la flûte l'Étoile ; Louis de Bougainville, en 1766, 1767, 1768 et 1769.*
**Paris: Saillant & Nyon, 1771.**

### Pierre Marie François de Pagès

*Pagès, Pierre-Marie-François de. Voyages autour du monde, et vers les deux pôles, par terre et par mer, pendant les années 1767, 1768, 1769, 1770, 1771, 1773, 1774 et 1776. Par M. de Pagès, capitaine des vaisseaux du roi, chevalier de l'ordre royal et militaire de Saint-Louis, correspondant de l'Académie des sciences de Paris.*
**2 vols. Paris: Moutard, 1782.**

### Jean-François de Galaup de Lapérouse

*Milet de Mureau, Louis Antoine. Voyage de La Pérouse autour du monde, publié conformément au décret du 22 avril 1791, et rédigé par M. L. A. Milet-Mureau, général de brigade dans le corps du génie, directeur des Fortifications, ex-constituant, membre de plusieurs sociétés littéraires de Paris.*
**4 vols. and atlas. Paris: Imprimerie de la République, 1797.**

### Étienne Marchand

*Marchand, Etienne. Voyage autour du monde, pendant les années 1790, 1791, et 1792, par Étienne Marchand, précédé d'une introduction historique ; auquel on a joint des recherches sur les terres australes de Drake, et un examen critique du voyage de Roggeween; avec cartes et figures : par C. P. Claret Fleurieu, de l'Institut national des sciences et des arts, et du Bureau des longitudes.*
**4 vols. Paris: Imprimerie de la République, 1797–1798.**

### Camille de Roquefeuil

*Roquefeuil, Camille-Joseph de. Journal d'un voyage autour du monde, pendant les années 1816, 1817, 1818 et 1819, par M. Camille de Roquefeuil, lieutenant de vaisseau, chevalier de Saint-Louis et de la Légion d'honneur, commandant le navire le Bordelais, armé par M. Balguerie Junior, de Bordeaux.*
**2 vols. Paris: Lesage-Gide-Ponthieu, 1823.**

### Louis Claude de Saulces de Freycinet

*Freycinet, Louis de. Voyage autour du monde, exécuté sur les corvettes de S. M. l'Uranie et la Physicienne pendant les années 1817, 1818, 1819, 1820 .... par M. Louis de Freycinet.*
**12 vols., 4 atlases and one historical atlas. Paris: Pillet Ainé, Paris, 1824–1829.**

*Freycinet, Rose de Saulces de. Campagne de l'Uranie (1817–1820). Journal de madame Rose de Saulces de Freycinet, d'après le manuscrit original accompagné de notes par Charles Duplomb.*
**Paris: Société d'Editions Géographiques, Maritimes et Coloniales, 1927.**

### Louis-Isidore Duperrey

*Duperrey, Louis Isidore. Voyage autour du Monde, exécuté par ordre du Roi, sur la corvette de Sa Majesté, la Coquille, pendant les années 1822, 1823, 1824 et 1825 ... par M. L. I. Duperrey ...*
**11 vols. and an atlas. Paris: Arthus Bertrand, Librairie-Éditeur, 1826–1830.**

*Lesson, René Primevere. Voyage autour du monde entrepris par ordre du gouvernement sur La Corvette La Coquille ; par P. Lesson, membre correspondant de l'institut.*
**2 vols. Paris: P. Pourrat Frères, 1839**

### Hyacinthe de Bougainville

*Bougainville, Hyacinthe de. Journal de la navigation autour du globe de la frégate La Thétis et de la corvette L'Espérance pendant les années 1824, 1825 et 1826, publié sous les auspices du département de la marine par M. le baron de Bougainville, capitaine de vaisseau, chevalier de Saint-Louis et de l'ordre des Cincinnati, combattant de la Légion d'honneur, chef de l'expédition.*
**2 vols. and an atlas. Paris: Arthus Bertrand, 1837.**

### Auguste Bernard Duhaut-Cilly

*Duhaut-Cilly, Auguste. Voyage autour du monde, principalement à la Californie et aux Îles Sandwich, pendant les années 1826, 1827, 1828, et 1829 ; par A. Duhaut-Cilly capitaine au long-cours, chevalier de la Légion d'honneur, membre de l'Académie d'industrie manufacturière, agricole et commerciale de Paris.*
**2 vols. Paris and Saint-Servan: Arthus-Bertrand and D. Lemarchand, 1834–1835.**

### Cyrille Pierre Théodore Laplace
**(First voyage, aboard the *Favorite*)**

*Laplace, Cyrille Pierre Théodore. Voyage autour du monde par les mers de l'Inde et de la Chine, exécuté sur la corvette de l'État la Favorite pendant les années 1830, 1831 et 1832 sous le commandement de M. Laplace, capitaine de frégate; publié par ordre de M. le vice-amiral comte de Rigny, ministre de la Marine et des Colonies.*
**4 vols. and historical album. Paris: Imprimerie Royale, 1833–1835.**

### Auguste Nicolas Vaillant

*Vaillant, Auguste, Nicolas. Voyage autour du monde exécuté pendant les années 1836 et 1837 sur la corvette la Bonite commandée par M. Vaillant capitaine de vaisseau publié par ordre du roi sous les auspices du département de la Marine, relation du voyage par A. de La Salle.*
**11 vols., 2 atlases. Paris: Arthus Bertrand, 1841–1852.**

### Abel Aubert du Petit-Thouars

*Petit-Thouars, Abel du. Voyage Autour de Monde sur la Frégate la Vénus, pendant les années 1836-1839, publié par ordre du roi, sous les auspices du ministre de la marine, par Abel du Petit-Thouars, capitaine de vaisseau, commandeur de la Légion d'honneur.*
**11 vols., 5 atlases. Paris: Gide, 1840–1855.**

### Cyrille Pierre Théodore Laplace
**(Second voyage, aboard the *Artémise*)**

*Laplace, Cyrille Pierre Théodore. Campagne de Circumnavigation de la frégate L'Artémise, pendant les années 1837, 1838, 1839, 1840, sous le commandement de M. Laplace, capitaine de vaisseau, publié par ordre du roi, sous les auspices du Ministre de la Marine.*
**6 vols. Paris: Arthus Bertrand, 1841–1854.**

### Jules Sébastien César Dumont d'Urville

*Dumont d'Urville, Jules Sebastien César. Voyage au pôle Sud et dans l'Océanie sur les corvettes l'Astrolabe et la Zélée, exécuté par ordre du roi pendant les années 1837–1840, sous le commandement de M. J. Dumont d'Urville, capitaine de vaisseau, publié par ordonnance de sa majesté ...*
**23 vols. and 7 atlases. Paris: Gide, 1841– 1854.**

*Le Guillou, Elie. Voyage autour du monde de l'Astrolabe et de la Zélée sous les ordres du contre-amiral Dumont-d'Urville pendant les années 1837, 1838, 1839 et 1840, par Elie Le Guillou ...*
**Paris: Berquet et Pétion, 1844.**

### Gaston de Roquemaurel

*Bergasse du Petit-Thouars, Abel. Le Vice-amiral Bergasse du Petit-Thouars d'après ses notes et sa correspondance 1832–1890.*
**Paris: Perrin et Cie, 1906.**

*Dumont d'Urville, Jules Sebastien César. Voyage au pôle Sud et dans l'Océanie sur les corvettes l'Astrolabe et la Zélée, exécuté par ordre du roi pendant les années 1837–1838–1839–1840, sous le commandement de M. J. Dumont d'Urville, capitaine de vaisseau, publié par ordonnance de sa majesté ...*
**23 vols. and 7 atlases. Paris: Gide, 1841–1854.**


Certain extracts and chapters dedicated to Gaston de Roquemaurel are drawn from published documents, letters and articles entrusted to the author by Jacques de Roquemaurel, and from the writings of Jean-Philippe Zanco, Gaston de Roquemaurel's biographer.

# ADDITIONNAL REFERENCES


- **Angelier, François.** *Dictionnaire des voyageurs et explorateurs occidentaux du XIIIe au XXe siècle.*
  **Paris: Pygmalion, 2011.**

- **Beriot, Agnes.** *Grands voiliers autour du monde, les voyages scientifiques 1760–1850.*
  **Paris: Editions du Pont Royal, 1962.**

- **Broc, Numa.** *Autour du monde, voyage de François de Pagès par terre et par mer 1767–1771.*
  **Paris: Editions Imprimerie Nationale, 1991.**

- **Broc, Numa.** *Dictionnaire illustré des explorateurs et grands voyageurs français du XIXe siècle.*
  **Aubervilliers: Editions de C.T.H.S., 1999.**

- **Brosse, Jacques.** *Les tours du monde des explorateurs, les grands voyages maritimes, 1764–1843.*
  **Paris: Bordas, 1983.**

- **Conan, Alain.** *Le Mystère Lapérouse, ou le rêve inachevé d'un roi.*
  **Paris: Edition de Conti, 2008.**

- **Cruchet, René.** *Le Tour du monde en 37 mois de Camille de Roquefeuil.*
  **Bordeaux: Imprimerie Delmas, 1952.**

- **Dahlgren, Erik Wilhelm.** *Voyages français à destination de la mer du Sud avant Bougainville (1695–1749).*
  **Paris: Imprimerie Nationale, 1907.**

- **Dunmore, John.** *Les Explorateurs français dans le Pacifique, XVIIIe siècle.*
  **Vol. 1. Paris: Les Editions du Pacifique, 1978.**

- **Guerin, Léon.** *Les navigateurs français, histoire des navigations, découvertes et colonisations françaises.*
  **Paris: Belin-Leprieur et Morizot, 1847.**

- **Hatin, L. E.** *Histoire pittoresque des voyages autour du monde.*
  **Vol. 1. Paris: Dutertre, 1848.**

- **Jacques, Guy.** *Jacques Arago... Ce frère inattendu.*
  **Paris: Editions François de Galice (Ixcéa), 2008.**

- **Lutteroth, Henri.** *O-Taïti histoire et enquête.*
  **Paris: Paulin L.-R. Delay, 1845.**

- **Morgat, Alain.** *Le tour du monde de la Coquille 1822–1825.*
  **Paris: Editions du Gerfaut, 2005.**

- **Polak, Michèle and Dugrand, Alain.** *Des livres à la découverte du monde.*
  **Paris: Hoëbeke, 2012.**

- **Polak, Michèle and Dugrand, Alain.** *Trésors des livres de mer, de Christophe Colomb à Marin Marie.*
  **Paris: Hoëbeke, 2011.**

- **Taillemite, Etienne.** *Bougainville et ses compagnons autour du monde 1766–1769, Journaux de navigations ...*
  **Paris: Imprimerie Nationale, 1977.**

- **Valode, Philippe.** *Les Grands explorateurs français de Jacques Cartier à nos jours.*
  **Paris: l'Archipel, 2008.**

- *Dictionnaire illustré des explorateurs et grands voyageurs français du XIXe siècle, Afrique.*
  **Paris: Editions du C.T.H.S., 1988.**

- *The Hill Collection of Pacific Voyages at the University of California, San Diego.*
  **2nd ed. New Haven and Sydney: William Reese Company and Hordern House Ltd., 2004.**

- *Académie du Var. Les Grandes expéditions autour du monde, Toulon, 1817–1840.*
  **Marseille: Editions Campanile, 2017.**

## ACKNOWLEDGMENTS

To those who helped me design this book: Juliane Cordes and Emmanuelle Halkin in Paris, for their expertise, their smiles and their availability.

To my wife Anne, for her artistic vision of the book, and her contribution to the passages about the women explorers.

To Mélissa Bilodeau (Montreal, Canada), whose skill and ingenuity, combined with her passion for the subject, enabled me to digitize thousands of documents and find technical solutions to the challenges I faced.

To Matthieu Ricard (Nepal), who gave me inspiration.

To Michèle Polak in Paris, for her advice and the loan of her wonderful books.

To Donald Heald in New York, whom I met for the first time on my return from a trip to Bhutan, and who advised me on my choice of books.

To Edward Duyker (Australia), one of the greatest Anglo-Saxon connoisseurs of French explorers in Oceania, for his precise foreword to the English edition of this book.

To Olivier Poivre d'Arvor (France), for his moving and vibrant choice of words in his foreword to the French version of this book.

To my bookseller friends, collectors, and lovers of antiquarian books, especially Jean-François Letenneur, Stephan Feldman, Ghislain de La Hitte, Jean-Louis Ceccarini, Christian Gasch (France), Yves Azemar (Hong Kong), Anne McCormick (Australia), and Patrick McGahern (Canada).

To those rare-book dealers who continue to make magnificent finds: Isabelle Bilbao and Guy Neplaz (Koegui bookshop, France), Gert Jan Bestebreurtje, Allard Schierenberg (Netherlands), and Peter Harrington (UK).

To Jacques de Roquemaurel, Jean-Philipe Zanco (France), and Ralph Kingston (USA), who helped me with the chapter on Gaston de Roquemaurel.

To Yves Bourgeois (France), director of numerous documentaries on Lapérouse, for his vast knowledge of the subject.

To my friends Françoise and Philippe Cazalis de Fondouce, Hetty and Bruno Schricke, Dominique and Laurent Solomon, Philippe Damas, and Jean-Baptiste Oudéa (Singapore), for their support and enthusiasm for this project.

To Georges Adam (Canada), captain of the *Spirit of St. Exupéry*, for his around-the-world voyages, which provided me with thousands of pages to read.

To Dr. Hoa Phong Le (Montreal) and Dr. Tan Ken Jin (Singapore), whose expertise and advice keep me in shape to continue my explorations around the world.

To Kristal Hale and Thomas Murray (USA), for their infectious passion.

To the team at Trigonix (Canada), in particular Dessie Radomirova, Saïd Abella, and André Cavanagh, who digitized thousands of illustrations from my book collection.

To Dennis Moren (Bangladesh), for his expertise with maps.

Many thanks to the Flammarion team in Paris: Emmanuelle Rolland, Henri Julien, Marion Cipriani, and Élodie Conjat.

To Charlène and the entire Artisans du Regard team in Paris.

Also to Anne Jouve (France), for her insightful comments, and to Florence Brutton for her wonderful translation work.

To Jeffrey Tan (Singapore), Kevin Page (Canada), Jean-Louis Diogo, and Linda Duarte (France), for their unfailing contribution to the project logistics.

And to all those others who helped and supported me.

MIX
Paper from responsible sources
FSC® C111592
www.fsc.org

Printed in Spain by Indice

English-language edition


Simultaneously published in French
as *Routes nouvelles, côtes inconnues*


editions.flammarion.com
@flammarioninternational
24 25 26 3 2 1
ISBN: 978-2-08-042845-5
L.01EBTN001045
Legal deposit: February 2024

ÉDITIONS FLAMMARION

**Director of Editorial Partnerships**
Henri Julien

**Project Manager**
Emmanuelle Rolland
assisted by Marion Cipriani

**Editor**
Emmanuelle Halkin

**Design and Typesetting**
Juliane Cordes and Corinne Dury

**Digitization**
Trigonix (Montreal), Desislava Radomirova,
André Cananagh, Saïd Abella

**Translation from the French**
Florence Brutton

**English copy editing**
Mark Brutton

**Production**
Élodie Conjat

**Color Separation**
Les Artisans du Regard

**Marketing Coordinator**
CDRATIER (Montreal),
Mélissa Bilodeau